Lecture Notes in Physics

Edited by J. Ehlers, München, K. Hepp, Zürich, H. A. Weidenmüller, Heidelberg, and J. Zittartz, Köln
Managing Editor: W. Beiglböck, Heidelberg

40

Effective Interactions and Operators in Nuclei

Proceedings of the Tucson International Topical Conference on Nuclear Physics Held at the University of Arizona, Tucson, June 2–6, 1975

Edited by B. R. Barrett

Springer-Verlag
Berlin · Heidelberg · New York 1975

Editor
Prof. Bruce R. Barrett
Department of Physics
University of Arizona
Tucson, Arizona 85721
USA

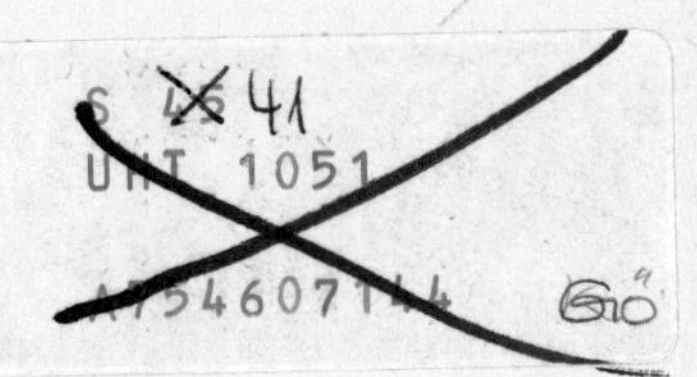

Library of Congress Cataloging in Publication Data

Tucson International Topical Conference on Nuclear
Physics, University of Arizona, 1975.
Effective interactions and operators in nuclei.

(Lecture notes in physics, 40)
Bibliography: p.
Includes index.
1. Effective interactions (Nuclear physics)--
Congresses. 2. Operator theory--Congresses.
I. Barrett, Bruce Richard, 1939- II. Title.
III. Series.
QC794.8.E35T82 1975 539.7'54 75-25688

ISBN 3-540-07400-7 Springer-Verlag Berlin Heidelberg New York
ISBN 0-387-07400-7 Springer-Verlag New York Heidelberg Berlin

Offsetprinting and bookbinding: Julius Beltz, Hemsbach/Bergstr.

PREFACE

The International Topical Conference on Effective Interactions and Operators in Nuclei, held at the University of Arizona in Tucson, Arizona, June 2 - 6, 1975, was the first international meeting on this subject. It brought together essentially all the principal physicists working in this field plus many other interested physicists. This volume contains the invited talks presented at the conference along with the discussions which followed each talk. There is also a list of participants and a copy of the conference program. This volume constitutes the second volume of the Proceedings of the Tucson Conference; the first volume, published by the University of Arizona, contained the abstracts of the contributed papers. Copies of volume one can be obtained by writing to the Chairman of the Organizing Committee for the conference: Dr. Bruce R. Barrett, Department of Physics, University of Arizona, Tucson, Arizona 85721, USA.

This volume has been reproduced by photo-offset, and most submitted manuscripts were included without any editing. Only obvious typographical errors were corrected, where it was reasonable and convenient to do so. The ideas and opinions are those of the invited speakers. The discussions were typed from hand-written copies submitted by the questioners and speakers and have been reproduced to the best of our ability to read the original copies.

I would like to thank Prof. Dr. H.A. Weidenmüller, Prof. J.D. McCullen and Dr. W.T. Weng for their help in proof-reading the manuscripts and the discussions. Special thanks go to Marijke Oskam-Tambaesser for typing the discussions and making corrections to the invited papers, to Frau Karl-Kratky for typing the final corrections, and to Prof. Weidenmüller for his help and encouragement and for the hospitality of the Max-Planck-Institut für Kernphysik in Heidelberg where the final editing was performed.

The Tucson Conference helped to define the present status of effective interaction and operator theory and calculations and to indicate the directions for future research and progress in this field. It was an exciting and stimulating conference. I hope that this volume will increase the understanding of this subject by both active researchers and non-workers in the field plus facilitating future work by bringing together in one volume all relevant and up-to-date information on the subject of effective interactions and operators in nuclei.

Finally, the Conference was not without its light moments, and several of these have been included in the Proceedings, such as Dr. Landé's comment during the discussion following Dr. Ratcliff's talk. A graduate student's comment on core-polarization is given before Dr. Barrett's talk (original idea and drawing by Margaret Sandel), and Dr. Negele's original drawing regarding Dr. Kirson's philosophy on density-dependence is presented before Dr. Negele's talk.

Heidelberg, July 1975 Bruce R. Barrett

Table of Contents

CONFERENCE PROGRAM

Monday, June 2

OPENING SESSION

9:00 a.m. Opening address by:
Dr. A.B. Weaver, Executive Vice President, Un. of Ariz.

I. Opening Talk

Session Chairman: F. Coester (Argonne)

9:15 a.m. Comments on and Overview of the Conference:
B.R. Barrett, Chairman, Organizing Committee

II. General Theory Session

Session Chairman: F. Coester (Argonne)

9:30 a.m. Time Independent Approach: B.H. Brandow (Los Alamos)

11:00 a.m. Time Dependent Approach (Johnson and Baranger Approach):
M.B. Johnson (Los Alamos)

3:00 p.m. Time Dependent Approach (Kuo, Lee and Ratcliff and Kuo and Krenciglowa Approaches): K.F. Ratcliff (Albany)

3:50 p.m. Determination of Effective Matrix Elements from Experimental Data: I. Talmi (Weizmann Institute)

5:20 p.m. Contributed Paper

Two-Body Propagators with Dynamic Effective Interactions:
C.A. Engelbrecht, F.J.H. Hahne, W.D. Heiss (Pretoria)

Tuesday, June 3

Session Chairman: P. Signell (Michigan State)

9:00 a.m. Contributed Papers

A New Method for Computing Effective Interactions:
H. Kümmel (Bochum)

What do we want to know about the Nucleon-Nucleon Potential?:F. Coester (Argonne)

III. Choice of the Unperturbed Hamiltonian

10:00 a.m. How do we Decide Which Unperturbed Basis to Use?
What is the Role of Self-Consistency?:
P.U. Sauer (Hannover)

IV. Calculation of the Effective Interaction

Session Chairman: A. Landé (Groningen)

11:15 a.m. Computation of the Reaction Matrix G:
R.L. Becker (Oak Ridge)

3:00 p.m. Perturbation Calculation in a Double-Partitioned Hilbert Space: B.R. Barrett (Tucson)

3:45 p.m. Behavior of the G-matrix Expansion for the Average Effective Interaction: P. Goode (Rutgers)

5:00 p.m. Contributed Papers

Single-Nucleon Potential Leading to Vanishing Rearrangement Energies in Knock-Out Reactions: T. Berggren (Lund)

Equivalence of Deformed Hartree-Fock (HF) and the Renormalized Spherical Shell Model (SSM) in the Intrinsic Frame for a Schematic Hamiltonian (I): M. Harvey (Chalk River)

Improved Effective Shell-Model Interaction in the s-d Shell: J.P. Vary (Brookhaven), S.N. Yang (Univ. of Washington)

Wednesday, June 4

V. Problems in Computing the Effective Interaction

Session Chairman: M. Harvey (Chalk River)

9:00 a.m. Convergence Properties of the Perturbation Expansion for the Effective Interaction: H.A. Weidenmüller (Heidelberg)

10:00 a.m. Is There an Universal Relationship Connecting All Two-Body Effective Interactions?: J.P. Schiffer (Argonne)

11:00 a.m. Contributed Papers

Comparison of Approximations to the Effective Interaction: S. Pittel (Bartol), J.D. Vergados (Univ. of Pennsylvania), C.M. Vincent (Univ. of Pittsburgh)

Self Consistency Corrections in Various Orders in the Linked Cluster Expansion (L.C.E.): Y. Starkand and M.W. Kirson (Weizmann Institute)

Thursday, June 5

VI. Alternate Approaches to Perturbation Theory

Session Chairman: D. Koltun (Rochester)

9:00 a.m. What Features of Nuclear Spectra Are Different from What We Expect Based on Statistics? J.B. French (Rochester)

10:00 a.m. Infinite Partial Summation: D.W.L. Sprung (McMaster)

11:15 a.m. Padé Approximations: T.H. Schucan (SIN)

3:00 p.m. Shell Model Diagonalizations in an Expanded Space: D.J. Rowe (Toronto)

3:45 p.m. Density-Dependent Interactions: J.W. Negele (MIT)

5:00 p.m. Contributed Papers

Construction of Variational Bounds for the N-Body Eigenstate Problem by the Method of Padé Approximations: D. Bessis (Saclay)

Convergence of Continued Fractions and Padé Approximations to the Energy-Independent Effective Interaction: H.M. Hofmann (Weizmann Institute)

Effective Delta Interactions in Nuclei: C.W. Wong (UCLA)

Friday, June 6

VII. Calculations, Tests and Features of Other Effective Operators

Session Chairman: P. Brussaard (Utrecht)

9:00 a.m. Calculations of Other Effective Operators: P.J. Ellis (Minnesota)

10:00 a.m. Contributed Paper

Core Polarization in Inelastic Scattering: F. Petrovich (Florida State University)

10:15 a.m. Other Tests of Effective Interactions and Operators: L. Zamick (Rutgers)

VIII. Summary Talk

Session Chairman: S. Moszkowski (UCLA)

11:30 a.m. Where Do We Stand at the Present Time Regarding the Microscopic Theory of Effective Interactions and Operators?: M.W. Kirson (Weizmann Institute)

LIST OF PARTICIPANTS

B.R. Barrett
Un. of Arizona

W. Bassichis
Texas A & M

F. Beck
Darmstadt, Germany

R.L. Becker
Oak Ridge

B. Brandow
Los Alamos

P.J. Brussaard
Utrecht, Holland

G.G. Calderon
Mexico

R. Chestnut
Darmstadt, Germany

F. Coester
Argonne

M. de Llano
Mexico

A.E.L. Dieperink
Amsterdam, Holland

D. Donahue
Un. of Arizona

M. Dworzecka
Un. of Maryland

J. Eisenberg
Hebrew Un., Israel

P. Ellis
Un. of Minnesota

H. Flocard
Orsay, France

H. Floyd
Un. of Arizona

B. French
Un. of Rochester

W.A. Friedman
Un. of Wisconsin

A. Gal
Hebrew Un., Israel

M. Golin
Rutgers

P. Goode
Rutgers

F.J.W. Hahne
Pretoria, South Africa

M. Harvey
Chalk River, Canada

H. Hofmann
Weizmann Institute, Israel

M.G. Huber
Erlangen, Germany

P. Huddy
Un. of Arizona

M. Johnson
Los Alamos

A. Kallio
Oulu, Finland

M.W. Kirson
Weizmann Institute, Israel

W. Knüpfer
Erlangen, Germany

S. Köhler
Un. of Arizona

D.S. Koltun
Rochester

H.G. Kümmel
Bochum, Germany

A. Landé
Groningen, Holland

S.Y. Lee
National Central Un., Taiwan

J. Le Tourneux
Un. of Montreal

M.A.K. Lohdi
Texas Tech. Un.

N. Lo Iudice
Un. of Naples, Italy

C.P. Malta
Sao Paulo, Brasil

P. Manakos
Darmstadt, Germany

R. McCarthy
Un. of Arizona

J. McCullen
Un. of Arizona

P. McNamee
Un. of Arizona

E. Middlesworth
Un. of Arizona

J. Miller
Carnegie-Mellon Un.

S. Moszkowski
UCLA

O. Mott
Un. of Arizona

J.W. Negele
MIT

G. Oberlechner
Strasbourg, France

G. Ohlen
Lund, Sweden

R. Padjen
Un. of Montreal, Canada

R. Parmenter
Un. of Arizona

F. Petrovich
Florida State Un.

S. Pittel
Bartol

M.A. Preston
McMaster Un., Canada

K. Ramavataram
Laval, Canada

V. Ramavataram
Laval, Canada

K.F. Ratcliff
Un. of New York, Albany

J. Richert
Strasbourg, France

D.J. Rowe
Un. of Toronto, Canada

M. Sandel
Un. of Arizona

P.U. Sauer
Hannover, Germany

R.R. Scheerbaum
Texas A & M

J. Schiffer
Argonne

T. Schucan
SIN, Switzerland

S. Shlomo
Michigan State Un.

P. Signell
Michigan State Un.

D. Sprung
McMaster Un., Canada

A. Stamp
Auckland, New Zealand

I. Talmi
Weizmann Institute, Israel

I.S. Towner
Chalk River, Canada

S. Tsai
Michigan State Un.

J. Vary
Brookhaven

J.D. Vergados
Un. of Pennsylvania

G. Vichniac
Saclay, France

C.M. Vincent
Un. of Pittsburgh

H.A. Weidenmüller
Heidelberg, Germany

R.R. Whitehead
Glasgow, Scotland

B.H. Wildenthal
Michigan State Un.

W.T. Weng
Un. of Arizona

Chun Wa Wong
UCLA

L. Zamick
Rutgers

PERTURBATION THEORY OF EFFECTIVE HAMILTONIANS *

B. H. Brandow

Theoretical Division
Los Alamos Scientific Laboratory, University of California
Los Alamos, New Mexico 87545

I. Introduction

As most of you probably know, I have not been working in the field of nuclear structure for several years. So I will not attempt to tell you how to deal with the very difficult problems of convergence, intruder states, strong core polarizability, tensor force correlations, etc. This talk will be mainly a review of the many papers which have used perturbation theory to derive "effective" or "model' hamiltonians. During the several years my Rev. Mod. Phys. article[1] was developing, I made a diligent search of the literature to trace the history of the techniques I was using. I was really quite surprised to find that the subject of degenerate perturbation theory was so obscure. If there was any significant body of literature on this subject, it was certainly not known to the nuclear physics community. The only significant papers I could find were those of Bloch, des Cloizeaux, Bloch and Horowitz, Morita, Löwdin, and Dawson, Talmi, and Walecka, as quoted in my article. Since then I have been exposed to the field of quantum chemistry and also the magnetic materials area of solid state theory, with the result that I have come across a considerable literature extending back as far as 1929. And of course there are a number of more recent works I want to comment on.

If we merely want to introduce the concept of an effective hamiltonian, surely the easiest way to do this is by means of the partitioning technique of Löwdin[2] and Feshbach[3]. They partition the Hilbert space into two subspaces, "P" and "Q", such that the Schroedinger equation becomes a 2 x 2 block matrix equation,

$$\begin{pmatrix} H_{PP} & H_{PQ} \\ H_{QP} & H_{QQ} \end{pmatrix} \begin{pmatrix} \Psi_P \\ \Psi_Q \end{pmatrix} = E \begin{pmatrix} \Psi_P \\ \Psi_Q \end{pmatrix} \tag{1.1}$$

The Ψ_Q variable is easily eliminated to produce the "projected" Schroedinger equation

$$\left[H_{PP} + H_{PQ}\,(E-H_{QQ})^{-1}\,H_{QP}\right]\Psi_P = E\Psi_P \tag{1.2}$$

These manipulations are obviously independent of perturbation theory. Nevertheless, there are two reasons why we must be concerned with perturbation theory. In the first place, this is the only method to date which has been able to produce general linked cluster results for open-shell systems. Secondly, perturbation theory is the tried and proven cornerstone of nuclear many-body theory. Most of the worthwhile developments in this field have come about because someone took the trouble to systematically calculate some of the higher-order terms of this or that rearrangement of the basic perturbation series, to show us which effects are big and which are small. Armed with this vital information one can usually find some more efficient means for treating the "big" terms, but as a method for generating the initial insights, and for doing the necessary bookkeeping, perturbation theory is very hard to beat. I want to take this opportunity to emphasize that I have never claimed or assumed that perturbation theory should converge, in the elementary sense that a straightforward evaluation of the series should give successively smaller terms, at least not for the open-shell problem where long range correlations are important. Nevertheless I believe that perturbation theory is going to retain its position as the backbone of our subject, because it is the most flexible and efficient bookkeeping system that we have.

I'm going to start off with a brief review of non-degenerate and non-many-body perturbation theory, and then spend quite a bit of time on the degenerate but non-many-body problem. We know that, at least for the hard-core part of the nuclear problem, it is necessary to formally sum selected parts of the perturbation series out to infinite order. To do this we need to have some understanding of the general structure of the expansion. I shall therefore be concentrating on this "structural" aspect of the various formal approaches. If one merely wants to evaluate the first two or three orders of the series, then any formalism which can generate these terms correctly may be considered a good one. But if we also ask for some general insights, we shall see that some formalisms are clearly superior. It turns out that the degenerate perturbation problem is not uniquely defined, and we shall also see that there are some practical criteria for choosing among the various possibilities. Finally, I shall review the literature dealing with the linked-cluster aspects of open-shell many-body systems.

II. Non-Degenerate Perturbation Theory

We start with an ordinary (non-many-body) quantum system, and the usual notation,

$$H = H_o + V \tag{2.1}$$

$$H_o \Phi_i = E_i \Phi_i \tag{2.2}$$

$$H\Psi = E\Psi = (E_o + \Delta E)\Psi \tag{2.3}$$

$$\Psi = \sum_i a_i \Phi_i \quad . \tag{2.4}$$

By introducing the projection operators

$$P_o = |\Phi_o\rangle\langle\Phi_o| \ , \ Q_o = 1 - P_o \ , \tag{2.5}$$

and adopting the so-called intermediate normalization convention

$$\langle \Phi_o | \Phi_o \rangle = \langle \Phi_o | \Psi \rangle = 1, \tag{2.6}$$

we can define a wave operator Ω with the properties

$$\Psi = \Omega \Phi_o, \tag{2.7}$$

$$P_o \Omega = P_o, \; \Omega P_o = \Omega, \; \Omega Q_o = 0, \tag{2.8}$$

and thus obtain

$$\Delta E = \langle \Phi_o | V | \Psi \rangle = \langle \Phi_o | V\Omega | \Phi_o \rangle \quad . \tag{2.9}$$

This last expression suggests the introduction of a reaction matrix or energy-shift operator

$$\mathcal{V} = V\Omega \quad . \tag{2.10}$$

Brillouin-Wigner Theory

The standard Brillouin-Wigner (BW) perturbation expressions are

$$\Omega = \left[1 + \sum_{n=1}^{\infty} \left(\frac{Q_o}{E_o + \Delta E - H_o} V \right)^n \right] P_o \tag{2.11}$$

and

$$\Delta E = \langle \Phi_o | V \left[1 + \sum_{n=1}^{\infty} \left(\frac{Q_o}{E_o + \Delta E - H_o} V \right)^n \right] | \Phi_o \rangle \quad . \tag{2.12}$$

Note that the BW form of perturbation theory is characterized by two general features: (i) it has the simple structure of a geometric series; (ii) the unknown energy shift ΔE appears within the energy denominators. The series (2.11) is formally equivalent to the closed-form or "integral" operator equation

$$\begin{aligned} \Omega &= P_o + \frac{Q_o}{E_o + \Delta E - H_o} V\Omega \\ &= \left[1 - \frac{Q_o}{E_o + \Delta E - H_o} V \right]^{-1} P_o \quad , \end{aligned} \tag{2.13}$$

and similarly

$$\begin{aligned} \mathcal{V} &= VP_o + V \frac{Q_o}{E_o + \Delta E - H_o} \mathcal{V} \\ &= \left[1 - V \frac{Q_o}{E_o + \Delta E - H_o} \right]^{-1} VP_o \; , \end{aligned} \tag{2.14}$$

from which the corresponding perturbation expansions can be obtained by iteration. We also note that in view of the convention (2.6), the norm of the perturbed wave-function can be expressed formally as

$$\begin{aligned} \langle \Psi | \Psi \rangle &= \langle \Phi_o | \Omega^\dagger \Omega | \Phi_o \rangle \\ &= 1 - \frac{d}{d(\Delta E)} \langle \Phi_o | \mathcal{V} | \Phi_o \rangle \quad . \end{aligned} \tag{2.15}$$

Rayleigh-Schroedinger Theory

The essence of the Rayleigh-Schroedinger (RS) perturbation theory is to get rid of the dependence on the unknown quantity ΔE, which characterizes the BW theory. There are a number of ways of doing this. The standard textbook method is to formally expand E and Ψ in powers of V, substitute in the Schroedinger equation, and equate terms of the same order in V. This method is tedious, and it fails to provide any insight into the general structure of the series.

Another method is to start with the BW expressions (2.11), (2.12), and expand each of the BW energy denominators into a geometric series in ΔE:

$$\frac{1}{E_o+\Delta E-H_o} = \frac{1}{E_o - H_o} \sum_{n=o}^{\infty} \left(\frac{-\Delta E}{E_o - H_o}\right)^n \quad . \qquad (2.16)$$

This leads to energy denominators of the desired form, but we still have to use the BW expression (2.12) to evaluate the numerator $(-\Delta E)$'s. This means that one has to proceed back and forth between (2.12) and (2.16). To obtain the RS series up to any finite order in V, this process can be terminated after a finite number of cycles (roughly $\frac{n}{2}$ cycles for order n). Although somewhat tedious, this is faster than the textbook method, and it gives considerable insight into the general structure of the RS series. A diagrammatic representation for this procedure was discussed in Ref. 1.

Another method is based on the observation that the two types of geometric series expansion, seen respectively in (2.12) and (2.16), can be generated simultaneously by means of the integral expression

$$\Omega = P_o + \frac{Q_o}{E_o - H_o} (V - \Delta E)\Omega \quad , \qquad (2.17)$$

which follows from (2.11) upon making the substitutions

$$H_o \rightarrow H_o + Q_o \Delta E Q_o \quad , \qquad V \rightarrow V - Q_o \Delta E Q_o \qquad (2.18)$$

Inserting (2.17) in (2.9) and equating like powers of V leads to the useful recursion formulas

$$\Delta E^{(n)} = \langle \Phi_o | V\Omega^{(n-1)} | \Phi_o \rangle \qquad (2.19)$$

$$\Omega^{(n)} = \frac{Q_o}{E_o - H_o}\left[V\Omega^{(n-1)} - \sum_{j=1}^{n-1} E^{(j)}\Omega^{(n-j)}\right] \quad . \qquad (2.20)$$

These formulas were used by Brueckner,[4] in the classic paper where he introduced the concept of a linked-cluster perturbation expansion. This approach has also been used by Baker[5] to derive the Goldstone linked-cluster result, in a manner closely paralleling my derivation[1] based on (2.16). (An outline of Baker's derivation was, in fact, given in Goldstone's paper.[6])

Finally, we note that the approach of (2.16) can be systematized in an elegant way. Let the right-hand side of (2.12) define the function $F(E_o+\Delta E)$. The combined result of the expansions (2.16) for all of the denominators appearing in (2.12) can then be expressed as a Taylor expansion,

$$\Delta E = \sum_{n=o}^{\infty} \frac{(\Delta E)^n}{n!} \frac{d^n F(E_o)}{d E_o^n} \quad . \qquad (2.21)$$

The result of repeatedly substituting this equation into itself will evidently be to express ΔE as a series of products of $F(E_o)$ and its various derivatives. The general form of this series is given by a formula of Lagrange[7], namely

$$\Delta E = \sum_{n=1}^{\infty} \frac{1}{n!}\left(\frac{d}{dE_o}\right)^{n-1} \left[F(E_o)\right]^n , \qquad (2.22)$$

as was pointed out in this connection by des Cloizeaux[8].

III. Degenerate Perturbation Theory

There is now a rather extensive literature on degenerate perturbation formalisms for ordinary quantum systems. In my opinion, however, most of the worthwhile developments may be found in just three papers, namely those of Kato[9], Bloch[10], and des Cloizeaux[8]. A number of the ideas of Bloch and des Cloizeaux were presented slightly earlier, in a far less readable form, by Speisman[11]. It is also quite interesting to compare with some Japanese works on the old Tamm-Dancoff theory of the two-nucleon interaction[12]. A survey of the various formalisms has recently been given by Klein[13], but we shall see that his observations must be taken with a grain of salt.

Degeneracy leads to the following new types of problems: (1) The proper definition for the "model" state vectors which should be associated with the desired exact eigenstates is not unique. (2) The analogs of the function $F(E_o)$ and its derivations are now matrices in the model subspace, thus the order of the various factors appearing in expressions such as that resulting from (2.22) is non-trivial. (3) It is possible to express the results in terms of recognizable closed-form expressions, such as (2.22) for example, in several different ways. The equivalence of these very-different-appearing expressions is non-trivial. (4) Most of the formal strategies are considerably easier to carry out when one assumes exact degeneracy for the model subspace

$$P_o = \sum_{i=1}^{d} |\Phi_i\rangle \langle \Phi_i| = 1 - Q_o \quad . \qquad (3.1)$$

We shall therefore assume this exact degeneracy for most of the present discussion.

Kato[9] was not particularly interested in developing a practical degenerate formalism; his main concern was the formal problem of convergence. Nevertheless, his resolvent-integral approach led to very compact formal expressions for the RS perturbation expansions of the operators

$$P = \sum_{\alpha=1}^{d} |\Psi_\alpha\rangle\langle\Psi_\alpha| \tag{3.2}$$

and

$$HP = PHP \quad , \tag{3.3}$$

where the Ψ_α's of P are the eigenstates which develop out of the model subspace P_o as the interaction is switched on adiabatically. Bloch[10] observed that the P_o projections of these operators follow trivially from Kato's expressions, and that the resulting operators lead to a model-space matrix equation

$$\left[P_o PHPP_o - E_\alpha P_o P P_o\right] |\bar{\alpha}\rangle_o \tag{3.4}$$

which generates the desired exact eigenvalues E_α of the states Ψ_α. Note that although the effective hamiltonian $P_o PHPP_o$ is manifestly hermitean, it requires the use of a non-diagonal metric $P_o PP_o$. This means that the Kato-Bloch model eigenvectors $|\bar{\alpha}\rangle_o$ are not mutually orthogonal. Des Cloizeaux[8] pointed out, however, that one can use the usual elementary metric by introducing what we shall call the rationalized Kato-Bloch hamiltonian

$$\mathcal{H}_{rKB} = (P_o PP_o)^{-1/2}\, P_o PHPP_o\, (P_o PP_o)^{-1/2} \quad . \tag{3.5}$$

Since this is still manifestly hermitean, its eigenvectors (denoted by $|\hat{\alpha}\rangle_o$) are mutually orthogonal.

In a separate development, Bloch[10] observed that the ordinary BW formulas have a very straightforward degenerate analog, namely

$$\left[H_o + P_o\, \mathcal{V}(E_o + \Delta E_\alpha) - (E_o + \Delta E_\alpha)\, I\right] |\alpha\rangle_o = 0, \tag{3.6}$$

where the matrix operator

$$\mathcal{V}(E) = VP_o + V \frac{Q_o}{E - H_o} \mathcal{V}(E) \tag{3.7}$$

is formally identical to (2.14) except for the definitions of P_o and Q_o. In this case the model eigenvectors $|\alpha\rangle_o$ are the degenerate projections of the exact eigenstates Ψ_α, which Bloch denotes by $|\alpha\rangle$:

$$|\alpha\rangle_o = P_o \Psi_\alpha \equiv P_o |\alpha\rangle \quad . \tag{3.8}$$

(These $|\alpha\rangle_o$'s differ only in normalization from my $\Psi_{D\alpha}$'s.) Since they are projections of orthogonal vectors, these $|\alpha\rangle_o$'s are generally not mutually orthogonal. Nevertheless they are (presumably) linearly independent, so one can define their bi-orthogonal complements $|\bar{\alpha}\rangle_o$ which have the properties

$${}_o\langle\alpha|\bar{\beta}\rangle_o = {}_o\langle\bar{\alpha}|\beta\rangle_o = \delta_{\alpha\beta} \tag{3.9}$$

and

$$P|\bar{\alpha}\rangle_o = \sum_\beta |\beta\rangle\langle\beta|\bar{\alpha}\rangle_o = \sum_\beta |\beta\rangle\ {}_o\langle\beta|\bar{\alpha}\rangle_o = |\alpha\rangle \quad . \tag{3.10}$$

The latter result demonstrates that these $|\bar{\alpha}\rangle_o$'s are the eigenvectors of the Kato-Bloch equation (3.4). From (3.8)-(3.10) we see that

$$|\alpha\rangle_o = P_o|\alpha\rangle = P_oP|\bar{\alpha}\rangle_o = P_oPP_o|\bar{\alpha}\rangle_o \quad , \tag{3.11}$$

and thus

$$P_oPP_o = \sum_\alpha |\alpha\rangle_o\ {}_o\langle\alpha| \quad , \quad (P_oPP_o)^{-1} = \sum_\alpha |\bar{\alpha}\rangle_o\ {}_o\langle\bar{\alpha}| \quad . \tag{3.12}$$

These relations enabled des Cloizeaux to define an orthonormal basis $|\hat{\alpha}\rangle_o$ which is formally just "half way" between the bi-orthogonal bases $|\alpha\rangle_o$ and $|\bar{\alpha}\rangle_o$:

$$|\hat{\alpha}\rangle_o = (P_oPP_o)^{1/2}\ |\bar{\alpha}\rangle_o = (P_oPP_o)^{-1/2}\ |\alpha\rangle_o \quad . \tag{3.13}$$

The eigenvectors of the "rationalized" hamiltonian (3.5) are just these $|\hat{\alpha}\rangle_o$'s. [It turns out that the $|\hat{\alpha}\rangle_o$'s of (3.13) depend on the norms of the $|\alpha\rangle_o$'s. The last-mentioned result follows, via (3.8)-(3.12), from the convention that the $|\alpha\rangle$'s (Ψ_α's) have unit norms.]

To obtain an RS analog of the energy-dependent operator (3.7), Bloch derived the operator equation

$$\Omega = P_o + \frac{Q_o}{E_o - H_o}\,[V\Omega - \Omega P_oV\Omega] \quad , \tag{3.14}$$

which is the appropriate analog of (2.17). This can be solved recursively, as in (2.19)-(2.20), to generate an RS expansion for the energy-independent operator

$$\mathcal{W} = P_oV\Omega \tag{3.15}$$

(my $\mathcal{W}$ = Bloch's $\mathcal{A}$ = des Cloizeaux's h) which replaces (3.6) by

$$[H_o + \mathcal{W} - E_\alpha I]\ |\alpha\rangle_o = 0 \quad . \tag{3.16}$$

Des Cloizeaux has pointed out that $\mathcal{W}$ can also be calculated recursively by a matrix analog of the Taylor expansion method (2.21), based upon

$$\mathcal{W} = \sum_{n=o}^{\infty} \frac{1}{n!}\left[\frac{d^n}{dE_o{}^n} P_o\,\mathcal{V}\,(E_o)\right](\mathcal{W})^n \quad . \tag{3.17}$$

This equation deserves careful study, as it indicates in a very concise manner the origin of the "folded diagrams" of the many-body RS theory. The basic topological structure of the diagrams with repeated folds can easily be worked out from this equation.

The $\mathcal{W}$ operator cannot be hermitean, since the $|\alpha\rangle_o$'s are not orthogonal. Des Cloizeaux showed, however, by means of (3.13), that the transformed model hamiltonian

$$K_{dC} = H_{dC} - H_o = (P_o P P_o)^{-1/2} \, W (P_o P P_o)^{1/2} \qquad (3.18)$$

does have orthogonal eigenvectors, which are again the $|\hat{\alpha}\rangle_o$'s. This means that K_{dC} is actually a hermitean operator, although this is not at all apparent from (3.18).

Unfortunately, the expressions (3.14)-(3.17) do not provide any direct way to calculate $P_o P P_o$. We have resolved this practical difficulty in the following manner[1,14,15]. Note that

$$\langle\alpha|\beta\rangle = {}_o\langle\alpha|\Omega^\dagger \Omega|\beta\rangle_o = N_\alpha \delta_{\alpha\beta} \quad , \; N_\alpha > 1, \qquad (3.19)$$

where, merely for consistency with our previous papers, we have now reverted to the intermediate normalization convention

$${}_o\langle\alpha|\alpha\rangle_o = {}_o\langle\alpha|\alpha\rangle = 1 \quad . \qquad (3.20)$$

The orthogonality feature of (3.19) enables us to construct an orthonormal basis,

$$|\hat{\alpha}\rangle_o = (\Omega^\dagger \Omega)^{1/2} \, |\alpha\rangle_o \, N_\alpha^{-1/2} \quad , \qquad (3.21)$$

and thus a hermitized effective interaction,

$$K_B = H_B - H_o = (\Omega^\dagger \Omega)^{1/2} \, W \, (\Omega^\dagger \Omega)^{-1/2} \quad . \qquad (3.22)$$

Allowing for the different norm conventions for the $|\alpha\rangle_o$'s in (3.13) and (3.21), it turns out that the $|\hat{\alpha}\rangle_o$'s from these two expressions are identical. Furthermore, the model H's in (3.5), (3.18), and (3.22) are all formally identical.

Since these last statements are not at all obvious, we shall now give a short demonstration.[16] Using (3.19)-(3.21), we see that

$$P_o P P_o = \sum_{\gamma=1}^{d} P_o|\gamma\rangle N_\gamma^{-1} \langle\gamma|P_o = \sum_{\gamma=1}^{d} |\gamma\rangle_o \, N_\gamma^{-1} \, {}_o\langle\gamma|$$

$$= \sum_{\gamma=1}^{d} (\Omega^\dagger \Omega)^{-1/2} \, |\hat{\gamma}\rangle_o \, {}_o\langle\hat{\gamma}| (\Omega^\dagger \Omega)^{-1/2}$$

$$= (\Omega^\dagger \Omega)^{-1} \quad . \qquad (3.23)$$

From (3.19) it also follows that the bi-orthogonal complements of the present $|\alpha\rangle_o$'s (our $\Psi_{D\alpha}$'s) are

$$|\bar{\alpha}\rangle_o = \Omega^\dagger \Omega \, |\alpha\rangle_o \, N_\alpha^{-1} \quad , \qquad (3.24)$$

and thus

$$P_o P H P P_o = \sum_{\gamma=1}^{d} P_o H|\gamma\rangle \, N_\gamma^{-1} \, {}_o\langle\gamma|$$

$$= \sum_{\gamma=1}^{d} P_o H|\gamma\rangle \, {}_o\langle\bar{\gamma}| (\Omega^\dagger \Omega)^{-1}$$

$$= P_o H\Omega\, (\Omega^\dagger\Omega)^{-1} \quad . \qquad (3.25)$$

The identities (3.23) and (3.25) confirm the full formal equivalence of H_{rKB}, H_{dC}, and H_B.

Critique

We now have a variety of final RS expressions to choose between, namely (3.4), (3.5), (3.16), (3.18), and (3.22). How do these compare from the practical standpoint? Bloch[10] has given explicit formulas for the number of terms of nth order in the RS expansions of the operators $P_o PHPP_o$, $P_o PP_o$, and W, showing that W is considerably simpler than either of these other operators. Rather surprisingly, the most complicated of these operators is $P_o PP_o$. For example, in 4th order the RS expansions involve 10, 15, and 5 terms respectively. Bloch also applied the corresponding eigenvalue recipes, (3.4) and (3.16), truncated to the same order, to a numerically solvable problem (the Mathieu equation). The W form (3.16) was found to be more accurate for all orders and parameter choices studied. Furthermore this approach is "more physical", in view of (3.8). It therefore appears that the method (3.4) does not merit serious consideration as a practical tool. It appears to us that the lack of hermiticity of W is a small price to pay for the great increase in simplicity, as compared to (3.22) for example.

On the other hand, (3.22) has one redeeming virtue. By means of its orthonormal eigenvectors $|\hat{\alpha}\rangle_o$, which satisfy (3.21), the transition matrix elements of an arbitrary operator O can be expressed as

$$\langle\alpha|O|\beta\rangle = \frac{{}_o\langle\alpha|\Omega^\dagger O\Omega|\beta\rangle_o}{({}_o\langle\alpha|\Omega^\dagger\Omega|\alpha\rangle_o\ {}_o\langle\beta|\Omega^\dagger\Omega|\beta\rangle_o)^{1/2}}$$

$$= {}_o\langle\hat{\alpha}|m(O)|\hat{\beta}\rangle_o \qquad (3.26)$$

where the effective operator has the relatively convenient form

$$m(O) = (\Omega^\dagger\Omega)^{-1/2}\, \Omega^\dagger O\Omega(\Omega^\dagger\Omega)^{-1/2} \quad . \qquad (3.27)$$

This result was pointed out by Bulaevski[17] as well as myself, and I have demonstrated that its many-body expansion is fully linked[1]. To choose between the formally equivalent expressions (3.5), (3.18), and (3.22), one should examine whether $\Omega^\dagger\Omega$ or $P_o PP_o$ has the simplest expansion. In view of the fact that

$$\Omega^\dagger\Omega = P_o + \Omega^\dagger Q_o\Omega \equiv P_o + \Theta, \qquad (3.28)$$

whereas $P_o PP_o = (\Omega^\dagger\Omega)^{-1}$, it is clear that the former operator is simplest, and thus that our (3.22) is the preferred form. For actual calculation, however, we have suggested that it may be more satisfactory to replace (3.22) by its explicitly hermitized average, and then carry out a formal expansion in powers of the operator Θ, thus

$$1/2 \left[(\Omega^\dagger\Omega)^{1/2} \mathcal{W}(\Omega^\dagger\Omega)^{-1/2} + (\Omega^\dagger\Omega)^{-1/2} \mathcal{W}^\dagger(\Omega^\dagger\Omega)^{1/2} \right] \quad (3.29)$$

$$= 1/2 \, [\mathcal{W} + \mathcal{W}^\dagger] + \text{terms of order } \Theta.$$

Truncation at order Θ^n then gives a manifestly hermitean result, even for n=o.

Incidentally, (3.26)-(3.27) leads to still another representation[14] for the hermitized effective interaction, namely

$$\mathcal{H} = m(H_o) + m(V), \quad (3.30)$$

whose eigenvectors are obviously the $|\hat{\alpha}\rangle_o$'s. The formal equivalence to the previous expressions can again be demonstrated by means of (3.23) and (3.25). This expression corresponds to a "true" rather than a "model" description[1,14]. We have argued elsewhere[14,18] that the "model" description [(3.16) or (3.18)] is more appropriate for systems with a hard-core interaction.

Other Approaches

Historically, the earliest paper which led to a systematic treatment of degenerate perturbation theory appears to be that of Van Vleck[19] in 1929. A more complete exposition of his method appeared in a paper by a student of his named Jordahl[20]. The basic idea is to carry out a unitary transformation on the full hamiltonian,

$$\mathcal{H}_{VV} = U^\dagger HU , \quad (3.31)$$

where this U is required to block-diagonalize $\mathcal{H}_{VV}$:

$$Q_o \mathcal{H} P_o = P_o \mathcal{H} Q_o = 0 \quad (3.32)$$

This requirement does not determine U uniquely, but it is natural to add the requirement that U should have as little effect within Q_o and P_o as possible, consistent with (3.32). Kemble[21] pointed out the convenience of expressing U in an exponential form, say

$$U = e^G, \quad (3.33)$$

so that the unitary condition becomes

$$G^\dagger = -G, \quad (3.34)$$

and the somewhat vague "minimal effect" requirement can be formulated as

$$P_o G P_o = Q_o G Q_o = 0 \ . \quad (3.35)$$

G is now fully defined. It can be obtained by writing G as a formal expansion in powers of V, and using (3.32) as a recursion relation. This general program has been pursued by Primas[22], Klein[13], and Kvasnicka[23]. This method is straightforward in principle, but it turns out to be tedious in practice, for the following reason.

Comparison with the preceding discussion suggests that the eigenvectors of (3.31) will be the $|\hat{\alpha}\rangle_o$'s and that the unitary transformation amounts to

$$U|\hat{\alpha}\rangle_o = \Omega(\Omega^\dagger\Omega)^{-1/2}\,|\hat{\alpha}\rangle_o \quad . \tag{3.36}$$

The known structure of the RS expansion for the right-hand side [see (3.15), (3.17), (3.28)] shows that it is "unnatural" to try to express this in an exponential form. Nevertheless, it is noteworthy that Klein was able to demonstrate that the $\mathcal{H}_{VV}$ of (3.31) is formally identical to $\mathcal{H}_{rKB}$ (and thus to the other $\mathcal{H}$'s discussed above). He did this by showing, in essence, that in both cases the set of orthonormal eigenvectors $|\hat{\alpha}\rangle_o$ maximizes the quantity $\Sigma_\alpha |\langle\alpha|\hat{\alpha}\rangle_o|^2$, and that this variational property leads to a <u>unique</u> orthonormal basis $|\hat{\alpha}\rangle_o$. This also confirms the identification (3.36).

We have now seen that the hermitean effective hamiltonian can be expressed in a number of formally equivalent ways. The equivalence of these different expressions is not at all obvious, although it can be surmised from the fact that none of these approaches have introduced any unnecessary unitary transformations U_o within the P_o subspace. The equivalence of the various expressions can be checked explicitly in low orders, by expanding everything in powers of V and collecting like terms, but this is very tedious. The resulting "bare" RS expansion in V does not exhibit any simple regularities (other than manifest hermiticity), and thus is rather unenlightening. We believe that the grouping or partial summation of terms into recognizable expressions, such as Ω, $\mathcal{W}$, $\Omega^\dagger\Omega$ etc., offers the only prospect for practical applications involving high (perhaps infinite) orders of perturbation theory. From this standpoint the above unitary transformation method does not appear to be useful, since no simple regularities are directly discernable within the various higher-order terms of G.

Another approach which should be mentioned is that of Hirschfelder and Silverstone and their collaborators[24]. Here the idea is to develop <u>each</u> of the eigenvalues and eigenfunctions as a <u>strict</u> Taylor series in V. They begin by solving the <u>first-order</u> secular equation

$$P_o\,[\,H_o + V\,]\,P_o\Psi_\alpha^{(1)} = E_\alpha^{(1)}\,\Psi_\alpha^{(1)} \quad , \tag{3.37}$$

and then making successive corrections to $E_\alpha^{(1)}$ and $\Psi_\alpha^{(1)}$. (This assumes that all of the degeneracy is lifted in first order. Their procedure becomes more complicated when some of the degeneracy is lifted in higher orders.) In other words they diagonalize first, and then perturb. We feel that this is an unwise way to proceed. The formalism becomes quite complicated, there is no simple relation between $\Psi_\alpha^{(1)}$ and the desired eigenstate Ψ_α, and one does not end up with any "effective hamiltonian". Furthermore, some test calculations by Gershgorn and

Shavitt[25] indicate that this approach gives lower accuracy than when one calculates an effective hamiltonian to the same order and then diagonalizes.

In a more positive vein, we would like to mention a very recent work of Lindgren[26]. He begins by left-multiplying the Schroedinger equation by Ω ,

$$\Omega(H_o + V)\Psi_\alpha = E_\alpha \Omega\Psi_\alpha = E_\alpha \Psi_\alpha \quad , \tag{3.38}$$

where $\Omega = \Omega P_o$ has been used, and then subtracting this from the original Schroedinger equation to eliminate the eigenvalue E_α. The result is the operator equation

$$[\Omega, H_o] = V\Omega - \Omega V\Omega, \tag{3.39}$$

which is similar to but more general than the Bloch equation (3.14). This can also be solved recursively. As compared to (3.14), the advantage is that this leads very directly to an expansion for systems where H_o has more than one eigenvalue within P_o, i.e. for quasi-degeneracy. A different but closely-related treatment of quasi-degeneracy has been presented by Kvasnicka[23]. Our approach[1] to the problem of quasi-degeneracy has been to add a degeneracy-breaking term

$$V_{db} = \sum_{i=1}^{d} |\Phi_i\rangle (E_i - \bar{E}_o) \langle \Phi_i| \tag{3.40}$$

to the "true" perturbation V, and then to formally sum out the effect of this V_{db} to all orders, using the exact-degeneracy results as the starting point.

Finally, we should mention a work of Bulaevski[17], based on the time-dependent approach of Morita[27]. In my opinion, his main accomplishment was to obtain an expression equivalent to (3.27). He tried but failed to obtain a linked-cluster form of (3.22), for magnetic insulator systems where it is appropriate to define H_o to explicitly include part of the two-body Coulomb interaction. (This contrasts with most many-body formalisms, where H_o is taken to be a purely one-body operator.) A solution of this particular problem is given in Ref. 16.

Applications

The idea of an effective hamiltonian has been found appealing in a number of other fields besides nuclear physics. For example, Jordahl[20], Pryce and Abragam[28], Griffith[29], and Soliverez[30], among others, have used degenerate perturbation theory to obtain effective spin hamiltonians for isolated magnetic ions within crystal lattices. The work of Pryce and Abragam, in particular, has made this a standard tool of paramagnetic resonance theory. Kramers[31] and Anderson[32] have also used degenerate perturbation theory (but not with formal precision; the terms with a "folded" structure were overlooked) to explain the long range "superexchange" type Heisenberg spin couplings between ions in magnetic insulator materials. I have recently used my degenerate many-body expansion to tidy up the weak points of their formalism[16]. (This has led to a general formal justification for the existence of effective spin hamiltonians in macroscopic magnetic insulator systems--a justification which, for the first time, does not run afoul of any aspect of the "non-orthogonality catastrophe".) Incidentally, the idea of a spin hamiltonian is probably the oldest example of the effective hamiltonian concept--this dates back to the original "molecular field" idea of Weiss[33]. The work of Feshbach[3] in nuclear reaction theory is well known, although this is not perturbative. Quantum chemists have applied Van Vleck's method quite extensively in studies of rotation-vibration coupling in molecules (see Klein for references). This "canonical transformation" method is also widely scattered throughout the physics literature in miscellaneous applications, whereever it has been found desirable to eliminate certain degrees of freedom. The Bohm-Pines[34] treatment of electron-gas correlations is merely one example, as is also the Tamm-Dancoff theory[12] of the two-nucleon interaction. Quantum chemists are now beginning to recognize the utility of the degenerate many-body perturbation formalism for studies of atomic and molecular properties.[35]

IV. Degenerate Many-Body Theory

Our main reason for focussing on the various RS versions of degenerate perturbation theory, in the previous section, is because the RS form of perturbation theory is needed to obtain linked-cluster results. This was first pointed out by Brueckner[4] in connection with the closed-shell problem, but it is equally true for open shells.

The first paper referring to the linked-cluster properties of a (quasi) degenerate system is probably that of Okubo (1954).[12] He showed, in connection with an energy-independent (RS) version of the Tamm-Dancoff theory, that the diagrams with "valence interactions" overlapped by vacuum fluctuations are cancelled by similar diagrams with different relative positions of these disconnected parts. This was shown explicitly in fourth order, and a sketchy argument was given to show that this "core-valence separation" should be a general phenomenon. It is not clear whether any of the later many-body investigators were influenced by this field-theoretic work.

The first complete proof of the core-valence separation (absence of diagrams containing both valence-particle interactions and disconnected core or "vacuum fluctuation" parts) was given by Bloch and Horowitz (1958),[36] by using an ingenious combination of a thermodynamic analog of the time-development technique of Goldstone[6], and the resolvent-integral technique of Hugenholtz.[37] They showed that the total interaction energy separates cleanly into two terms,

$$\Delta E = \Delta E_c + \Delta E_v \quad , \tag{4.1}$$

where the core part ΔE_c is given by Goldstone's result for the closed shells, evaluated exactly as if all the valence particles (or valence holes) had been physically removed. Their valence part of the interaction energy is given by a "valence" analog of the secular equation (3.6), where the diagrams defining the $P_o \mathcal{V}_v P_o$ matrix elements $\langle \Phi_i | \mathcal{V}_v | \Phi_j \rangle$ have all of their interactions connected (directly or indirectly) to the external valence lines. This is altogether a very elegant paper. It is especially remarkable when one considers that this appeared only a year after Goldstone's work. Their result does have some limitations, however. The valence portion of their final recipe has a BW instead of RS character: the valence interaction energy ΔE_v appears in all of the energy denominators within the $\mathcal{V}_v$ matrix elements, and, correspondingly, the valence diagrams (which make up $\mathcal{V}_v$) do not all form a single connected piece. There are unlinked or disconnected valence diagrams such as

(4.2)

Unfortunately, this diagram must be interpreted physically as an effective <u>four</u>-body interaction term, contrary to one's intuition, when using the Bloch-Horowitz formalism.

Proceeding in historical sequence, the next papers to attempt to derive a linked-cluster result for open-shell systems are those of Primas (1961-3)[22]. His argument was accepted uncritically and reproduced in somewhat more detail in the previously mentioned review by Klein.[13] Primas employed the unitary transformation technique, using the exp(G) representation (3.33) for U. His "proof" consists of the argument that the $\mathcal{H}$ of (3.31) must be fully linked because the G operator has this property. However, his expression for the second-order contribution to G can easily be seen to include the unlinked diagram

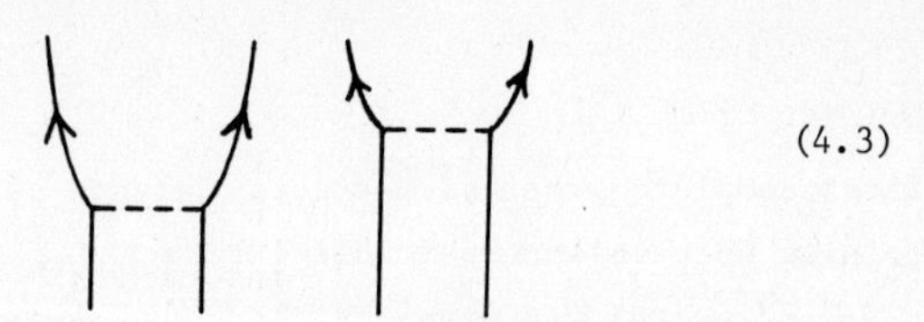

for a system with four valence particles, thus his "proof" fails already in the lowest non-trivial order. Primas relied entirely on what he referred to as the "Lie-group structure" of his algebraic expressions, and he claimed it to be an advantage that he did not use any diagrams. The moral of this story should be obvious.

For those with more knowledge of many-body theory, it must be acknowledged that there is some temptation to look for an exponential form for the transformation U. This is because it is known that the exact wavefunction for a closed-shell system can be expressed in the form

$$\Psi_o = e^W \Phi_o \quad , \tag{4.4}$$

where W consists entirely of linked diagrams.[1,37,38] But there are some important differences here. This W is a second-quantized operator, in the sense that every particle or hole line eminating from the top of the diagrams carries a one-body creation or annihilation operator. The G of (3.33) is merely an ordinary quantum-mechanical operator. Besides, the above transformation exp(W) cannot be unitary because it is not norm-preserving.

Returning to the Bloch-Horowitz result, one may reasonably surmise that the removal of ΔE_v from all of the valence diagram energy denominators should somehow eliminate the unlinked valence diagrams such as (4.2). The first attempt to demonstrate this was by Morita (1963).[27] He used an adiabatic switching technique which differed from Goldstone by starting at a finite negative time ($t < 0$), such that

$$[H_o, U_\alpha(0,t)] = U_\alpha(0,t)V_\alpha(t) - V_\alpha(0)U_\alpha(0,t) - i\alpha\lambda \frac{\partial}{\partial\lambda} U_\alpha(0,t) \quad , \tag{4.5}$$

where

$$V_\alpha(t) = e^{iH_o t} \lambda V e^{-iH_o t} e^{\alpha t} \quad , \quad \lambda=1 \quad . \tag{4.6}$$

Actually, he substituted into (4.5) the relation

$$U_\alpha(0,t)\, V_\alpha(t) = i \frac{\partial U_\alpha(0,t)}{\partial t} \quad . \tag{4.7}$$

[It may be worth noting that, with only a trivial modification of his basic strategy, his algebraic steps could have been considerably simplified by setting $t = -\infty$ at the outset, so that all explicit time derivatives resulting from (4.7) could have been avoided. His notation is a further source of confusion, since he has translated his time origin by $-|t|$, thus $t_i \to t_i + |t|$ and $\exp(\alpha t_i) \to \exp\, \alpha(t_i - |t|)$.]

Morita's main result is the factorization

$$U|\Phi_i\rangle = \sum_j \{U|\Phi_i\rangle\}_Q \langle\Phi_i|U|\Phi_i\rangle \quad , \quad i \ \& \ j \text{ in } P_o \ , \tag{4.8}$$

where $\{U|\Phi_i\rangle\}_Q$ has no singular energy denominators as $\alpha \to 0$. Combining this with (4.5)-(4.7) and taking the limits $|t| \to \infty$, $\alpha \to 0$ leads to a secular equation of the form (3.16), with the core terms removed as in Bloch and Horowitz, and with no unlinked valence diagrams such as (4.2). Unfortunately, Morita did not fully prove (4.8). His argument involved a diagrammatic analysis in which the concept of folded diagrams was introduced (although not by that name). His topological characterization of these folded diagrams was, however, far too vague to serve either as a proof or as a basis for calculations.

The first correct diagrammatic representation (and thereby a justification) for the result (4.8) was given in Ref. 1, most of which actually dates from 1965[39]. We showed there (pp. 789-91, also topological rules on pp. 792-3) a proper and systematic way to introduce the folded diagrams within the adiabatic context of Morita, even though we did not actually use the adiabatic approach to derive our main results. [Within our time-independent context, the folded diagrams were generated by a method equivalent to (3.17).] The topological properties and diagrammatic rules for the folded diagrams were spelled out in full detail, and a general proof was given for the cancellation of unlinked valence diagrams. Fully-linked results were obtained for the model-space operators (3.15), (3.22), and (3.27). (It should be noted, however, that the operator Θ, by itself, is not fully linked.)

The next publication in this field is that of Sandars (1969).[40] The basic strategy of his derivation is identical to that of Baker.[5] Sandars' discussion of the closed-shell case is fine, but it is seriously inadequate for open shells. He introduces the concept of "backwards" (= folded) diagrams, but he does not seem to be aware of the need for diagrams containing more than one fold. There is no discussion of the existence or nonexistence of unlinked valence diagrams such as (4.2). In common with some other authors, he uses the term "linked" to mean simply the absence of "vacuum fluctuations," thus (4.2) would be described as "linked but disconnected." (One should be aware that the claim of a "completely linked result"[41] sometimes means no more than the core-valence separation obtained by Bloch and Horowitz.)

The next paper is that of Oberlechner, Owono-N'-Guema, and Richert (1970).[42] These authors follow Morita quite closely as to general strategy (though with clearer notation), again keeping t finite until the end. Their diagrammatic justification of (4.8) follows the procedure described in Ref. 1. This paper can be characterized most simply as "Morita done right."

We come now to Johnson and Baranger (1971).[43] I'm rather at a loss to characterize this paper, because everything about it is so different from the rest of the literature. It is certainly a highly original work. Whereas in the RS context we encountered a non-hermitean W form, and a hermitean (H or K) form appearing in several disguises, one now finds an infinity of possibilities. The "time bases" of their diagrams are not determined by the formalism itself, with the consequence that these may be specified afresh in each order of the perturbation expansion. This amorphous nature of the formalism is somewhat disturbing as it currently stands, although it holds forth the prospect of a number of new and potentially useful general structures. Unfortunately, there has not yet been a thorough analysis of any of these new general structures.

The Kuo-Lee-Ratcliff work (1971)[44] follows the general adiabatic approach of Morita and Oberlechner et al., with the following differences: (1) In common with Goldstone, "t" is set equal to $-\infty$ at the outset. This is a welcome simplification. (2) Their justification of a secular equation analogous to (3.16) is very similar to the argument used by Bloch and Horowitz. This is considerably less direct than the procedure of Morita, which is based on the observation that, as long as the adiabatic parameter α remains finite, the matrix $P_o U_\alpha P_o$ exists and has a well-defined inverse. A nice feature of Kuo et al. is their demonstration that the entire expansion for $U(0, -\infty)|\Phi_i\rangle$, i in P_o, can be cleanly decomposed into three factors, thus

$$UP_o = (\text{valence diagrams}) \times (\text{core vacuum} + \text{core excitations}) \times (1 + \text{vacuum fluctuations}) \quad . \qquad (4.9)$$

The first of these factors is then decomposed as in (4.8), following the folded diagram procedure of Refs. 1 and 42. Kuo et al. claim to have shown also that the folded diagrams eliminate all unlinked valence terms such as (4.2).

In some related papers, Krenciglowa et al.[45] have investigated a possible partial summation for the folded-diagram series, namely, the "Q-box summation". This amounts to running my time-independent derivation backwards [see (7.2-(7.8) of Ref. 1], thus converting the fully-linked valence expansion back into the Bloch-Horowitz result. We regard this as a step backwards, because of the presence of diagrams such as (4.2). Even in the case of just two valence particles--the famous ^{18}O example--this has the unfortunate effect of entangling the 1-body (^{17}O removal) and 2-body (effective interaction) energies, so all that the formalism allows one to do now is to compute the differences between the ^{18}O levels and the ^{16}O ground state. The clean identification of the two-body-interaction component has been lost. Nevertheless, there is a compromise approach by which one can salvage the good aspects of the "Q-box summation" idea. One can sum selected infinite subsets of the fully-linked $\mathcal{V}$ diagrams into BW-like expressions for appropriately-chosen subsystems of the valence-particle system. We have discussed this idea in (8.4)-(8.6) and page 811 of Ref. 1.

The latest paper claiming to derive a linked-cluster result for open-shell systems is that of Lindgren (1974).[26] This is notable mainly for the innovation (3.39). The discussion of the many-body aspects is closely parallel to that of Sandars. Again there is no mention of the existence of diagrams containing more than one fold. The effective interaction is, however, shown to be fully linked, thus eliminating (4.2).

Valence Diagram Reduction

I'd like to close this survey with a discussion of what is perhaps the most obscure and difficult part of Ref. 1, the business of "reducing" the Bloch-Horowitz (BH) diagrams before folding them together to generate the fully-linked RS expansion for $\mathcal{W}$. The basic problem is that the separation of the core and valence aspects is not completed by the elimination of the core vacuum fluctuations. The set of BH diagrams contains many instances of core-particle excitations which ought, according to nuclear-matter experience, to be factorized and rendered independent of the ΔE_v which enters in all of the BH energy denominators. In addition, there are "projecting core excitations" which tie together two or more BH-like valence-excitation blocks, thus forming a composit diagram which fulfills the BH requirement of having no intermediate states within P_o. Some examples are

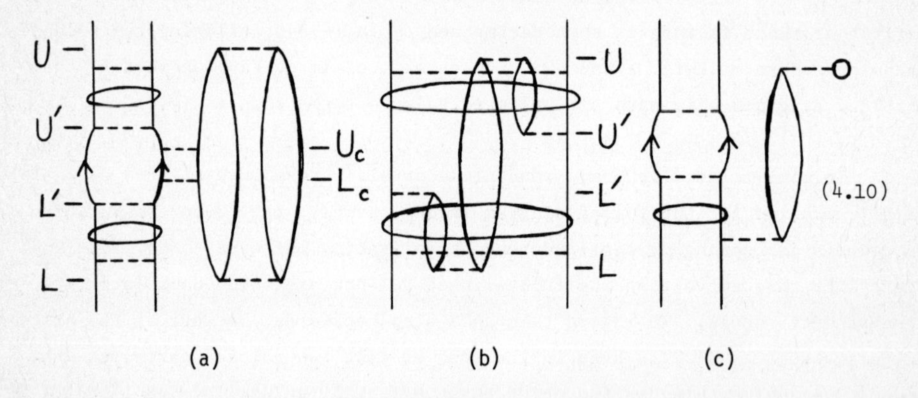

where the horizontal loops indicate levels which would be in P_o, were it not for the presence of overlapping core excitations. We became concerned about this problem because of our interest in the true single-particle occupation probabilities for the core orbitals,[14,15,18] and also through an attempt (initially unsuccessful) to demonstrate the full formal equivalence of the one-valence-particle diagrams to the usual field-theoretic formula for the "quasi-particle" energy. This problem is also of vital concern for the core-phonon-exchange diagrams, as seen in (4.10b).

The valence-interaction "insertions" seen near the bottoms of diagrams (a) and (b) will obviously be cancelled when one forms the usual folded-diagram series. That is, the removal of ΔE_v from the energy denominators between the levels L and L′ just compensates the effect of all possible valence-interaction insertions between these levels. This cancellation is easily demonstrated in both the time-independent and time-dependent approaches, as has been recognized also by Oberlechner et al. and Kuo et al. Thus the "downwards projecting core excitations" are easily dealt with.

In our "reduced BH expansion", which is intermediate in character between the original BH expansion and the RS expansion for $\mathcal{W}$, the rule is that between the levels L and L′ there are no valence insertions and no ΔE_v's in the denominators. This simplification can be carried a step further: the entire downwards projecting core excitation, which begins at L_c in (4.10a), can now be factorized and thus rendered completely independent of the valence-particle system. By carrying out this reduction before folding the BH diagrams to generate $\mathcal{W}$, these simplifications lead to the convenient rule that the "bottom folding level" for this diagram is now L′ instead of L. When constructing the folded diagrams, each downwards projecting core excitation should be thought of as merely a numerical factor which acts "instantaneously" (i.e., it has no vertical extension) at the level L_c. This new aspect of the core-valence separability is not at all surprising--indeed it is almost intuitively obvious--but it is nice to have this clearly formulated.

As you may expect by now, it is possible to obtain a completely analogous set of results for the "upwards projecting core excitations", where the characteristics of levels U, U′, and U_c are entirely similar to those of L, L′, and L_c. We were able to demonstrate this by reducing the tops (as well as the bottoms) of the BH diagrams before folding these "$\mathcal{V}_v$ blocks" to generate the RS diagrams for $\mathcal{W}$. This point is quite non-trivial, because it does not seem possible to achieve this "top reduction" in any simple manner within the usual time-dependent adiabatic context. To test this, I worked out the adiabatic integral for diagram (4.10c) and its generalized time orderings, and found a messy result with no obvious interpretation. (I recommend this exercise to the devotees of the Goldstone approach.) The problem is that the time limits of U, namely 0 and $-\infty$, are not equivalent. Tastes may differ about which type of derivation is simplest and most straightforward, but the fact remains that none of the other approaches have led to as complete an analysis of the valence diagrams (with the resulting simplifications) as in Ref. 1.

There was, however, one inadequacy in the treatment of reduction given in Ref. 1 (see end of page 784). This relates to the above-mentioned problem of demonstrating the full formal equivalence of the single-valence-particle diagrams to the usual mass-operator prescription for the quasi-particle energy--the Hugenholtz-Van Hove theorem. This equivalence has now been worked out in detail.[46] Furthermore, some connections with the "strength of the quasi-particle pole" of the one-body causal Green's function have also been worked out.[14]

. . . .

References

* Work performed under the auspices of the United States E.R.D.A.

1) B. H. Brandow, Rev. Mod. Phys. 39, 771 (1967).
2) P.-O. Löwdin, J. Math. Phys. 3, 969 (1962), and references therein.
3) H. Feshbach, Ann. Phys. (N.Y.) 19, 287 (1962).
4) K. A. Brueckner, Phys. Rev. 100, 36 (1955).
5) G. A. Baker, Jr., Rev. Mod. Phys. 43, 479 (1971).
6) J. Goldstone, Proc. Roy. Soc. (London) A239, 267 (1957).
7) E. T. Whittaker and G. N. Watson, Modern Analysis (Cambridge University Press, 1952), p. 133.
8) J. des Cloizeaux, Nuclear Phys. 20, 321 (1960).
9) T. Kato, Prog. Theoret. Phys. 4, 514 (1949).
10) C. Bloch, Nuclear Phys. 6, 329 (1958).
11) G. Speisman, Phys. Rev. 107, 1180 (1957).
12) N. Fukuda, K. Sawada, and M. Taketani, Prog. Theoret. Phys. 12, 156 (1954); S. Okubo, ibid. 12, 603 (1954).
13) D. J. Klein, J. Chem. Phys. 61, 786 (1974).
14) B. H. Brandow, in Lectures in Theoretical Physics, Vol. XI B (Boulder, 1968) (Gordon and Breach, N. Y., 1969).
15) B. H. Brandow, Ann. Phys. (N.Y.) 57, 214 (1970); see Sec. IV and Appendices A - C.
16) B. H. Brandow, submitted to Phys. Rev. B.
17) L. N. Bulaevski, Soviet Phys.-JETP 24, 154 (1967).
18) B. H. Brandow, Phys. Rev. 152, 863 (1966).
19) J. H. Van Vleck, Phys. Rev. 33, 467 (1929).
20) O. M. Jordahl, Phys. Rev. 45, 87 (1934).
21) E. C. Kemble, The Fundamental Principles of Quantum Mechanics (McGraw-Hill, New York, 1937) p. 394.
22) H. Primas, Helv. Phys. Acta 34, 331 (1961); H. Primas, Rev. Mod. Phys. 35, 710 (1963).
23) V. Kvasnička, Czech. J. Phys. B 24, 605 (1974).
24) J. O. Hirschfelder, Int. J. Quantum Chem. 3, 731 (1969); H. J. Silverstone, J. Chem. Phys. 54, 2325 (1971); H. J. Silverstone and T. T. Holloway, Phys. Rev. A 4, 2191 (1971); J. O. Hirschfelder and P. R. Certain, J. Chem. Phys. 60, 1118 (1974).
25) Z, Gershgorn and I. Shavitt, Int. J. Quantum Chem. 2, 751 (1968).
26) I. Lindgren, J. Phys. B 7, 2441 (1974).
27) T. Morita, Prog. Theoret. Phys. 29, 351 (1963).
28) M. H. L. Pryce, Proc. Phys. Soc. (London) A63, 25 (1950); A. Abragam and M. H. L. Pryce, Proc. Roy. Soc. A205, 135 (1951).
29) J. S. Griffith, The Theory of Transition Metal Ions (Cambridge U. Press, 1961).
30) C. E. Soliverez, J. Phys. C 2, 2161 (1961).
31) H. A. Kramers, Physica 1, 182 (1934).
32) P. W. Anderson, Solid State Physics 14, 99 (1963).
33) P. Weiss, J. de Physique (4) 6, 661 (1907).
34) D. Bohm and D. Pines, Phys. Rev. 92, 609 (1953).
35) U. Kaldor, Phys. Rev. Letters 31, 1338 (1973); S. Garpman, I. Lindgren, J. Lindgren, and J. Morrison, Phys. Rev. A 11, 758 (1975).
36) C. Bloch and J. Horowitz, Nucl. Phys. 8, 91 (1958).
37) N. M. Hugenholtz, Physica 23, 481 (1957).

38) J. Hubbard, Proc. Roy. Soc. (London) A240, 539 (1957).
39) B. H. Brandow, in Proceedings of the International School of Physics "Enrico Fermi", Course 36 (Varenna, 1965) edited by C. Bloch (Academic Press, New York 1966). Preprints containing most of Ref. 1 were widely circulated in 1965.
40) P. G. H. Sandars, Adv. Chem. Phys. 14, 365 (1969).
41) See for example E. M. Krenciglowa and T. T. S. Kuo, Nuclear Phys. A240, 195 (1975).
42) G. Oberlechner, F. Owono-N'-Guema, and J. Richert, Nuovo Cimento B68, 23 (1970).
43) M. B. Johnson and M. Baranger, Ann. Phys. (N.Y.) 62, 172 (1971).
44) T. T. S. Kuo, S. Y. Lee, and K. F. Ratcliff, Nuclear Phys. A176, 65 (1971); Ratcliff, Lee, and Kuo (unpublished).
45) E. M. Krenciglowa, T. T. S. Kuo, E. Osnes, and B. Giraud, Phys. Letters 47B, 322 (1973); E. M. Krenciglowa and T. T. S. Kuo, Nuclear Phys. A235, 171 (1974).
46) B. H. Brandow, Ann. Phys. (N. Y.) 64, 21 (1971); Appendix C.

B.H. BRANDOW: TIME INDEPENDENT APPROACH

Negele: You mentioned Bloch had checked the three forms of hermitized and non-hermitian expansions on some problem, and found the form you prefer the most accurate. What was the test problem?

Brandow: Bloch only tested the two forms discussed in his article, namely the Kato-Bloch form, which is hermitian but with a non-diagonal metric, and what I call the $\mathcal{W}$ form, which is non-hermitian with a simple metric. The test was actually a non-degenerate (d=1) problem, namely the Mathieu equation. This showed that the $\mathcal{W}$ method was more accurate for all orders and interaction strengths studied.

Kümmel (to Negele's question): Is it not true that usually the non-hermiticity is quite small? Or do you know of a case where this is not true?

Brandow: I agree that this is generally true for the eigenstates that one is interested in. But shell model calculations frequently use model spaces much larger than the number of interesting low-lying eigenstates. The convergence of the general scheme is undoubtedly much worse for some of the higher eigenstates obtained from these calculations. In the $\mathcal{W}$ scheme these questionable eigenstates are effectively decoupled from the "good" eigenstates, but in the $\mathcal{K}$ scheme which involves the entire matrix Θ, this is no longer true. I feel that this question deserves some study. It is certainly conceivable that this mixing could cause problems. However, my main reason for contrasting the hermitian and non-hermitian formulations was to emphasize that their perturbative expansions are quite different. To answer your question more specifically, I'd like to mention that the RPA theory can be expressed in the $\mathcal{W}$ context -- this is shown in my Rev. Mod. Phys. article. There the eigenvectors are just the "X" vectors, and one knows that these are not orthogonal.

Koltun: What is the specific problem with the non-factorable, reducible diagrams you mentioned? They can, in fact, be calculated in each order. Also, reducibility does not guarantee better convergence, does it?

Brandow: Nothing is guaranteed in this messy problem of convergence, but whenever there are fewer diagrams that need to be calculated, I think this constitutes a step in the right direction. My "valence diagram reduction" is a natural extension of the generalized time order factorization idea, and I think everyone would agree that the latter is a good thing to do.

Vincent: Can you please explain why unlinked valence diagrams are a bad thing? Or perhaps this is such an old question that you did not think of mentioning it?

Brandow: Sorry I overlooked this. There are several reasons why it is nice to have a fully linked result. There is the practical matter of having far fewer diagrams to calculate. Formally, this leads to much more straightforward physical interpretations. And in cases where there are a large number of valence particles, it saves one from the sort of pathology found in the unlinked BW treatment of the closed-shell problem -- denominators spuriously large because they contain the energy shift ΔE which is proportional to the number of particles.

Kümmel: If I remember correctly there is a paper by the chairman of this session dating back to 1958 where the linked cluster theorem has been proven without use of perturbation theory.

Brandow: Yes, of course, but that was only for the non-degenerate case.

Kümmel: I claim that it can be proven as well for the degenerate cases. One more remark: I believe that bookkeeping in perturbation theory is not simpler than in some other theories.

Brandow: Your remark gives me the opportunity to expound on one of my favourite prejudices. I feel that many-body theory is not just a science, it is an art. There are many different formalisms on the market which are rigorous and in some sense systematic. The art lies in selecting the most appropriate formalism and adapting it to the essential physics of the problem at hand. There is a psychological problem here, because every formalism carries some aesthetic appeal, at least among its devotees. The structure of each formalism tends to suggest that certain types of approximations are most reasonable or "natural", regardless of the actual quantitative characteristics of the physical system. The many-body literature is full of papers where people have tried to make the physics fit into their preconceived approximation schemes, instead of vice versa. One of the reasons I like perturbation theory is that it seems to be the "least biased" of all the many-body techniques. When you run out of insight, you can always start drawing and calculating diagrams -- this is a fully systematic procedure. In this way one can discover what are the dominant physical considerations for the particular system. This approach has frequently been very fruitful. Some good examples are Bethe's unpublished estimates of certain 4th order diagrams which led him to apply Faddeev's three-body scattering theory to nuclear matter, and the Barrett-Kirson calculations which have led to some quite unexpected insights into the effects of core polarization. Of course, once you have discovered the dominant effects, you will want to treat them by more efficient methods. But after you have managed to do this somehow, you can always go back to perturbation theory to discover what needs to be done next. History has demonstrated that perturbation theory is the most versatile bookkeeping system we have.

FOLDED DIAGRAM THEORY, TIME-DEPENDENT APPROACH OF JOHNSON AND BARANGER

Mikkel B. Johnson*
Meson Physics Division
University of California
Los Alamos Scientific Laboratory
Los Alamos, New Mexico 87544

The folded diagram expansion found by Brandow[1] and extensively developed by him using time-independent methods has been subsequently explained from several points of view[2,3,4] on the basis of time-dependent quantum mechanics, the framework in which the expansion was originally conceived by Morita.[5] This talk is intended to be an advertisement for the point of view taken in Ref. 2, and to be a review of the methodology found therein.

The basic goal is the same in both time-independent and time-dependent approaches. Folded diagrams may be regarded as providing an answer to the following question, in perturbation theory. Let

$$H = H_o + H_1 \tag{1}$$

be an arbitrary many-fermion Hamiltonian operating between all possible configurations of a complete set of one-body orbitals, eigenvectors of H_o. Pick a set of <u>active orbitals</u> in the vicinity of the Fermi surface of the unperturbed system, as in Fig. 1. The set of particles and holes in active orbitals only, forms the <u>model space</u> of states. Is it possible to replace the exact many-body problem, for reasonably low excitations, by a problem stated entirely in the model space? In other words, can we find a <u>model Hamiltonian</u> (or <u>effective Hamiltonian</u>) $\overline{H}$ and a set of <u>effective</u> operators $\overline{A}$, $\overline{B}$,... such that the eigenvalues of $\overline{H}$ inside the model space are the same as some of the eigenvalues of the true H in the entire space; and the matrix elements of $\overline{A}$, $\overline{B}$,... between eigenstates of $\overline{H}$ are the same as matrix elements of true operators A, B,... between corresponding eigenstates of the true H?

In time-independent quantum mechanics folded diagrams is considered to be a method for eliminating the energy dependence of the effective interaction. In time-dependent quantum mechanics one would more aptly say that folded diagrams is

*Work performed under the auspices of the U.S. Energy Res. and Dev. Admin.

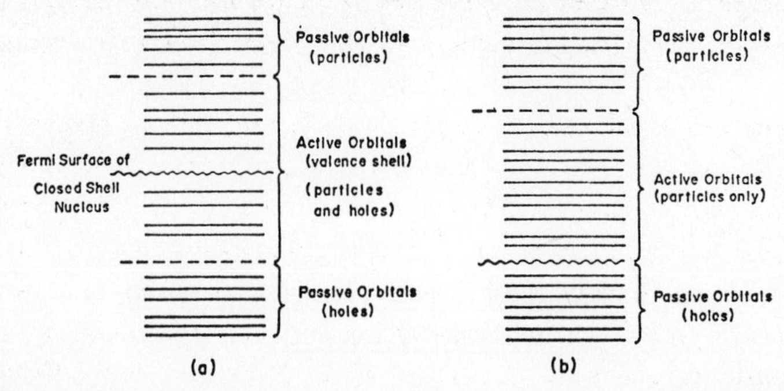

Fig. 1. Unperturbed orbital energies. The model space is defined by choosing a set of "active" orbitals in the vicinity of the Fermi surface. Example (a) has both active particles and active holes; example (b) has no active holes.

a theory for representing a time-delayed interaction by an instantaneous effective interaction. Whichever description one chooses for the dynamics, the utility of folded diagrams is commonly agreed to be the diagrammatic linked cluster character of the result and the simplification attendant in having decoupled the problem of calculating the energy eigenvalue and wave-function from the problem of calculating the effective Hamiltonian and other observables.

Brandow's folded diagram expansion may be derived from diverse approaches to many-body theory. For example, the time-dependent theories of Refs. 3 and 4 lead to a result no different from that found by Brandow, and Brandow's expansion may also be derived from the methods of Ref. 2. Having recognized this one may be tempted to ask, what is the value of learning the time-dependent method?

Quantum mechanics is of course equivalently formulated in time-dependent and time-independent language. However, one's intuition often functions differently in each, and the technical details necessary to come to a particular result can also be significantly different. It is actually the case that Brandow's folded diagrams can be not only reproduced in a time-dependent approach, but also simplified and extended by taking full advantage of the special flexibility of the time-dependent language.

It is worth emphasizing at this point that it is common throughout all physics to formulate theories in terms of effective instantaneous interactions in a model space. The subject of this conference, effective interactions in nuclei, is just one possible application of folded diagrams. Because the time-dependent formulation works through the time-evolution operator it would not be surprising to find this approach better suited to a particular problem than the time-independent approach. For example, the problem of determining the nucleon-nucleon potential from a meson exchange model is naturally solved in terms of folded diagrams[6] using time-dependent theory of Ref. 2 because of the close connection between the scattering S matrix and the time-evolution operator. But it is also true that for the problem of calculating the effective interaction in nuclei the time-dependent theory has certain advantages, some of a useful practical nature, over the time-independent theory. For the theoretician who calculates these diagrams it is worth recognizing the differences.

One technical advantage of using time-dependent methods to evaluate diagrams, be they Feynman or folded, is that each particle propagator naturally makes its own contribution independent of what happens elsewhere in the diagram. The so-called "individual particle propagator"[7] description used in Ref. 2 is exceedingly efficient for going into the middle of a large diagram to see how an instantaneous interaction replaces a particular piece of it. This contrasts especially to time-independent theory where "Global" Feynman-Goldstone energy denominators are often used; in general, these denominators couple propagators of the diagram piece under consideration to propagators in other parts of the diagram. Furthermore, diagrams written in terms of individual propagators are generally more highly summed than when written in terms of Global propagators. Thus, often one comes directly to a simple result using the individual particle propagators which can be obtained in other approaches only after summing a large class of diagrams and making use of the so-called "factorization theorem," which is a relation among energy denominators.

The time-dependent theory provides additional advantages. By exploiting them as in Ref. 2, one is lead <u>directly</u> to an Hermitian effective interaction; in contrast, the other methods[1,3,4] must introduce separate considerations to Hermitize $\bar{H}$. The methods of Ref. 2 also permit one to find the correct diagrams when the model space includes orbitals which are normally occupied; this is a problem which has not yet been solved within the time-independent framework or within the time-dependent framework of Refs. 3 and 4.

It is my hope to put across the simplicity of folded diagrams and to convince you that for practical problems the methods of Ref. 2 are especially useful.

I. Feynman-Goldstone Diagrams

The language of the time-dependent approach to folded diagrams is time-dependent perturbation theory, and the basic element is the Feynman-Goldstone diagram. We briefly review the rules, in order to establish conventions used in the remainder of the talk.

Consider a many-fermion system whose Hamiltonian (1) is treated in perturbation theory. One usually thinks of H as a two-body interaction, but it is often convenient to subtract a one-body interaction from H and to include it in H_o, which is always pure one-body. Arbitrary states of the system are specified by comparing them to the reference state $|\phi_0\rangle$, called the Fermi sea, which is an eigenstate of H_o. Orbitals that are empty in $|\phi_0\rangle$ will be designated by a lower case Roman letter a, b,... and orbitals normally occupied will be designated by a capital Roman letter A, B,... .

We define the time-evolution operator T(t,t') by

$$T(t,t') \, |\psi(t')\rangle e^{iE_ot'} = |\psi(t)\rangle e^{iE_ot} \tag{2}$$

for any state $|\psi\rangle$, with E_o being the unperturbed energy of the Fermi sea. T obeys the Schroedinger equation

$$i \, d\,T(t,t')/dt = (H-E_o)T(t,t') \quad (\text{with } T(t,t) = 1) \tag{3}$$

A Feynman-Goldstone diagram is a term in the perturbation expansion for a matrix element of T. Figure 2 shows one such term for the matrix element

$$\langle abA|T(t,t')|cdD\rangle, \tag{4}$$

which is the probability that if the system is prepared in the state $|cdD\rangle$ at time t', it will be analyzed in the state $|abA\rangle$ at time t. The diagrams are drawn generally in terms of particle lines (designated by upgoing arrows and labeled by the lower case Roman letter of the corresponding single particle state), hole lines (designated by downgoing arrows and labeled by the corresponding upper case Roman letter), and vertices (labeled by the times at which they occur). Diagrams may be arbitrarily complicated, but to be valid they must of course obey conservation laws such as conservation of the total number of bodies, and they must be entirely contained between times t and t'. The Pauli principle may be ignored in intermediate states.

To each diagram there corresponds a number, found as follows:

(1) Every vertex contributes an antisymmetrized matrix element of H;

(2) Every line α (α can refer to either a particle or a hole) contributes $e^{-i\varepsilon_\alpha \Delta t}$, where ε_α is an unperturbed orbital energy and Δt is the time-difference between the two ends of the line, counted in the direction of the arrow. In addition, every hole line contributes a factor -1;

(3) Every closed loop contributes an additional factor -1.

To find the value of the matrix element of T, all valid diagrams evaluated in this fashion must be summed together. Each distinct diagram must be counted

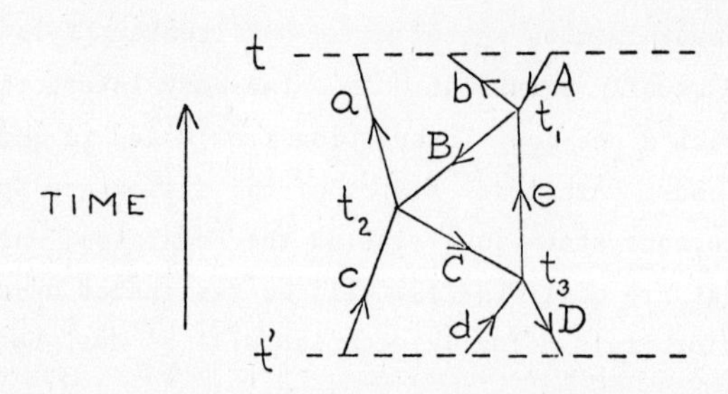

Fig. 2. A Feynman-Goldstone diagram.

once and only once. We consider two diagrams distinct if they have different topology, or have the same topology but differ by the label on any line or the time associated with any vertex. Thus, the sum includes an integration over times; a factor (-i) is associated with each time variable. One has to be a bit careful in specifying exactly which matrix element of T is evaluated by these rules owing to the fact that for fermions the sign of the diagram changes when lines are exchanged. See Ref. 2.

II. Folded Diagrams for the Case of No Active Holes

We want to define the effective interaction and the effective operators so that diagrams in the true problem are equal to corresponding diagrams in the model problem, and vice versa. In terms of equations, we want to define $\overline{H} = H_o + \overline{H}_1$ systematically so that

$$\overline{T}(+\infty,-\infty) = T(+\infty,-\infty) \tag{5}$$

and define the effective operator $\overline{A}$ so that with $\overline{H}$ defined by Eq. (5)

$$\overline{T}(+\infty, t)\overline{A}\,\overline{T}(t, -\infty) = T(+\infty,t)A\,T(t,-\infty) \tag{6}$$

It is of course true that Eqs. (5) and (6) can be satisfied only when the initial and final states are configurations in the model space; equality of these matrix elements is all that is necessary to establish the desired properties of $\overline{H}$ and $\overline{A}$, $\overline{B}$,... . This result is relatively plausible and will not be proved here. In Ref. 2 it is shown to be a direct consequence of Eqs. (5) and (6), provided that H_1 and $\overline{H}_1$ are appropriately switched on and off at $t = -\infty$ and $+\infty$. We shall also not give any detail on folded diagrams for effective operators; the considerations are almost identical to those for the effective interaction, however, and the expansion is completely linked.

The other methods[1,3,4] impose at this point an additional requirement, namely that the model eigenstates be the projections of the true eigenstates onto the model space. These theories are thus led to define $\overline{H}$ by a transformation on $T(o,-\infty)$. We impose no such requirements and prefer setting the problem up with $t = +\infty$ and $t = -\infty$ playing a symmetrical role. This constitutes a significant point of departure from the other approaches and also constitutes a simplification.

We have seen how to draw diagrams that contribute to matrix elements of T; these consist of active lines (propagators for particles in active or valence states), passive lines (corresponding to states not in the valence space) and matrix elements of the interaction. We can group all passive lines together into "boxes" so that diagrams for T can be equivalently drawn as boxes joined by active particle lines. Diagrams for $\overline{T}$ are drawn as matrix elements of $\overline{H}_1$, represented by small circles, and active particle lines. Drawn in this way, the diagrams in the true and model problems are very similar, and one might guess that the effective interaction may be thought of as the sum of many small circles, one corresponding to each box. Actually, because the boxes are time-delayed, there are also circles corresponding to two and more boxes, as we shall see.

Before going into the details of the relation between boxes and circles it is useful to give examples of boxes. Precisely defined, a box is a connected set of passive lines (particles or holes), together with the vertices they join, plus active particle lines drawn between vertices already belonging to the box. Some examples are given in Fig. 3. We have assumed in drawing these that the interaction is sufficiently weak that perturbation theory in powers of H_1 is possible.

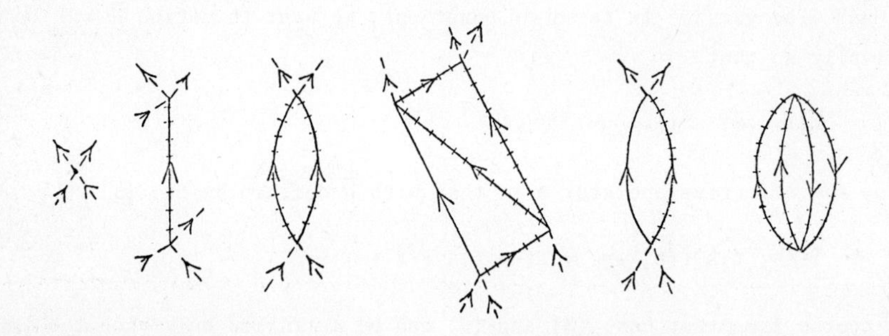

Fig. 3. Examples of "boxes." Cross-hatched lines are passive; smooth lines are active; dashed lines are active lines which are not included in the definition of the box.

If not, it sometimes makes sense to do perturbation theory in terms of the G-matrix, defined diagrammatically in Fig. 4. The G-matrix is another example of a box; more complicated boxes may be built up from several G-matrices in a very obvious way.

For those diagrams in which boxes and circles are sufficiently separated in time, then equality between these true and model diagrams may be achieved when circles correspond to single boxes only. This may be seen in Fig. 5. Figure 5a is

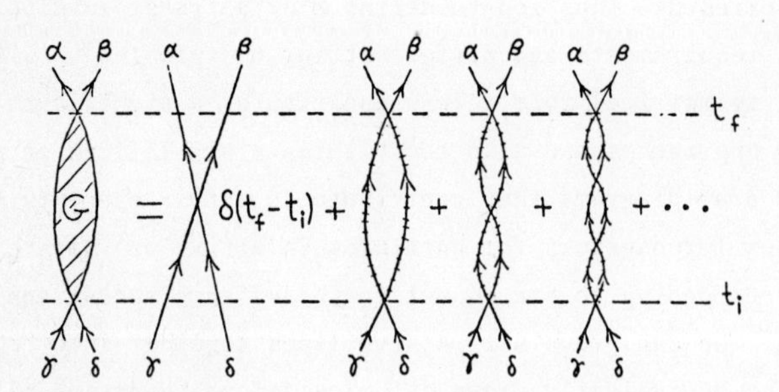

Fig. 4. Definition of the G matrix. At least one internal line between each interaction is passive.

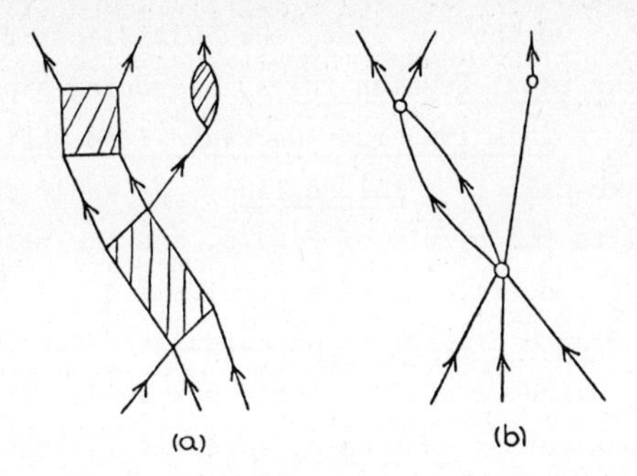

Fig. 5. (a) A true diagram, made of boxes and active lines. We often represent boxes by hatched areas, without specifying the details of the interior. (b) The equivalent model diagram, made of circles and active lines.

a diagram of the true problem drawn in terms of boxes and Fig. 5b is a model diagram of the same value drawn in terms of corresponding circles. It is easy to see how the circles are defined in terms of boxes because, as emphasized earlier, time-dependent theory permits us to consider an individual diagram piece independently of what happens elsewhere in the diagram. We have therefore merely to equate corresponding pieces of the two diagrams extending over the same time interval, as shown in Fig. 6. Remember that the times in the diagram have specific values; therefore, diagrams having different times associated with the interactions are treated separately. In a similar vein, we consider two topologically similar boxes as being different if the relative times are different, so that the sum over boxes, which occurs later to get $\overline{H}_1$, will also involve an integration over times.

Actually it is the value of the circle which we want, so we must take the external lines from both sides from Fig. 6. Because line a corresponds to a numerical factor $e^{-i\varepsilon_a(t_1-t_0)}$ we can take it out by multiplying both sides of the

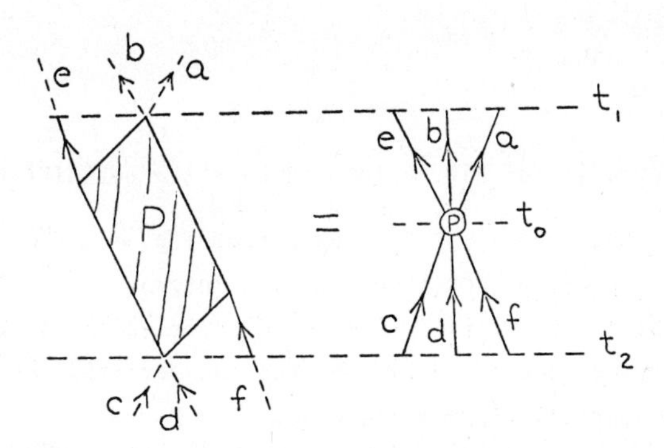

Fig. 6. The correct equation relating box to circle. Horizontal dashed lines are lines of constant t and are drawn to help the eye. Other dashed lines indicate active lines not included in the numerical value of the picture.

equation by $e^{-i\varepsilon_a(t_0-t_1)}$. On the box side, the multiplication can be effected, without departing from the usual Feynman rules, by adding a piece of line labeled a and running from t_1 to t_0. This line runs backward from what a particle line should normally do, and we call it a folded line. We apply the same treatment to all other lines external to the circle of Fig. 6, and the net result is Fig. 7, an equation giving an explicit definition of a circle, i.e., an element of the effective interaction, as a Feynman diagram to be calculated according to the usual rules. In this diagram, which we call a single box folded diagram, all external lines are folded back to a common time-base, which is indicated by a horizontal dashed line. This time, which is the time at which the circle acts in the model description, is so far completely arbitrary; by choosing it appropriately the diagrammatic expansion may be made to correspond to the non-Hermitian prescription which has been commonly employed in calculations, or with no additional effort to an Hermitian prescription. However, it is a consequence of the theory that predictions are the same for all choices, which is not surprising because the same time-base chosen for the single box is used whenever the box occurs as part of a multiple-box diagram. The freedom available here can and should be used to simplify the folded diagram series; see Sec. IV.

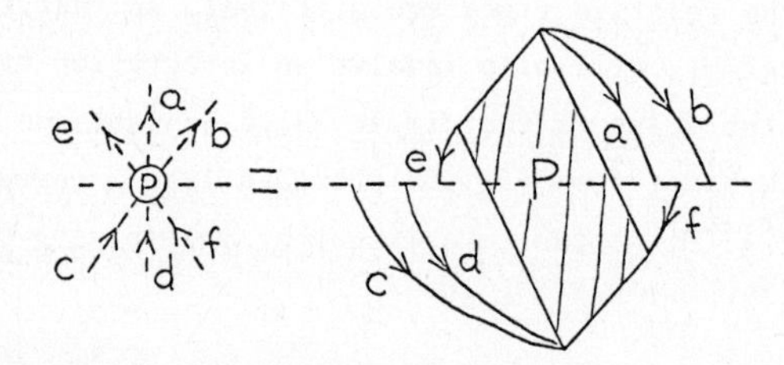

Fig. 7. Transformation of Fig. 6 into an equation defining the circle P.

Because the other theories[3,4] begin from T(o,-∞), the time-base must then always be the last time of the box in order that interaction be completely contained below the boundary at t = 0. This is why these interactions always come out non-Hermitian. There may be some advantage for the non-Hermitian prescription, but so far it has not been convincingly demonstrated.

When boxes are not sufficiently separated in time, then single box folded diagrams are insufficient to guarantee that each diagram for the true T is equal to a corresponding diagram of the same value for the model $\bar{T}$. Figure 8 gives an example of how this can happen. Because Fig. 8a is a true diagram which does not exist as a model diagram when $\overline{H}_1$ consists of single boxes only, we construct a new circle called a "true-correcting double-box folded diagram" which is <u>added</u> to $\overline{H}_1$. Note that a new time-base must be introduced for the double-box diagram; it is again arbitrary.

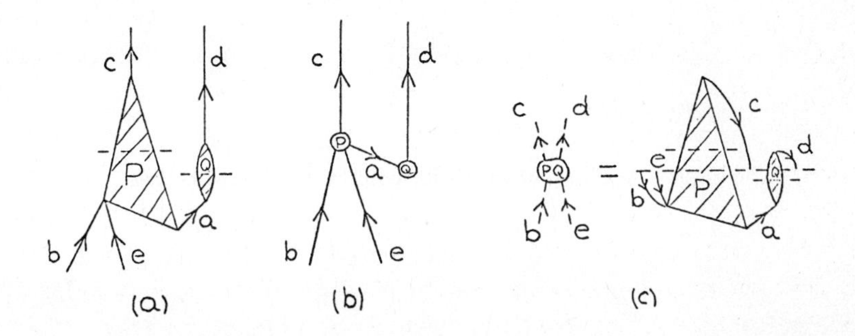

Fig. 8. (a) An allowed true diagram. (b) The corresponding model diagram, which is not allowed because particle <u>a</u> runs backwards. (c) Definition of the double box contribution PQ to $\overline{H}_1$.

<u>Prescription</u>. Whenever two boxes are connected by active lines in such a way, and have their time-bases chosen in such a way, that you cannot draw an equivalent model diagram without reversing the directions of some of the connecting lines, these two boxes should be considered as a single unit and their contribution to $\overline{H}_1$ should be calculated directly.

True correcting <u>multiple</u>-box diagrams also occur; their identification is straightforward.

When <u>circles</u> are not sufficiently separated in time, then single-box diagrams are insufficient to guarantee that each diagram for the model $\overline{T}$ is equal to a diagram of the same value for the true T. This gives rise to a second type of double-box folded diagram, the "model-correcting" diagram. Figure 9a gives an example of a model diagram without a true equivalent; the circles P and Q are single box diagrams defined in Fig. 9b, c. Figure 10 shows explicitly that Fig. 9a has no true equivalent; when expressed in terms of boxes some active lines run in the wrong direction. Because Fig. 10e is a contribution which is among the model but not the true diagrams, we make a new circle for it and <u>subtract</u> it from $\overline{H}_1$. This is shown in Fig. 11. It is of course necessary to respect the time-ordering of the original circle diagram, so the time-base of P is always above that of Q.

<u>Prescription</u>. Whenever two circles, connected by active lines, succeed each other so rapidly that it is impossible to replace them by equivalent boxes, with time-bases coinciding with the times of circles, without reversing the direction of some of the connections, the whole contribution should be subtracted from $\overline{H}_1$.

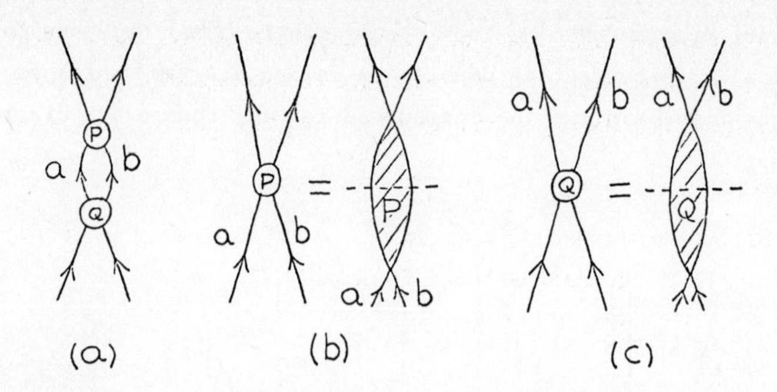

Fig. 9. A model diagram without a true equivalent.

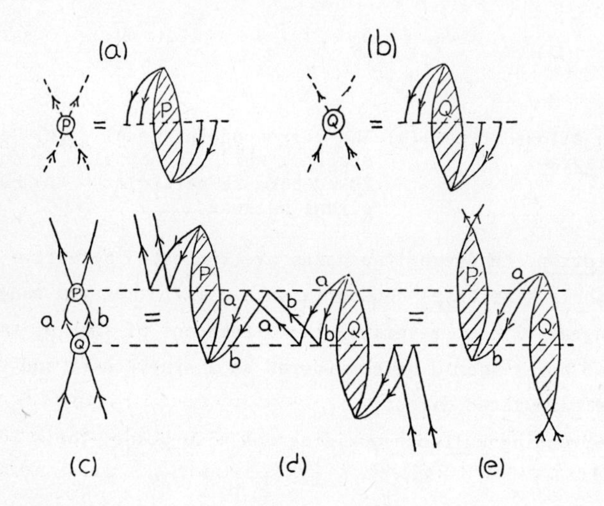

Fig. 10. Calculating Fig. 9a in terms of boxes.

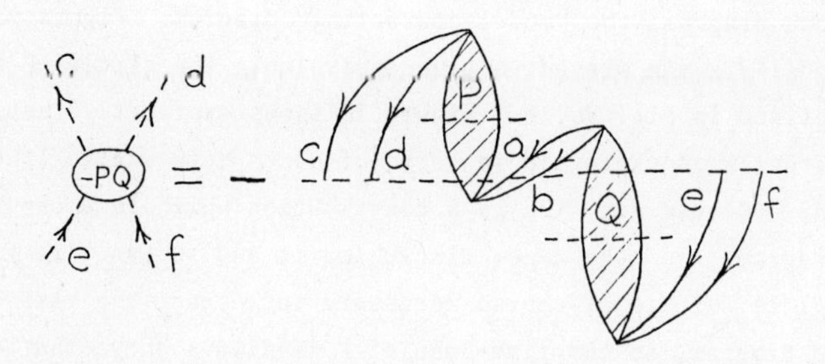

Fig. 11. The model-correcting double-box contribution to $\overline{H}_1$ whose effect is to remove the unwanted process of Fig. 9a.

Model-correcting multiple-box diagrams also occur, as do multiple-box diagrams which have a mixed character. We shall not consider these here; see Ref. 2 for detail on model-correcting multiple-box diagrams.

III. An Example

Consider the problem of calculating the effective interaction for nuclei beyond ^{208}Pb. Let the unperturbed Hamiltonian be

$$H_o = K + U \tag{7}$$

and the perturbation

$$H_1 = V - U \tag{8}$$

where V and U are a two-body and one-body potential, respectively. Let there be no active holes and only one active particle orbital for each charge and each set of angular momentum quantum numbers ℓ,j,m. We next show diagrams for zero, one- and two-body contributions to the effective interaction as illustrations of the considerations of the last section. We draw the boxes in terms of G-matrices as defined in Fig. 4. Occasionally we also show the three-body analog of G, the Bethe F-matrix, which we denote by a rectangle. In addition we use a dot to represent the vertex -U.

We shall not attempt to give in each case a complete list of all diagrams out to a specified order. We simply draw a few representative ones and try to avoid duplication. The external lines are always active; some of the internal lines could have been drawn as active, and when a doubly-partitioned model space[1,8] is used for the intermediate states of G some internal active particle lines may also be passive.

Figure 12 shows diagrams contributing to the zero-body part of the effective interaction. The sum of this series is the perturbation expression for the ground state energy of ^{208}Pb.

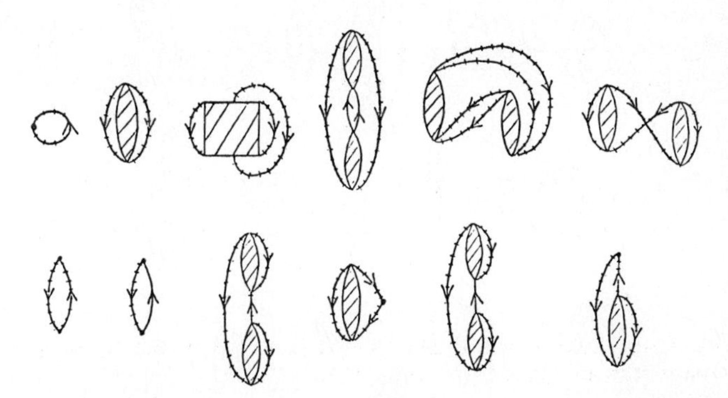

Fig. 12. Some zero-body diagrams of the effective interaction.

Figure 13 shows the one-body part, which is necessarily diagonal because of our assumption about conserved quantum numbers. This series yields the exact low-lying energies of ^{209}Pb and ^{209}Bi; in a realistic shell-model calculation, it is perfectly appropriate to replace them by experimental energies.

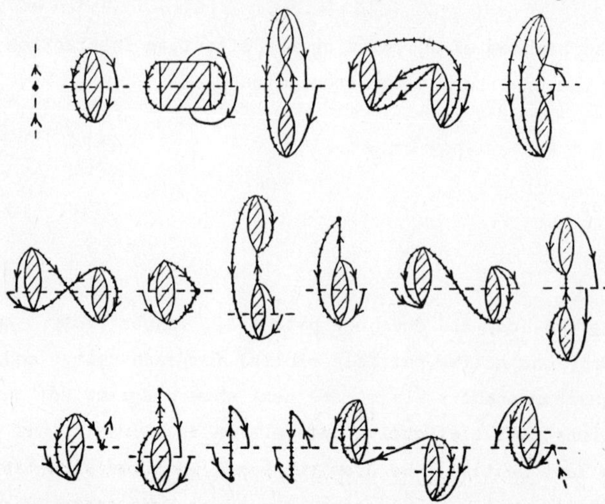

Fig. 13. Some one-body diagrams of the effective interaction. We recall that a dashed line is an active line whose contribution is not included in the value of the diagram.

The two-body part of the effective interaction is shown in Fig. 14. The Bertsch-Brown-Kuo cone polarization diagram is the third one. The present series is different from the usual perturbation series, since we have used the individual particle propagators. It may happen, although we have no reason to expect so, that this form of the series will provide better convergence than the usual form. In any case, it is easy enough to return to the usual form by breaking up the G matrices into pieces such that no more than one G occurs at any one time. For instance, when the last diagram of Fig. 14 is broken up, the lowest order part is Fig. 15,

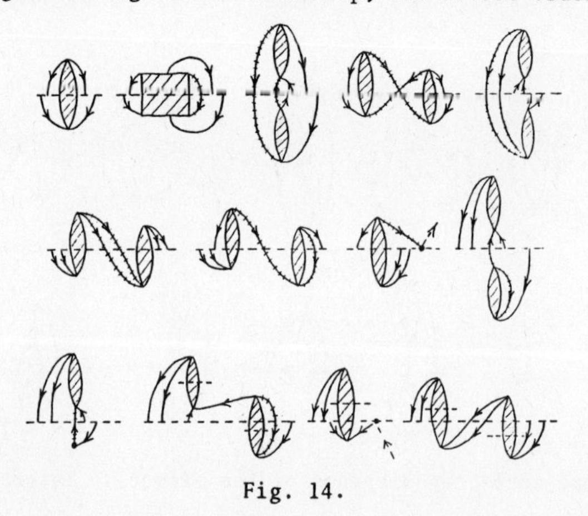

Fig. 14.

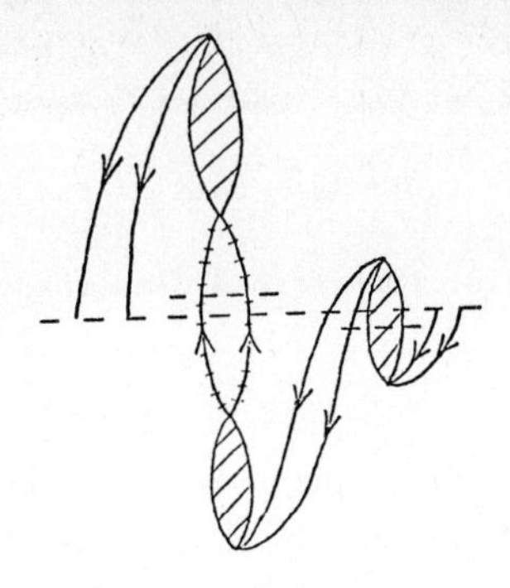

Fig. 15.

and it is more closely related to the usual Goldstone diagram. We shall next give examples which show how to do the time integrations.

IV. Calculation of Folded Diagrams

To obtain $\overline{H}$, we must at some point sum over all folded diagrams. As remarked, the sum includes an integration over times, which we show how to do now.

The simplest folded diagram for the two-body effective interaction of Fig. 14 is the G-matrix itself, redrawn in Fig. 16.

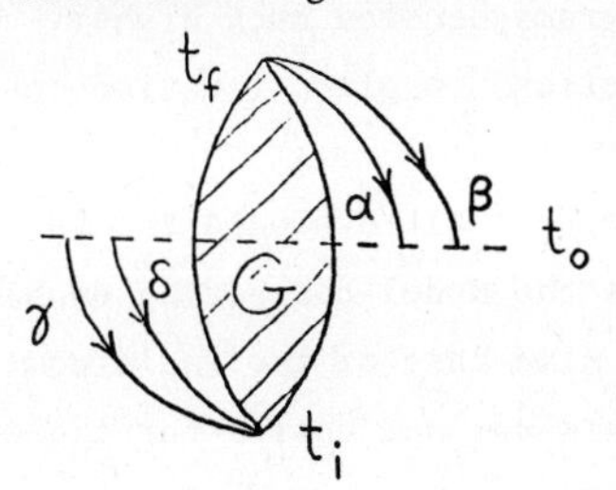

Fig. 16.

Let the time-base be located at

$$t_0 = (1 - \rho)t_f + \rho t_i \tag{9}$$

where ρ is an arbitrary parameter. The value of this diagram is

$$G(t_f - t_i)e^{-i(\varepsilon_\alpha+\varepsilon_\beta)(t_0-t_f)}\ e^{-i(\varepsilon_\gamma+\varepsilon_\delta)(t_i-t_0)} \tag{10}$$

Because of the constraint of Eq. (9) there is only one independent time to be integrated. Choose this to be

$$T = t_f - t_i \tag{11}$$

In terms of T, Eq. (10) becomes, integrated over times

$$(-i)\int_0^{\infty} dT\, e^{+i(\varepsilon_\alpha+\varepsilon_\beta)\rho T}\, e^{+i(\varepsilon_\gamma+\varepsilon_\delta)(1-\rho)T}(\alpha\beta|G(T)|\gamma\delta) \tag{12}$$

$$= (\alpha\beta|g[\rho(\varepsilon_\alpha + \varepsilon_\beta) + (1 - \rho)(\varepsilon_\gamma + \varepsilon_\delta)]|\gamma\delta) \tag{13}$$

where we have introduced the Fourier transformed G-matrix

$$g(\omega) = -i\int_0^{\infty} dt\; G(t)e^{i\omega t} \tag{14}$$

We are now faced with specifying the time-base. Brandow[1] and the authors of Refs. 3 and 4 must choose the time-base to be the last interaction for reasons already discussed; this choice means $\rho = 0$. To guarantee Hermiticity, the time-base is placed in such a way to preserve the past-future symmetry originally in the problem. There are many ways to do this. For instance we could put the time-base at the center of the G-matrix (ρ=1/2) which gives

$$(\alpha\beta|g\, \tfrac{1}{2}(\varepsilon_\alpha + \varepsilon_\beta + \varepsilon_\delta + \varepsilon_\gamma)]|\gamma\delta) \tag{15}$$

or we could average over the choice $\rho = 1$ and $\rho = -1$ which is equivalent to putting the time-base at the vertices. The advantage of the latter choice is that each diagram is given exactly by the Goldstone rules; the disadvantage is that the number of distinct multiple-box diagrams becomes much higher. It is therefore more convenient for practical calculations to place the time-base at the center of the diagram.

As our second example we shall evaluate the last diagram of Fig. 14, which is redrawn in Fig. 17. It is the model-correcting double-box diagram discussed in Sec. II. We must choose the time-base of the individual G-matrices to be half-way between the vertices since this was our choice for the one-box diagram, Eq. (15).

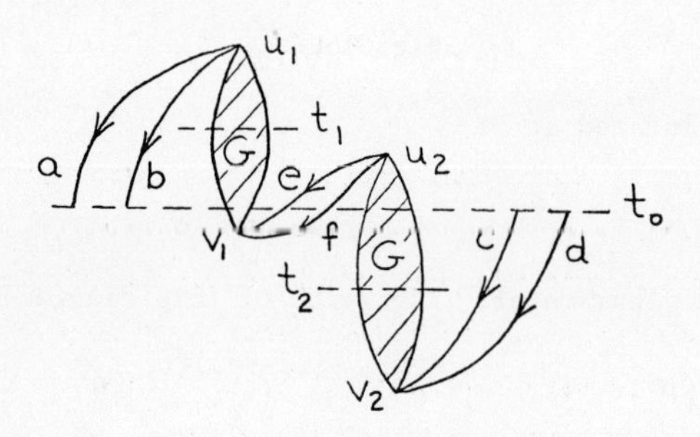

Fig. 17.

Thus $t_i = 1/2(u_1+v_1)$ and $t_2 = 1/2(u_2+v_2)$. We choose the time-base for the diagram as a whole to lie half-way between t_1 and t_2, $t_o = 1/2(t_1+t_2)$. There are four independent times, say u_1, v_1, u_2, v_2, but the time-base t_o must stay fixed, which leaves three variables of integration. These are most conveniently chosen to be the basic time-intervals in the picture, $T_1 = u_1 - v_1$, $T_2 = u_2 - v_2$, $T_m = u_2 - v_1$. The restrictions on the time variations are

$$u_1 > v_1,\ u_2 > v_2,\ u_2 > v_1,\ t_1 > t_2$$

or

$$T_1 > 0,\ T_2 > 0,\ T_m > 0,\ T_m < 1/2(T_1 + T_2) \tag{16}$$

Therefore, the contribution of this diagram to $(ab|\overline{H}_1|cd)$ is

$$-(-i)^3 \int_0^\infty dT_1 \int_0^\infty dT_2 \int_0^{1/2(T_1 + T_2)} dT_m \ (ab|G(T_1)|ef)\ (ef|G(T_2)|cd) \tag{17}$$

$$\exp i[(\varepsilon_a + \varepsilon_b)(t_0 - u_1) + (\varepsilon_e + \varepsilon_f)T_m + (\varepsilon_c + \varepsilon_d)(v_2 - t_0)]$$

One easily finds

$$u_1 - t_o = \frac{3}{4} T_1 + \frac{1}{4} T_2 - \frac{1}{2} T_m \tag{18}$$

$$t_o - v_2 = \frac{1}{4} T_1 + \frac{3}{4} T_2 - \frac{1}{2} T_m$$

The integrals can then be performed, beginning with T_m, and the result is

$$\left[g_1(\tfrac{1}{2} E_f + \tfrac{1}{2} E_m) g_2(\tfrac{1}{2} E_i + \tfrac{1}{2} E_m) - g_1(\tfrac{3}{4} E_f + \tfrac{1}{4} E_i) \times g_2(\tfrac{1}{4} E_f + \tfrac{3}{4} E_i) \right] (E_m - \tfrac{1}{2} E_f - \tfrac{1}{2} E_i)^{-1} , \tag{19}$$

with the definitions

$$E_f = \varepsilon_a + \varepsilon_b,\ E_i = \varepsilon_c + \varepsilon_d,\ E_m = \varepsilon_e + \varepsilon_f \tag{20}$$

$$\begin{aligned} g_1(\omega) &= (ab|g(\omega)|ef) \\ g_2(\omega) &= (ef|g(\omega)|cd) \end{aligned} \tag{21}$$

where $g(\omega)$ is given by Eq. (14). Expression (19) is Hermitian and has no singularity when the denominator vanishes, because the numerator vanishes then too.

This is the end of the examples. I hope that you have seen how easy it can be to enumerate and evaluate folded diagrams. If perturbation theory can be used to calculate the effective interaction, then the approach of Ref. 2 provides a very simple conceptual and calculational framework.

REFERENCES

(1) B. H. Brandow, Rev. Mod. Phys. 39 (1967), 771; B. H. Brandow, in "Lectures in Theoretical Physics (K. T. Mahanthappa, Ed.), Vol. 11, Gordon and Breach, New York, 1969.

(2) M. B. Johnson and M. Baranger, Ann. Phys. 62 (1971), 172.

(3) G. Oberlechner, F. Oweno-N'-Guema, and J. Richert, Nuovo Cimento B68 (1970), 23.

(4) T. T. S. Kuo, S. Y. Lee and K. F. Ratcliff, Nucl. Phys. A176 (1971) 65.

(5) T. Morita, Prog. Theor. Phys. 29 (1963), 351.

(6) M. B. Johnson, "Theory of Meson Exchange Potentials for Nuclear Physics," Los Alamos Preprint LA-UR-74-1927.

(7) M. Baranger, Recent Progress in the Understanding of Finite Nuclei, in "Proc. Int. School of Physics Enrico Fermi," Course 40, Varenna 1967 (M. Jean, director), Academic Press, New York, 1969.

(8) G. E. Brown, Rev. Mod. Phys. 43 (1971) 1.

M.B. JOHNSON: TIME DEPENDENT APPROACH (JOHNSON AND BARANGER APPROACH)

Zamick: I wonder if anybody actually used your method in calculating the effective interaction?

Johnson: No.

Sauer: You suggest to add the model-correcting folded diagrams to the effective interaction with which you solve the model space Schrödinger equation. Would it not be theoretically cleaner first to commit all crimes in solving the model space Schrödinger equation and then to repair the mistake afterwards?

Brandow: (remark) My first approximation to the hermitized model hamiltonian involved the average of the Brueckner G matrix with the right-hand starting energy, and the corresponding G with the left-hand starting energy. I think it is very interesting that your corresponding leading terms involve just a single G, evaluated with the average of the left and right starting energies. This looks like a simpler result. Nevertheless I have analyzed the formal structure of the higher-order corrections to my "naively hermitized" result --- this is just my θ expansion. I think it would be interesting to explore the corresponding general formal structure of the corrections to your form of "naive hermitization".

A TIME-DEPENDENT, DIAGRAMMATIC ANALYSIS OF EFFECTIVE INTERACTIONS AND OPERATORS

Keith F. Ratcliff
Department of Physics
State University of New York at Albany
Albany, New York 12222

I. Introduction

An important motivation of our particular approach[1-3] is to realize degenerate perturbation theory as a natural extension of the well-known and successful linked-cluster expansion of non-degenerate perturbation theory. Gell-Mann and Low[4] showed that the true ground state $|\psi_c^Q\rangle$ of a non-degenerate system could be developed adiabatically from the finite ratio of two singular terms

$$|\psi_c^Q\rangle \equiv \frac{U(0,-\infty)|o\rangle}{\langle o|U(o,-\infty)|o\rangle} \tag{1}$$

where $U(0,-\infty)$ is the time development operator $U(t,t')$ in interaction representation and $|o\rangle$ is the particle-hole vacuum supplied by the one-body unperturbed Hamiltonian, H_o. The projection of the correlated ground state $|\psi_c^Q\rangle$ onto the one-dimensional model space $|o\rangle$ clearly has unit norm

$$\langle o|\psi_c^Q\rangle = 1. \tag{2}$$

The true ground state energy E_c and ground state energy shift $\Delta E_c = E_c - \varepsilon_o$ are then given by

$$E_c = \frac{\langle o|HU(o,-\infty)|o\rangle}{\langle o|U(o,-\infty)|o\rangle}$$
$$\Delta E_c = \frac{\langle o|H_1 U(o,-\infty)|o\rangle}{\langle o|U(o,-\infty)|o\rangle} \tag{3}$$

where

$$H_1 = H - H_o$$
$$H_o|o\rangle = \varepsilon_o|o\rangle. \tag{4}$$

Six years later, Goldstone[5] gave a diagrammatic representation of these terms. His work can be summarized by three equations

$$U(0,-\infty)|o\rangle = U_L(o,-\infty)|o\rangle\langle o|U(0,-\infty)|o\rangle \quad (5)$$

$$\langle o|U_L(0,-\infty)|o\rangle = 1 \quad (6)$$

$$\langle o|HU_L(0,-\infty)|o\rangle = E_c. \quad (7)$$

His factorization theorem is expressed in Eqn. (5). The term $U_L(0,-\infty)|o\rangle$ is defined diagrammatically as the set of all diagrams which are free from vacuum fluctuations. This term is finite and survives the ratio formed by Gell-Mann and Low. We interpret Eqn. (6) as demonstrating that $U_L(0,-\infty)$ functions like a unit operator within the one dimensional model space while Eqn. (7) is merely the one-dimensional secular equation. Finally the energy shift ΔE_c is given by $\langle o|H_1U_L(0,-\infty)|o\rangle$ and is represented diagrammatically by the famous linked cluster expansion.

For the degenerate problem we shall denote with upper case letters $|A\rangle$, $|B\rangle$,... the orthonormal many-particle eigenstates of H_o which span our model space and which we refer to as active states. Eigenstates of H_o which lie outside the model space are called passive states and are labeled with Greek letters $|\alpha\rangle$, $|\beta\rangle$,... . We further define the projection operator P onto the model space

$$P = \sum_A |A\rangle\langle A|. \quad (8)$$

In parallel with the Gell-Mann, Low, Goldstone theory which is summarized by equations (5-7) the present theory[1] is summarized by

$$U(0,-\infty)|A\rangle = \sum_B U_Q(0,-\infty)|B\rangle\langle B|U(0,-\infty)|A\rangle \quad (9)$$

$$\langle B|U_Q(0,-\infty)|A\rangle = \delta_{BA} \quad (10)$$

$$\sum_B \langle A|HU_Q(0,-\infty)|B\rangle b_B^\lambda = E_\lambda b_A^\lambda \quad (11)$$

Eqn. (9) is the more generalized statement of factorization appropriate to a degenerate model space. The term $U_Q(0,-\infty)|B\rangle$ will be defined diagrammatically. Eqn. (10) tells us that $U_Q(0,-\infty)$ functions like the unit operator within the model space. The model space secular equation, Eqn. (11), yields true eigenvalues E_λ of H with the corresponding eigenvectors $|\phi_\lambda\rangle$ proportional to the projection of the true eigenvector $|\psi_\lambda\rangle$ onto the model space, i.e. $|\phi_\lambda\rangle \propto |P\psi_\lambda\rangle$.

II. Analysis of Diagrams and First Factorization

In Fig. 1 we show a typical term or history from $U(0,-\infty)|A>$ which arises in the adiabatic evolution of a model space state. The elements of such a diagram fall into three classes. We designate as a vacuum fluctuation (VF) those elements, disconnected from the rest of the diagram, which are distinguished by having no fermion lines reach either time boundary. A core excitation (CE) is an element, disconnected from the rest of the diagram, which has fermion lines reaching only one time boundary. The valence diagram (VD) is that element with fermion lines reaching both time boundaries.

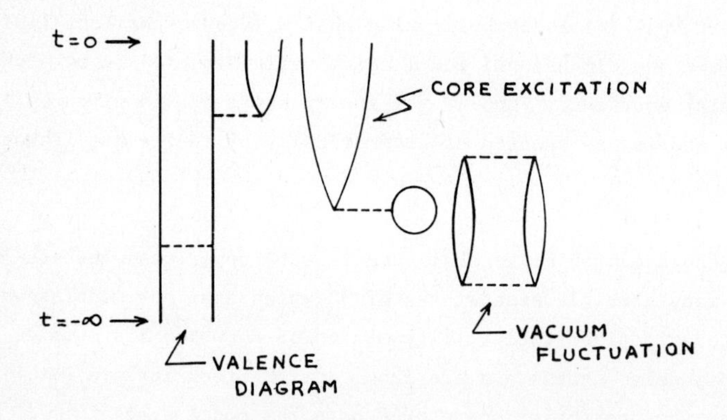

Fig. 1 The three elements of a diagram

Factorization of the first kind arises from the fact that an entire group of actual histories are related by generalized time ordering (GTO) the vertices of each of the three classes of elements relative to each other. A simple example is shown in Fig. 2 where the sum of six diagrams is seen to factorize into the product, (VD) x (CE) x (VF), of the three elements.

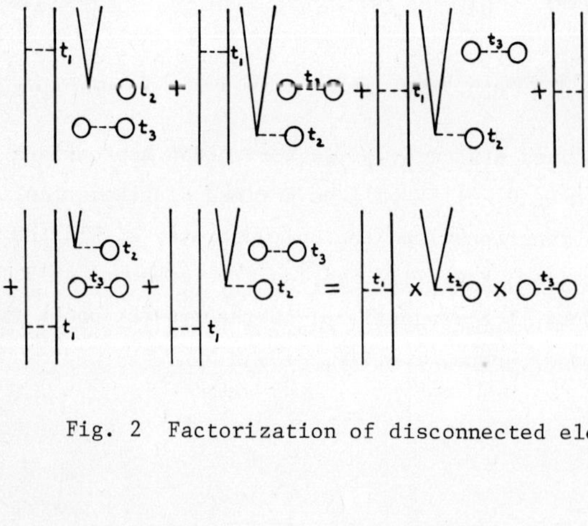

Fig. 2 Factorization of disconnected elements

The immediate generalization of this is that the sum of all histories factorizes into the product of the three classes of diagrammatic elements. This is shown in Fig. 3. The unit factor included with Σ(VF) accounts for histories which have no vacuum fluctuations. Likewise the incorporation of $|o\rangle$ with Σ(CE) accounts for histories with no core excitation. Histories in which no vertex attaches to the valence fermion lines are accounted for by the inclusion of inert valence lines as the leading term in Σ(VD). We recognize $[1+\Sigma(VF)]$ as identical to the divergent factor $\langle o|U(0,-\infty)|o\rangle$ in Goldstone's analysis, Eqn. (9). Likewise $[|o\rangle+\Sigma(CE)]$ is Goldstone's correlated ground state $|\psi_c^Q\rangle$ whose unit projection onto $|o\rangle$ follows immediately from the orthogonality of $|o\rangle$ to every term in Σ(CE). The term $U_L(0,-\infty)|o\rangle$ designates the set of all linked diagrams which evolve adiabatically from the unperturbed core. (The term "linked" is being used in its original sense to denote the absense of any vacuum fluctuations. The property that all the fermion lines of a diagram are attached to each other through vertices will be referred to as "completely connected".)

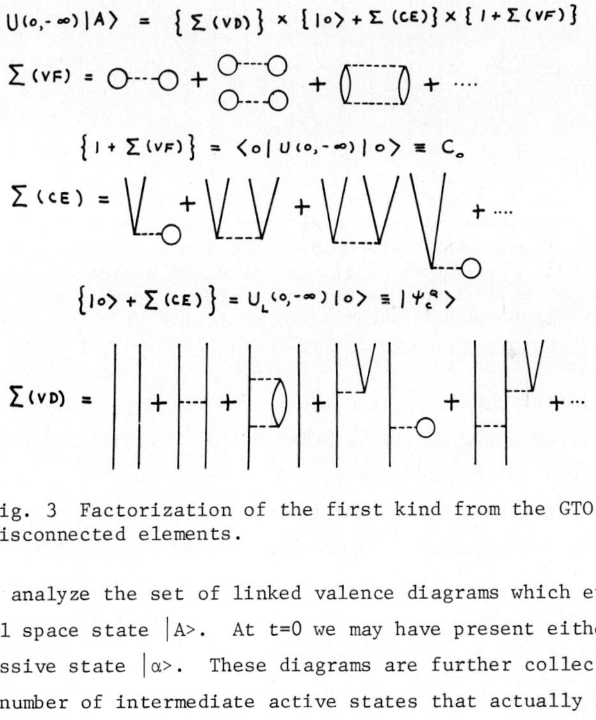

$$U(o,-\infty)|A\rangle = \{\Sigma(VD)\} \times \{|o\rangle + \Sigma(CE)\} \times \{1+\Sigma(VF)\}$$

$\Sigma(VF) =$ + + + ...

$$\{1+\Sigma(VF)\} = \langle o|U(o,-\infty)|o\rangle \equiv C_o$$

$\Sigma(CE) =$ + + + ...

$$\{|o\rangle + \Sigma(CE)\} = U_L(o,-\infty)|o\rangle \equiv |\psi_c^Q\rangle$$

$\Sigma(VD) =$ + + + + + ...

Fig. 3 Factorization of the first kind from the GTO of disconnected elements.

In Fig. 4 we analyze the set of linked valence diagrams which evolve adiabatically from a model space state $|A\rangle$. At t=0 we may have present either an active state $|B\rangle$ or a passive state $|\alpha\rangle$. These diagrams are further collected into groups according to the number of intermediate active states that actually appear in the history. (We emphasize number of intermediate active states that "actually appear" rather than that "can be made to appear by an operation of stretching"[6].) This grouping of valence diagrams is motivated by the fact that each intermediate active valence state gives rise to an energy denominator [(active energy) - (active energy)]$^{-1}$ which can vanish. Thus we achieve a grouping of the valence terms according to the order of this singularity. The small circles in our linked valence diagrams are called "Q-boxes" a name which reminds us that the Q-box is defined as one or more vertices with the intermediate fermion lines such that every intermediate state is orthogonal to the model space.

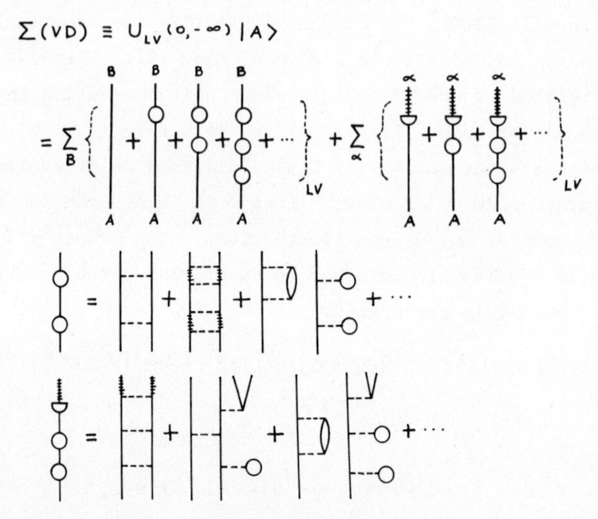

Fig. 4 Classification of linked valence diagrams and examples of 2 Q-box diagrams. The valence line represents any number of valence particle lines. The circle is the Q-box during which all intermediate states are passive. The semi-circle is the open Q-box, "open" because the system has not returned to an active state.

III. Folded Diagrams and Second Factorization

Factorization of the first kind arose when the GTO of disconnected elements of diagrams was seen to give rise to other actual histories (provided one ignores the Pauli principle in intermediate states). The resulting factorization isolated the singularities arising from vacuum fluctuations in a way that permits their removal from the secular equation. We retain however singularities arising from intermediate active states in valence diagrams. Factorization of the second kind will likewise isolate these singularities in a way convenient for their removal from the secular equation. The application of GTO to a set of vertices of a valence diagram must however involve diagrams in which fermion lines are reversed in direction. Such diagrams are the "folded diagrams" which do not correspond to actual histories but which are essential to achieve factorization (just as the Pauli violating diagrams were essential for factorization of the first kind).

In Fig. 5 we show an example of the GTO of vertices of a valence diagram. Here a pair of vertices is retained in definite order ($t_2 > t_3$) but the remaining vertex (t_1) is permitted to occur in any order relative to the other two. The sum

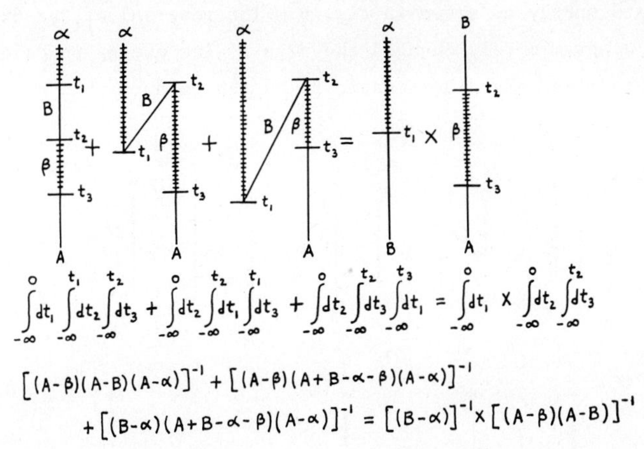

Fig. 5 Factorization of linked valence diagrams

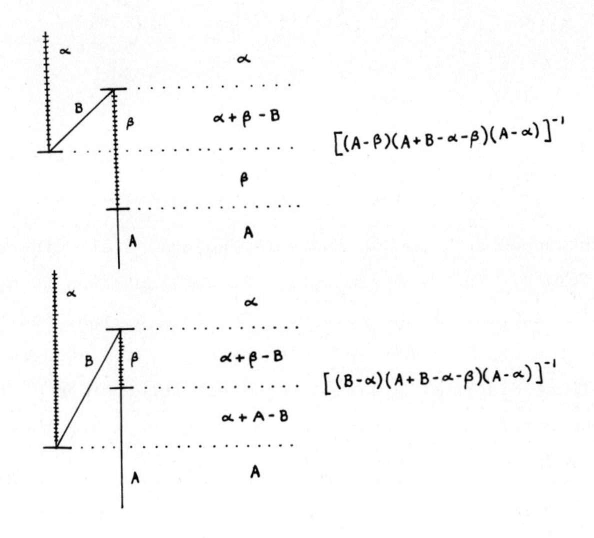

Fig. 6 Intermediate state energies for folded diagrams and resulting energy denominators.

of all three valence diagrams is identical with the product of a pair of valence diagrams. This factorization has been achieved at the expense of introducing two valence diagrams in which we reverse the direction of (or fold) the intermediate state $|B\rangle$ to achieve the order $t_1 < t_2$. The diagrammatic equality of Fig. 5 is easily seen as a relation among the time ordering operators. The equivalent relation among energy denominators is also shown. The energy denominators of folded diagrams are still calculated as the difference between the initial energy and each intermediate state energy as shown in Fig. 6. The reversal of the direction of $|B\rangle$ in a folded diagram merely changed the sign of its energy in calculating energy denominators. Otherwise the diagrammatic rules remain unchanged when applied to folded diagrams.

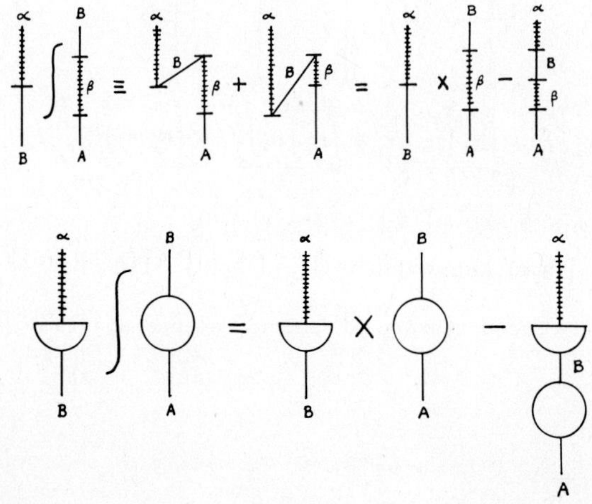

Fig. 7 Integral symbol for the sum of all folded diagrams.

In Fig. 7 we introduce a symbol to indicate the sum of all diagrams in which state $|B\rangle$ has been folded. The "integral sign" in this notation is directed to the common state $|B\rangle$ which appears at the bottom of the left element and top of the right element. Thus the equality at the bottom of Fig. 7 represented the set of all 1 Q-box open valence diagrams in which the latest interaction of the Q-box occurs later than the earliest interaction of the open Q-box and thus requires the folding of the intermediate active state between them.

In Fig. 8 we express each group of linked valence diagrams, which arrive at t=0 in a passive state, as a linear combination of products of multiply-folded valence diagrams and linked valence diagrams which arrive at t=0 in an active state. Summing the vertical columns of Fig. 8 then produces the factorization shown in Fig. 9. The inert valence line ($\sim|B\rangle$) which is combined with the folded diagrams in Fig. 9 accounts for the original set of linked valence diagrams which arrived at t=0 in an active state. The factorization has isolated a state $U_{QLV}(0,-\infty)|B\rangle$ in which every intermediate state in the linked valence diagrams is passive (hence the label QLV) due to the folding of any intermediate active state. This vector obviously has finite norm and clearly

$$PU_{QLV}(0,-\infty)|B\rangle = |B\rangle. \tag{12}$$

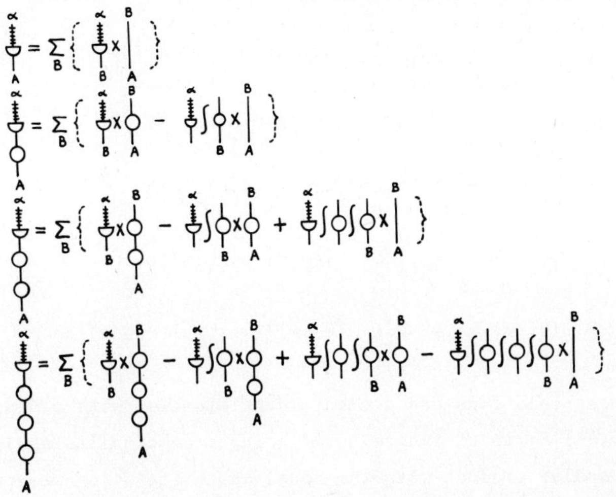

Fig. 8 Folded diagram expansion for linked valence diagrams terminating in a passive state. The inert valence line with A and B at opposite ends is the diagrammatic representation of the Kronecker delta. Summation of each column of terms gives the result shown in Fig. 9

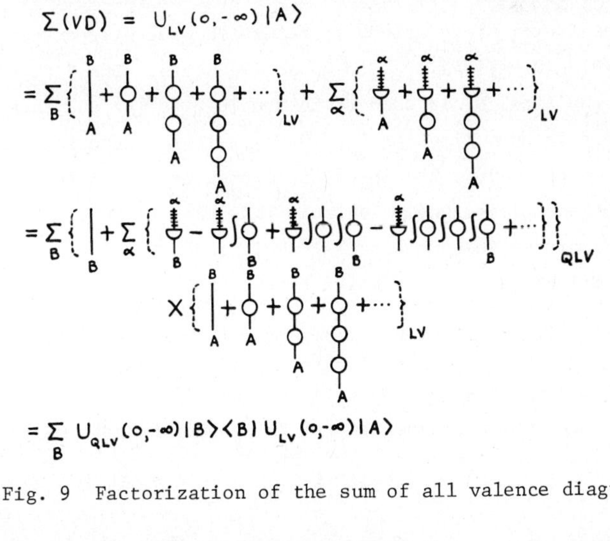

$$\Sigma(VD) = U_{LV}(0,-\infty)|A\rangle$$

$$= \sum_B U_{QLV}(0,-\infty)|B\rangle\langle B|U_{LV}(0,-\infty)|A\rangle$$

Fig. 9 Factorization of the sum of all valence diagrams.

The final result is achieved by combining the linked valence factorization with the earlier core factorization, Eqn. (5), to produce the decomposition theorem[1]

$$
\begin{aligned}
U(0,-\infty)|A\rangle &= \{\Sigma(VD)\} \times \{|o\rangle + \Sigma(CE)\} \times \{1+\Sigma(VF)\} \\
&= \sum_B U_{QLV}(0,-\infty)|B\rangle\langle B|U_{LV}(0,-\infty)|A\rangle \\
&\quad \times U_L(0,-\infty)|o\rangle\langle o|U(0,-\infty)|o\rangle \\
&= \sum_B U_Q(0,-\infty)|B\rangle\langle B|U(0,-\infty)|A\rangle
\end{aligned} \tag{13}
$$

$$U_Q(0,-\infty)|B\rangle = U_{QLV}(0,-\infty)|B\rangle \times U_L(0,\infty)|o\rangle \tag{14}$$

$$\langle B|U(0,-\infty)|A\rangle = \langle B|U_{LV}(0,-\infty)|A\rangle \times \langle o|U(0,-\infty)|o\rangle \tag{15}$$

Every term in these relations has a clear identification with a diagrammatic sum. The vector, $U_Q(0,-\infty)|B\rangle$ is of finite norm (being free of all singularities). In calculating its scalar product with the model space state $|A\rangle$ we get δ_{AB}, Eqn. (10), since $|A\rangle$ is obviously orthogonal to all but the product of inert valence line in $U_{QLV}(o,-\infty)|B\rangle$ and the $|o\rangle$ in $U_L(0,-\infty)|o\rangle$. The secular equation, Eqn. (11), then follows immediately[1,7] thus completing the extension of the Gell-Mann, Low, Goldstone approach to the degenerate model space.

IV. Analysis of the Secular Equation

The secular equation employs matrix elements of $PHU_Q(0,-\infty)P$ which are achieved by applying one factor of H_1 at t=0 to return the factorized vector diagram $U_Q(0,-\infty)|B\rangle$ $(=U_{QLV}(0,-\infty)|B\rangle x|\psi_c^Q\rangle)$ to an active state. The 0-body part of H_{eff} is produced when the final H_1 is used to return $|\psi_c^Q\rangle$ to $|o\rangle$ and thus utilizes the completely inert valence lines of $U_{QLV}(0,-\infty)|B\rangle$. The numerical value of $H_{eff}^{(0)}$ is thus $\langle o|H|\psi_c^Q\rangle$ which is just the Goldstone linked expansion for the true ground state energy of the closed shell system. This then is in complete agreement with the assumptions of the phenomenological shell model in that energies of 0d-1s shell nuclei will be measured relative to the true ground state of O^{16}.

The 1-body part of H_{eff} arises from diagrams of $U_Q(0,-\infty)|B\rangle$ in which the Q-boxes attach to only one valence fermion line, all others being inert. This folded fermion line which arrives at t=0 in a passive state may be returned to an active state either by applying H_1 to just the open Q-box in $U_{QLV}(0,-\infty)|B\rangle$ and utilizing $|o\rangle$ in $|\psi_c^Q\rangle$ or by using H_1 to connect the valence diagram to a core excitation. The result is just to produce a valence diagram sum

$$\langle i|H_1U_Q(0,-\infty)|j\rangle_V = \langle i|\ Q - Q'\int Q + Q'\int Q\int Q - \dots\ |j\rangle_V \qquad (16)$$

since these two operations merely serve to complete the final Q-box. The designation Q' in Eqn. (16) for the final Q-box reminds us that this final Q-box must have more than one vertex since it arose from an open Q-box which arrived at t=0 in a passive state. We note that the unfolded term in Eqn. (16) is Q not Q' because the one vertex term arises by applying the final H_1 to an inert valence line.

When the model space is so chosen that the single particle states of H_o, which define the model space, are uniquely characterized by exact symmetries of H(i.e. by J, J_z, T_z, and parity) then $PHU_Q(0,-\infty)P$ is a diagonal in the one-active-particle model space. This is the case for a shell model space consisting of one major shell. Under these circumstances the single particle energies of H_{eff} are unambiguously determined to be the empirical energies of the one-particle system measured relative to the true core energy (i.e. E_i-E_c). Thus for the 0d-1s shell nuclei we would use $O^{17}-O^{16}$ and $F^{17}-O^{16}$ single particle energies as is common practice in shell model phenomenology.

The 2-body part of H_{eff} arises from diagrams in which a pair of valence fermion lines are actively involved in the Q-boxes (all other valence fermion lines being inert). Again the final H_1 may be applied directly to the valence part or may couple the valence part to $|\psi_c^Q\rangle$. The result is the folded diagram series

$$\langle kl|H_1U_Q(0,-\infty)|ij\rangle_{CV} = \langle kl|\ Q - Q'\int Q + Q'\int Q\int Q - \dots\ |ij\rangle_{CV} \qquad (17)$$

Again the final box of folded diagrams is a Q'box with more than one vertex. Careful analysis(2) reveals that the only terms which survive the summation of this series are ones in which the pair of valence lines are connected by at least one vertex (which may be the vertex at t=0). The connectedness of the valence diagrams is reflected in the subscript C in Eqn. (17). This result generalizes to the statement that the matrix elements of the n-body part of H_{eff} are given by the sum of all folded, completely connected, valence diagrams of n valence fermion lines.(2)

It should be obvious that the number of folded diagrams is very much greater than the number of linked valence diagrams with normal time ordering (and hence not folded). Thus if the summing of folded diagrams depended on their evaluation one at a time as suggested in Figures 5-7, the usefulness of this theory would be bleak indeed. It should then be recorded here that the identity of the Q-box as a repeated diagrammatic element has permitted the development of rather powerful techniques for the summation of the folded diagram series as the limit of well defined sequences.[2,7-10] In these approaches evaluation only of the Q-box is required. The effect of repeated folding of this element is then automatically generated. To this end the identity of each Q-box is essential and it is precisely that identity which is destroyed by the "cancellation of all stretchable diagrams".[6]

The conjugate of the present model space eigenket equation will of course produce the identical eigenbra spectrum. The elements of the secular equation are shown in Fig. 10 and involve the adiabatic evolution of the system backwards in time from the remote future. We adopt a diagrammatic convention that the obvious folded diagram series in $\langle A|U_Q(\infty,0)$ is drawn with the folded elements to the left of the open Q-box that reaches t=0. The evaluation of these diagrams proceeds with the same diagrammatic rules as with $U_Q(0,-\infty)|A\rangle$, the only change being that the active state at $t=+\infty$ now plays the role of the initial state in evaluation energy denominators. Thus the specific example given at the bottom of Fig. 10 is just the conjugate of the relation in Fig. 7. The matrix elements for the effective interaction of the eigenbra equation are then simply $\langle ij|Q - Q\int Q' + Q\int Q\int Q' - ..|kl\rangle$.

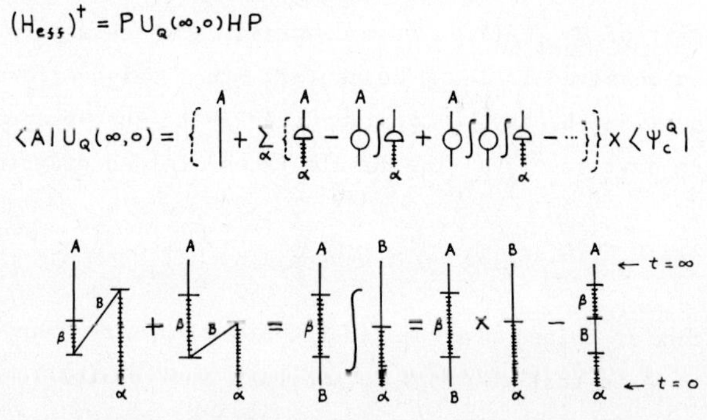

$(H_{eff})^{\dagger} = P U_Q(\infty,0) H P$

Fig. 10 The conjugate of H_{eff} and folded diagrams in the interval $(\infty,0)$

We denote by $|\phi_\lambda\rangle$ the finite set of d eigenkets of H_{eff} which correspond to projections of true eigenkets $|\psi_\lambda\rangle$ onto the model space (d=dimensionality of the model space)

$$PHU_Q(0,-\infty)P|\phi_\lambda\rangle = E_\lambda|\phi_\lambda\rangle$$
$$\langle\phi_\lambda|PU_Q(\infty,0)HP = \langle\phi_\lambda|E_\lambda . \qquad (18)$$

Since H_{eff} is not Hermitian i.e. in general

$$<A|HU_Q(0,-\infty)|B> \neq <B|HU_Q(0,-\infty)|A> \tag{19}$$

its eigenkets are not mutually orthogonal

$$<\phi_\lambda|\phi_\mu>\propto<P\psi_\lambda|P\psi_\mu> \neq \delta_{\lambda\mu} \tag{20}$$

We shall however assume that the set of d projections of true eigenvectors that we obtain are linearly independent so that the $|\phi_\lambda>$'s span the model space. Under these circumstances a unique set of d vectors $|\bar{\phi}_\lambda>$, which also span the model space, may be found[11] by solution of

$$<\phi_\lambda|\bar{\phi}_\mu> = \delta_{\lambda\mu} \tag{21}$$

which serves to define both the direction and norm of $|\bar{\phi}_\mu>$. The two sets of vectors $|\phi_\lambda>$ and $|\bar{\phi}_\lambda>$ are said to form a biorthogonal set. An immediate consequence of Eqn. (21) is that the model space projection operator may be expressed as

$$P = \sum_\lambda|\bar{\phi}_\lambda><\phi_\lambda| = \sum_\lambda|\phi_\lambda><\bar{\phi}_\lambda| \tag{22}$$

independent of any assumption as to the norm of the $|\phi_\lambda>$. It then follows that if $|\psi_\mu>$ is one of the true eigenvectors whose projection $|P\psi_\mu>$ is an eigenket of the secular equation, then

$$<\bar{\phi}_\lambda|\psi_\mu> = <\bar{\phi}_\lambda|P\psi_\mu> \propto <\bar{\phi}_\lambda|\phi_\mu> = \delta_{\lambda\mu} \tag{23}$$

This very important result suggests that since $|\bar{\phi}_\lambda>$ is orthogonal to all but one of the lowlying true eigenkets, these true eigenkets can be realized as the adiabatic evolution of a unique model space vector. Therefore in analogy to the original Gell-Mann, Low development Eqn. (1) we have

$$\frac{|\psi_\lambda>}{<\bar{\phi}_\lambda|\psi_\lambda>} = \frac{U(0,-\infty)\;|\bar{\phi}_\lambda>}{<\bar{\phi}_\lambda|U(0,-\infty)|\bar{\phi}_\lambda>}$$

$$\frac{<\psi_\lambda|}{<\psi_\lambda|\bar{\phi}_\lambda>} = \frac{<\bar{\phi}_\lambda|U(\infty,0)}{<\bar{\phi}_\lambda|U(\infty,0)|\bar{\phi}_\lambda>} \;. \tag{24}$$

These are precisely the relations that are needed to attack the problem of effective operators.

V. Effective Operators

We denote by X an operator such as an electric or magnetic multipole operator. The corresponding effective operator X_{eff} is commonly defined as that operator with model space matrix elements identical with those of X between true eigenstates

$$<\psi_\lambda|X|\psi_\mu> = \frac{<\phi_\lambda|X_{eff}|\phi_\mu>}{<\phi_\lambda|\phi_\lambda>^{\frac{1}{2}}<\phi_\mu|\phi_\mu>^{\frac{1}{2}}} \quad . \tag{25}$$

In Eqn. (25) it is assumed that the true eigenvectors ψ are normalized and clearly the definition of X_{eff} is independent of the norm of model space eigenvectors ϕ.

In Eqn. (24) we realized the adiabatic evolution of true eigenvectors as a ratio of singular quantities but this ratio does not yield a normalized true eigenvector. We therefore recast the expression for matrix elements of X in a normalization free manner and employ Eqn. (24) to give[3]

$$\begin{aligned} X_{\lambda\mu} &= \frac{<\psi_\lambda|X|\psi_\mu>}{[<\psi_\lambda|\psi_\lambda><\psi_\mu|\psi_\mu>]^{\frac{1}{2}}} \\ &= \frac{<\bar{\phi}_\lambda|U(\infty,0)XU(0,-\infty)|\bar{\phi}_\mu>}{[<\bar{\phi}_\lambda|U(\infty,0)U(0,-\infty)|\bar{\phi}_\lambda><\bar{\phi}_\mu|U(\infty,0)U(0,-\infty)|\bar{\phi}_\mu>]^{\frac{1}{2}}} \\ &= \frac{<\bar{\phi}_\lambda|U_L(\infty,0)XU_L(0,-\infty)|\bar{\phi}_\mu>}{[<\bar{\phi}_\lambda|U_L(\infty,0)U_L(0,-\infty)|\bar{\phi}_\lambda><\bar{\phi}_\mu|U_L(\infty,0)U_L(0,-\infty)|\bar{\phi}_\mu>]^{\frac{1}{2}}} \quad . \end{aligned} \tag{26}$$

The numerator of Eqn. (26) demands that we insert an X-vertex at t=0 and otherwise join at t=0 all fermion lines from histories which have evolved forward from the remote past under $U(0,-\infty)$ with those which evolved backwards from the remote future under $U(\infty,0)$. The denominator involves the same coupling but without the X-vertex. Factorization of the first kind immediately extracts a factor of C_o^2 (see Fig. 3) from both numerator and denominator since the vacuum fluctuations never reach t=0 and thus play no role in the application of the X-vertex. Removal of this common factor converts each U to a U_L as shown in the third line of Eqn. (26). In this last form both numerator and denominator retain singularities arising from intermediate active states in linked valence diagrams. This divergent nature will be removed below by folding and factorization of the second kind.

We turn now to an analysis of the terms involving the X-vertex. In Fig. 11 these terms are first collected into two groups, X^V_{AB} if the X-vertex is connected to valence lines which reach the time boundaries, and X^C_{AB} if the X-vertex is connected only to a core excitation. In either case, the remaining parts of core excitations are joined across t=0 and since they are disconnected we may apply GTO to their vertices relative to those of other parts of the diagrams and thereby extract a factor of the norm of the true core ground state, $\langle\psi^Q_c|\psi^Q_c\rangle$. This factor has been removed from the definition of X^V and X^C.

$$\langle A|U_L(\infty,0)XU_L(0,-\infty)|B\rangle = \{X^V_{AB} + X^C_{AB}\}\langle\psi^Q_c|\psi^Q_c\rangle$$

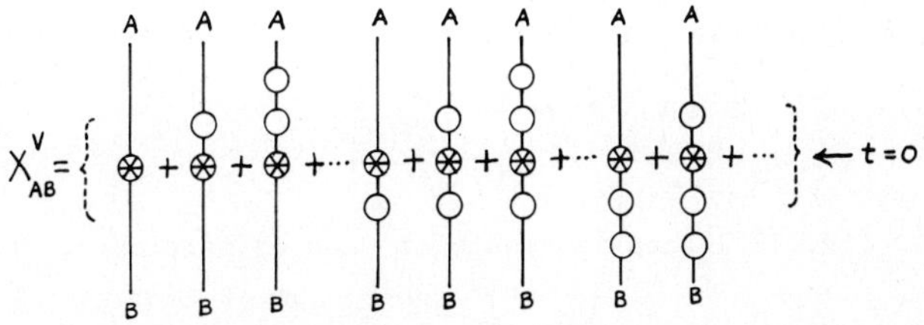

$$= \langle A|[U(\infty,0)XU(0,-\infty)]_{LV}|B\rangle$$

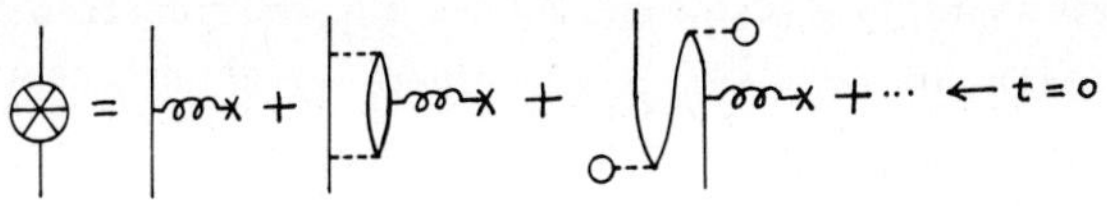

Fig. 11 Classification of diagrams in X^V and definition of the X-box.

Diagrams of the valence term X^V (see Fig. 11) are then grouped according to the number of intermediate active states present in the intervals $(\infty,0)$ and $(0,-\infty)$. We also define in Fig. 11 a new diagrammatic element, the X-box, consisting of the X-vertex and any number of associated H_1-vertices between which we have passive intermediate states on either side of t=0. From the examples given in Fig. 11, we see that the X-box is formed by the joining of linked valence diagrams which arrive at t=0 in active or passive states. In addition the third term shows how core excitations can be joined to valence diagrams at t=0.

In Fig. 12 the valence sum is factorized by the now familiar operation of folding intermediate active states and we extract the divergent factors $\langle A|U_{LV}(\infty,0)P$ and $PU_{LV}(0,-\infty)|B\rangle$ whereas the diagrammatic sum of folded diagrams, F(X), is well-defined and finite.

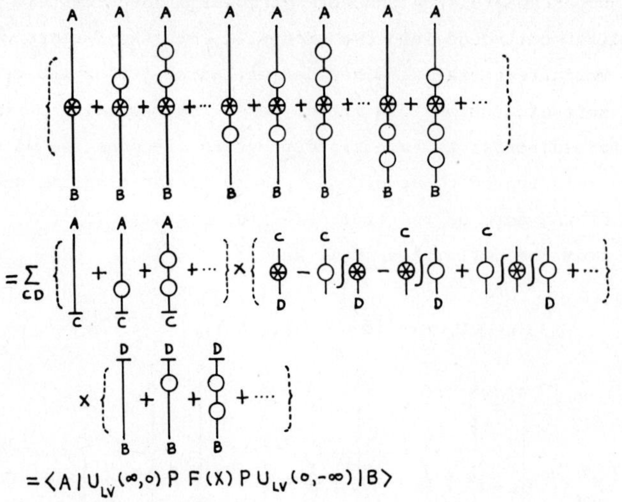

$$= \langle A | U_{LV}(\infty,0)\, P\, F(X)\, P\, U_{LV}(0,-\infty) | B \rangle$$

Fig. 12 Factorization of X^V and definition of F(X).

In Fig. 13 we turn to the analysis of diagrams of the core term X^C. The linked valence diagrams are of two forms. When we join at t=0 two valence diagrams which arrive at t=0 in active states we denote the t=0 junction by a horizontal line and realize their sum as $\langle A|U_{LV}(\infty,0)PU_{LV}(0,-\infty)|B\rangle$ as shown in Fig. 14. When we join at t=0 two valence diagrams which arrive at t=0 in a passive state or when we join core excitation and valence elements at t=0, we indicate this t=0 junction

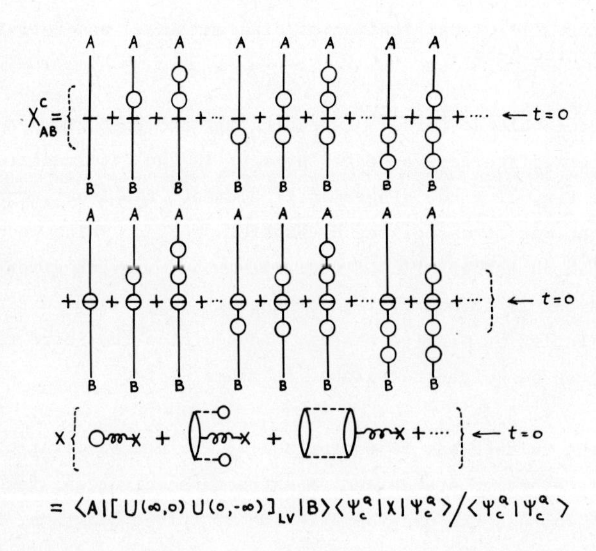

Fig. 13 Valence and core terms contained in X^C.

by a new diagrammatic element, the divided Q-box which is defined in Fig. 14. The horizontal line in the divided Q-box reminds us of the t=0 junction on either side of which is a common intermediate state to be accounted for in the application of the diagrammatic rules to the value of a divided Q-box (the divided Q-box is dimensionless whereas the Q-box has dimensions of an energy).

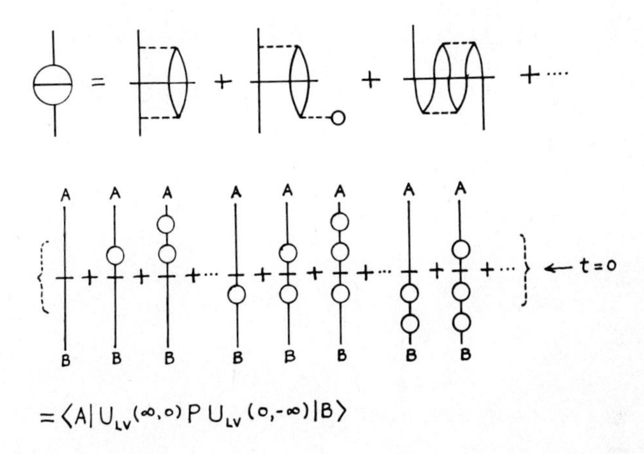

Fig. 14 Definition of the divided Q-box and factorization of valence diagrams which arrive at t=0 in active states.

The sum of core diagrams containing the X-vertex is the final factor in Fig. 13. These core diagrams are identical with the Goldstone linked cluster expansion for the true ground state expectation value of X in the non-degenerate case. Since the X-vertex here indicates time t=0, this set of diagrams is identical with $\langle\psi_c^Q|X|\psi_c^Q\rangle/\langle\psi_c^Q|\psi_c^Q\rangle$.

In Fig. 15 the set of divided Q-box diagrams is factorized by folding with obvious results. The notation F(I) has been introduced for the sum of folded diagrams involving divided Q-boxes since the creation of a pair of common intermediate states by the t=0 junction line is diagrammatically equivalent to applying a vertex of the unit operator I to such diagrams at t=0. (The reader should be warned to not interpret the sum F(I) literally as F(X) with the X-vertex merely replaced by the I-vertex. F(X) contains as its lowest order term an X-vertex unaccompanied by any H_1 vertices. Such a term is missing from our definition of F(I). In reality this lowest order I-vertex term is present and is represented by the factorized sum in Fig. 14. This difference in definition[3] of F(I) and F(X) is the reason why the final result in Fig. 16 contains F(X) in the numerator but I+F(I) in the denominator.)

When we turn to the denominator in Eqn. (26) we realize immediately that the diagrammatic representation of these normalization terms is just that of Fig. 13 for X^C if one merely omits the core diagram factor.

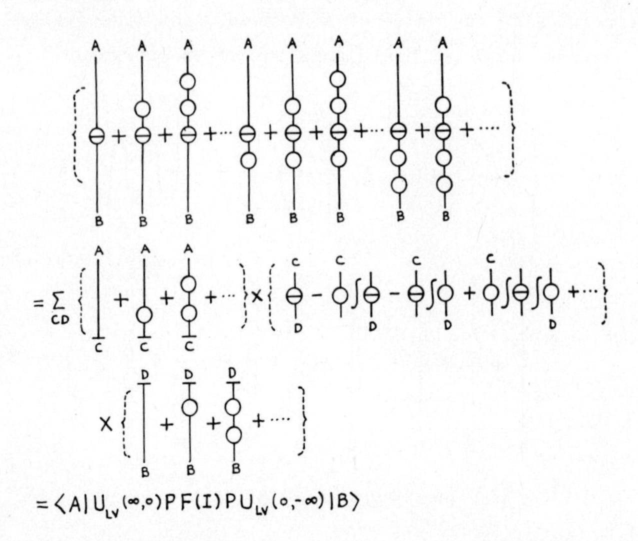

Fig. 15 Factorization of valence diagrams with a divided Q-box element.

The final element in the factorization arises from the application of P (in the form of Eqn. (22)) to Eqn. (24) to yield

$$|\phi_\lambda\rangle = \frac{P|\psi_\lambda\rangle}{\langle\bar{\phi}_\lambda|\psi_\lambda\rangle} = \frac{PU(0,-\infty)|\bar{\phi}_\lambda\rangle}{\langle\bar{\phi}_\lambda|U(0,-\infty)|\bar{\phi}_\lambda\rangle} = \frac{PU_{LV}(0,-\infty)|\bar{\phi}_\lambda\rangle}{\langle\bar{\phi}_\lambda|U_{LV}(0,-\infty)|\bar{\phi}_\lambda\rangle} \tag{27}$$

and its obvious conjugate relation. The replacement of U by U_{LV} in the final form of Eqn. (27) is possible only because the application of P yields a ratio of model space matrix elements out of which we can factor $\langle 0|\psi_c^\Omega\rangle$ as well as C_o. Thus we have displayed a factorization

$$PU_{LV}(0,-\infty)|\bar{\phi}_\lambda\rangle = |\phi_\lambda\rangle\langle\bar{\phi}_\lambda|U_{LV}(0,-\infty)|\bar{\phi}_\lambda\rangle \tag{28}$$

formally equivalent to the Goldstone relation, Eqn. (5), for the unperturbed core (which is its own biorthogonal element for the one-dimensional problem). We denote the divergent factor by U_λ

$$U_\lambda \equiv \langle\bar{\phi}_\lambda|U_{LV}(0,-\infty)|\bar{\phi}_\lambda\rangle = \langle\bar{\phi}_\lambda|U_{LV}(\infty,0)|\bar{\phi}_\lambda\rangle \tag{29}$$

In terms of this final factorization our matrix elements simplify

$$\begin{aligned}
&\langle\bar{\phi}_\lambda|U_{LV}(\infty,0)P[I+F(I)]PU_{LV}(0,-\infty)|\bar{\phi}_\lambda\rangle \\
&\qquad = \langle\phi_\lambda|I+F(I)|\phi_\lambda\rangle_{QLV}U_\lambda^2 \\
&\langle\bar{\phi}_\lambda|U_{LV}(\infty,0)PF(X)PU_{LV}(0,-\infty)|\bar{\phi}_\mu\rangle \\
&\qquad = \langle\phi_\lambda|F(X)|\phi_\mu\rangle_{QLV}U_\lambda U_\mu
\end{aligned} \tag{30}$$

where the subscript QLV reminds us that this term is now free of all singularities. In the ratios calculated for $X_{\lambda\mu}$ the divergent factors U_λ and U_μ cancel identically. As an example we consider the ratio which multiplies $\langle X\rangle_c$ and find

$$\begin{aligned}
&\frac{\langle\phi_\lambda|I+F(I)|\phi_\mu\rangle_{QLV}}{\langle\phi_\lambda|I+F(I)|\phi_\lambda\rangle_{QLV}^{\frac{1}{2}}\langle\phi_\mu|I+F(I)|\phi_\mu\rangle_{QLV}^{\frac{1}{2}}} \\
&= \frac{\langle\psi_\lambda|\psi_\mu\rangle}{\langle\psi_\lambda|\psi_\lambda\rangle^{\frac{1}{2}}\langle\psi_\mu|\psi_\mu\rangle^{\frac{1}{2}}} = \delta_{\lambda\mu}
\end{aligned} \tag{31}$$

which is thus seen to be merely a fancy representation of the Kronecker delta function.

$$X_{\lambda\mu} = \frac{\langle\psi_\lambda|X|\psi_\mu\rangle}{\langle\psi_\lambda|\psi_\lambda\rangle^{1/2}\langle\psi_\mu|\psi_\mu\rangle^{1/2}}$$

$$= \langle X\rangle_c\,\delta_{\lambda\mu} + \frac{\langle\phi_\lambda|F(X)|\phi_\mu\rangle_{QLV}}{\langle\phi_\lambda|I+F(I)|\phi_\lambda\rangle_{QLV}^{1/2}\,\langle\phi_\mu|I+F(I)|\phi_\mu\rangle_{QLV}^{1/2}}$$

$$\langle\phi_\lambda|F(X)|\phi_\mu\rangle_{QLV} = \sum_{AB}\langle\phi_\lambda|A\rangle\langle B|\phi_\mu\rangle$$

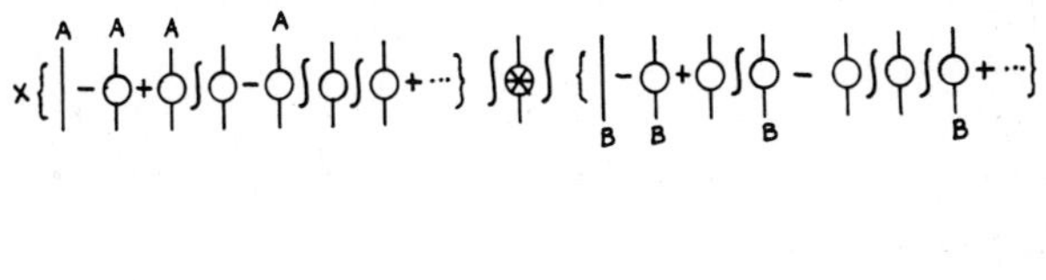

Fig. 16 True matrix elements of X expressed as well defined functions of model space matrix elements.

The final result[3] is shown in Fig. 16 in which every term is manifestly finite and unambiguously defined diagrammatically. We note that, as expected, the true core expectation value enters additively only for diagonal matrix elements. We also note that this final expression employs only the eigenstates of H_{eff}. The biorthogonal set $\bar{\phi}$ has proved invaluable as a means to an end but no longer appears in the final expression. We finally note that this expression is independent of the normalization chosen for ϕ.

I would like to express my appreciation to T. T. S. Kuo for a long and stimulating collaboration. I also thank Drs. Anastasio, Krenciglowa, Lee, Tsai, and Weng whose interaction helped clarify many of these ideas.

REFERENCES

1. T. T. S. Kuo, S. Y. Lee and K. F. Ratcliff, Nucl. Phys. A176, 65 (1971).
2. K. F. Ratcliff, S. Y. Lee and T. T. S. Kuo, unpublished.
3. E. M. Krenciglowa and T. T. S. Kuo, Nucl. Phys. A240, 195 (1975).
4. M. Gell-Mann and F. Low, Phys. Rev. 84, 350 (1951).
5. J. Goldstone, Proc. Phys. Soc. (London), A239, 267 (1957).
6. G. Oberlechner, F. Owono-N'-Güema and J. Richert, Nuovo Cimento 68B, 23 (1970).
7. T. T. S. Kuo in Dynamic Structure of Nuclear States, ed. D. J. Rowe et al (Univ. of Toronto Press, Toronto, 1972).
8. E. M. Krenciglowa and T. T. S. Kuo, Nucl. Phys. A235, 171 (1974).
9. M. R. Anastasio, J. W. Hockert and T. T. S. Kuo, Phys. Lett. 53B, 221 (1974).
10. M. R. Anastasio and T. T. S. Kuo, Nucl. Phys. A238, 79 (1975).
11. B. H. Brandow, Rev. Mod. Phys. 39, 771 (1967).

K.F. RATCLIFF: TIME DEPENDENT APPROACH (KUO, LEE AND RATCLIFF AND KUO AND KRENCIGLOWA APPROACHES)

Brandow: I'd like to remark that the square-root normalization factors in your expression for the effective operator matrix elements X_{AB} are not fully linked -- the valence norms involve disconnected valence diagrams. By using the orthogonal "hat" model vectors, however, it is possible to formally tie the normalization factors to the numerator of your X_{AB} so as to get a fully linked (or fully connected) result. I don't believe that a result of this character can be achieved without using the "hat" vectors of the hermitized model hamiltonian.

Ratcliff: The disconnected feature of these last diagrams arises by the joining of the valence elements at t = 0 without the requirements that the t = 0 vertex bring one back to a model space state. When this requirement is applied, as in the evaluation of H_{eff} which develops in the semi-infinite interval, then of course everything is completely connected. It is of course true that the sum of the subset of folded diagrams which are disconnected factorize into the product of the folded sum of the disconnected elements.

Kallio: I want to bring in some criticism of this whole approach. To my mind this theory of folded diagrams is reaching such a degree of complication that of the two possibilities

a) you can perform practical calculations

b) you cannot,

the case (b) would be preferable because just explaining what you have done in (a) is next to impossible. I would feel it preferable to break your initial symmetry and hence decrease its degeneracy so as to be able to stick to non-degenerate Bethe-Goldstone theory with deformation.

Ratcliff: In considering these three folded diagram theories, there are two criteria on which to focus. The first is whether or not the theory itself is correct and complete. I would hope that our approach is seen as so systematic that you are convinced that, beyond being a plausibility argument, everything really is included and is furthermore free from any form of double counting, etc. The second criterion is that of usefulness. Here again the theory must be systematic enough that it is unambiguous what is included in every term of the series and how to evaluate each such term. To this end the clean order of the series in our formulation should seem attractive. Finally, in this connection, let me rebut an observation Brandow made in his talk. The number of folded diagrams which enter these theories is very much larger than the number of diagrams which are not folded. If a particular formulation leaves you with the task of identifying and evaluating every folded diagram one at a time you have a right to be unhappy. In our formulation we emphasize the Q-box as the basic diagrammatic element and define it in such a way that the only thing that appears in our series is that same element repeated over and over in folded diagrams. In this way one has made considerable progress in summing the folded diagram series by expressing each term in the folded series as a definite function of the unfolded term. One's physical insight enters in determining what finite set of effects are most important to include in one's approximation to the Q-box. The simplicity of being able to sum the folded series of identically repeated elements is lost if one employs the reducibility of diagrams via a stretching operation as Brandow advocates. On balance it then seems to us that here is one case where the maximum possible cancellation of diagrams is not automatically a good thing to do.

Brandow: I disagree completely with that.

Ratcliff: There are no compensating folded diagrams which restore the original complexity to my "reduced diagram" result. This is because I have also redefined the "time boundaries" of the Q-boxes which mark the levels where the folds must be made. The final result is simpler in every respect.

Vary: In the previous talk there was considerable attention paid to the consequences of employing an effective interaction from Brueckner theory whereas this has not entered your presentations at all. Do you neglect this aspect and simply treat the G matrix as if it were H_1, i.e. some weak potential, without regard for its proper theoretical origin? This I am surmising based on the fact that you start from the Gell-Mann-Low theorem and never depart from the simple (i.e."weak" or perturbative) interaction picture.

Ratcliff: The first job one has is to take care of the short range correlations by performing the ladder sums in the Q-box to convert to G's. That this partial summation to give G's is not interfered with by the time ordering of folded elements is not obvious in the folded form. However, once one realizes that the sum of all such time orderings can be written as the difference between singular non-folded diagrams as in Fig. 7, then the outcome of the ladder sum becomes obvious. In principle there are no problems in the formalism with the proper handling of short range correlations.

Kümmel: Do you understand the relation of your theory to Baranger-Johnson?

Ratcliff: Of the two time dependent approaches, Baranger-Johnson stress the creation of an instantaneous model space interaction. We stress factorization in the direct calculation of the matrix elements of H_{eff}. The relationship of these approaches is not immediately obvious though the comparative examples I have worked out have given the same answers suggesting a relationship exists. In the Baranger-Johnson approach the instantaneous interaction is being formed for the model space evaluation operator everywhere in the full interval $(+\infty, -\infty)$. If I accept their basic approach, then it suggests to me that the model space evolution operator should give a result for an infinitesimal time interval which is independent of where that time interval is chosen. In our approach all the action takes place in a particular interval chosen at $t = 0$ for reasons I have explained. It may be argued that our factorization has the effect of generating an instantaneous interaction PHU_0P in an infinitesimal interval at $t = 0$, infinitesimal because the factor which multiplies PHU_0P is the full evolution operator U over the entire semi-infinite interval $(0, -\infty)$. I have not derived this relationship in detail but suspect these observations will prove to be the key to a formal understanding.

Landé: Your figures of Q-box-diagram series lead one to ask:

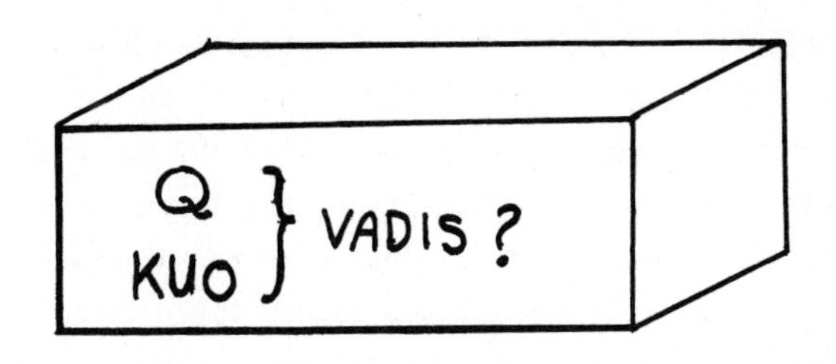

DETERMINATION OF EFFECTIVE MATRIX ELEMENTS FROM EXPERIMENTAL DATA

Igal Talmi

The Weizmann Institute of Science

Rehovot, Israel

Having listened to the previous lectures it seems to me safe to draw two simple conclusions. The first is that if the shell model really works, it must be some kind of a miracle. The second conclusion is that it will take us a long time to get the effective interaction in nuclei from many body calculations. I would therefore like first to assure you that the shell model does indeed work in certain cases remarkably well. I would then try to tell you what people have been doing in order to determine matrix elements of the effective interaction from experimental data in actual nuclei, these people being somewhat impatient to wait until all theoretical and computational difficulties are solved and the matrix elements handed over on a silver platter by many body theorists.

What do we mean by saying that the shell model works? The shell model could be considered just as a convenient basis of wave functions which can be used to express the various orders of the perturbation expansion. The shell model furnishes a complete set of states and in order to obtain results, the many-body problem should be solved. For most people, however, the meaning would be that reasonably simple shell model configurations with reasonably simple effective interactions give a reasonable explanation of the experimental data. It is clear that the larger the shell model space adopted the better the potential agreement with experiment. On the other hand, the bigger the space the more matrix elements of the effective interaction that should be determined. It is difficult to define a reasonable upper limit to the size of the shell model space which should be considered. On the other hand, the spherical shell model with the Pauli principle provide in every case the minimum configuration that must be considered. Such a configuration could serve as a starting point.

The simplest meaningful effective interaction is a two-body interaction. We can start by using it and trying to explain the data. If we get good agreement with the data, we may as well stop. What if we don't? We can enlarge the shell model space or we could increase the complexity of the interaction, including three-body forces etc. In principle, it is just a matter of taste. One could stick to a very simple model space and include rather large three-body terms and higher ones. In practice, however, there are definite criteria. Experimentally, we look most of the time at the low lying levels (certainly at the lowest levels of each spin and parity). If the simple configuration adopted cannot account for such levels, it makes more sense to include admixtures of other configurations.

It is true that such mixings may give rise to three-body terms in the Hamiltonian sub-matrix of the simplest configuration. Still, it would make sense to include these effects in the effective interaction only if corresponding states of the perturbing configurations lie much higher, their weights are small and do not change appreciably when going from one state to another.

If the perturbing states are low, their configurations should be included explicitly in the shell model space. This is justified not only by the requirement for a complete description of low lying levels. Large admixtures can actually be measured by stripping reactions, transition probabilities, etc. Moreover, if a larger shell model space is adopted, more states could be included in the analysis. If these also come out correctly from the calculation, there is more confidence in the procedure adopted.

Electromagnetic moments and transition probabilities usually require better wave functions than the ones obtained by a simple shell model calculation. The effect of admixtures of high configurations may be replaced by using effective single particle operators ("effective charge"). Still, transition probabilities are usually sensitive to small admixtures of nearby configurations. Therefore, the first step in a shell model program is to calculate energies which are stationary at the correct eigenstates. If these come out right, refinements should be carefully introduced to obtain also moments and transitions.

Even if we are trying to determine matrix elements of an effective two-body interaction, we cannot just look at systems of two nucleons outside closed shells. We have to know which states these nucleons occupy. We should therefore examine nuclei in which there are several valence nucleons and see whether the configurations adopted are sufficient to explain the data. It is always possible to take a sufficiently large shell model space and fit all measured experimental energy levels. It is true that at the same time many more levels are predicted, but these lie naturally higher and their energies are difficult to measure. Also at higher energies more configurations could be excited and these will have to be included in the analysis.

Some people believe that they know the exact interaction to be used in the shell model. In some cases this knowledge is based on experience, in other cases it is considered to be handed over directly by Nature ("it is well known that the effective interaction is"). A rather fashionable choice has been the Kuo-Brown matrix elements. Once such an interaction is adopted, a sufficiently large shell model space is chosen so that the particular feature of interest will emerge from the calculation.

The question that should be raised is whether such a procedure has any quantitative significance. Since agreement with experiment can always be achieved, we are unable to learn anything about nuclear wave functions or about effective interactions. The other approach which many people have been using may look at first equally unjustified. One assumes a simple configuration and two-body interactions and tries to see whether this could explain the data. The restriction to the simplest possible configuration is not a matter of faith or lack of imagination. The idea is to obtain consistency checks which should be satisfied by the experimental data. If we have a two-body effective interaction, matrix elements in the n-particle configuration are linear combinations of those in the two particle system. Thus, for simple configurations there are very simple relations that should be satisfied by the data before they can be accepted for the determination of the effective interaction.

These relations do not depend on the exact nature of the two-body effective interaction. The latter should be kept as general as possible. As we have learned from many-body theory, the effective interaction is a very complicated operator, it is non-local and should strongly depend on the configuration adopted. We therefore try to consider the most possible general two-body effective interaction, yet even such an interaction gives rise to many consistency relations which can be checked experimentally.

Let me illustrate this procedure by a rather old example[1,2], well known to many of us, which is still, however, the "best typical example". The spectra of ^{38}Cl and ^{40}K are shown in Fig. 1. We assume the simplest jj-coupling configurations, one $1f_{7/2}$ neutron and one and three $1d_{3/2}$ protons respectively, outside the closed shells of ^{36}S. We can then express energies in ^{40}K as linear combinations of energies of ^{38}Cl and vice versa. The agreement between experimental ^{38}Cl level spacings and those calculated from the ^{40}K levels is striking indeed. If this is not a conspiracy of Nature, we can deduce the matrix elements of the effective interaction in this case with some degree of certainty.

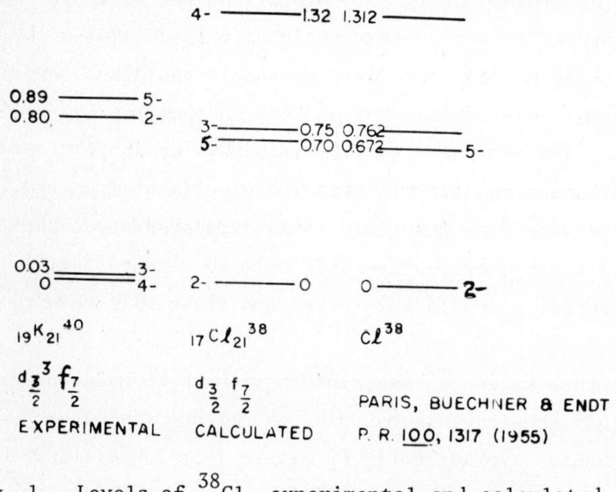

Fig. 1. Levels of ^{38}Cl, experimental and calculated from ^{40}K.

The binding energy of ^{40}K can also be accurately calculated from the binding energy of ^{38}Cl if we determine single proton and neutron energies from ^{37}Cl-^{36}S and ^{37}S-^{36}S binding energy differences. The agreement obtained indicates that differences in the various effective matrix elements do not change much between ^{38}Cl and ^{40}K. It is not clear, however, that they do not undergo a J-independent change, nor that single particle energies are exactly constant. This point will be further discussed in the following.

Another simple example is offered by configurations with $1g_{9/2}$ protons and $2d_{5/2}$ neutrons in the zirconium region[3,4,5]. The proton-neutron effective interaction gives rise to some isomeric states whose energies are very well reproduced by the shell model calculation. Details of the effective interaction in this case will be presented in the talk by J.P. Schiffer.

Let us now consider several nucleons in the same j-orbit. The necessary matrix elements can be determined from the spectrum of the two particle configuration. The energy differences in the n-nucleon configuration are linear combinations of these matrix elements with coefficients given by squares of fractional parentage coefficients. The energies calculated by using these two-body matrix elements should agree with the experimental ones if the configuration is pure. Let us first consider identical j-nucleons. In that case, for $j \leq 7/2$, some of the consistency conditions assume a very simple form. Nuclei with even number of j-nucleons should have the same level spacings as the nucleus with n=2. Also nuclei with n odd should have the same level spacings as in the n=3 nucleus. For $j > 7/2$ these simple consistency conditions would hold only for a certain class of two-body interactions for which seniority is a good quantum number[6].

A simple example can be seen in the $(2d_{5/2})^n$ neutron configurations of Zr isotopes[7]. Looking at even nuclei we see identical 0-2-4 levels in the two nucleon configuration of ^{92}Zr and the two hole configuration of ^{94}Zr (Fig. 2). We also see that in ^{96}Zr the 0-2 separation is much higher indicating possible closure of the $2d_{5/2}$ shell. In ^{93}Zr the 5/2 - 3/2 spacing (0.267 MeV) agrees very well with the one calculated from the spectra of even nuclei (0.263 or 0.254 MeV). Also the 3/2 - 5/2 M1 electromagnetic transition is strongly attenuated[8] as it should be within a j^n configuration of identical nucleons[9]. The J=9/2 level calculated to lie at 1.1 MeV may be identified with a level at 1.17 MeV[10] but this is still to be confirmed. This is another example where we can determine matrix elements of the effective interaction.

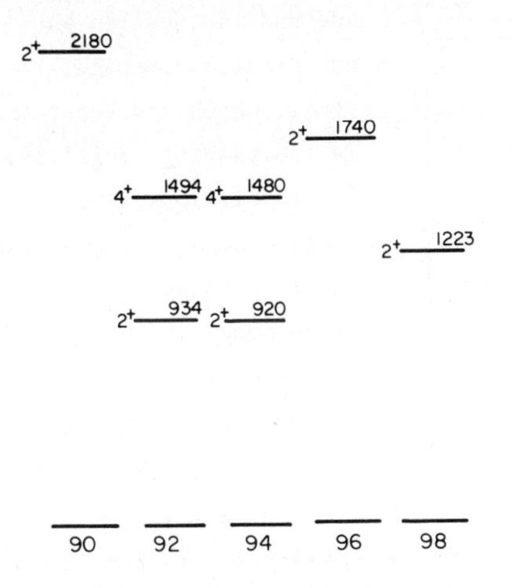

Fig. 2. Levels in even Zr isotopes (in keV)

I shall not discuss here more details of this interaction but it is of some interest to note that the spacings in this case are almost exactly those of a pure quadrupole interaction (0.9 and 1.5 respectively). If the effective interaction is expressed as a "pairing-plus quadrupole interaction", in this particular case the pairing term contributes less than 10% of the matrix elements. If a delta potential is used instead of the pairing interaction, its contribution is even smaller. The situation is similar in the $1d^n_{5/2}$ configurations in oxygen isotopes but there the rather close $2s_{1/2}$ orbit gives rise to considerable perturbations.

In addition to level spacings, we can consider also binding energies of j^n configurations. If seniority is a good quantum number, which is always the case up to $J \leq 7/2$, ground states for even n have J=0, v=0 and for odd n - J=j, v=1. Their energies are then given by[11)]

$$B.E.(j^n) = B.E.(n=0) + nC + \frac{n(n-1)}{2}\alpha + [\tfrac{n}{2}]\beta \tag{1}$$

In (1) the first term is the total energy of the closed shells, the second expresses the single nucleon energies and the last two terms are linear combinations of two-body interactions. The constants α and β are given by

$$\alpha = \frac{2(j+1)\bar{V}_2 - V_o}{2j+1} \qquad \beta = \frac{2(j+1)}{2j+1}(V_o - \bar{V}_2) \tag{2}$$

where

$$V_o = \langle j^2 J=0 | V | j^2 J=0 \rangle , \quad \bar{V}_2 = \sum_{J>0 \text{ even}} (2J+1)\langle j^2 J | V | j^2 J \rangle / \sum_{J>0 \text{ even}} (2J+1)$$

Single nucleon separation energies are then given by

$$B.E.(j^n) - B.E.(j^{n-1}) = C + (n-1)\alpha + \tfrac{1}{2}(1+(-1)^n)\beta \tag{3}$$

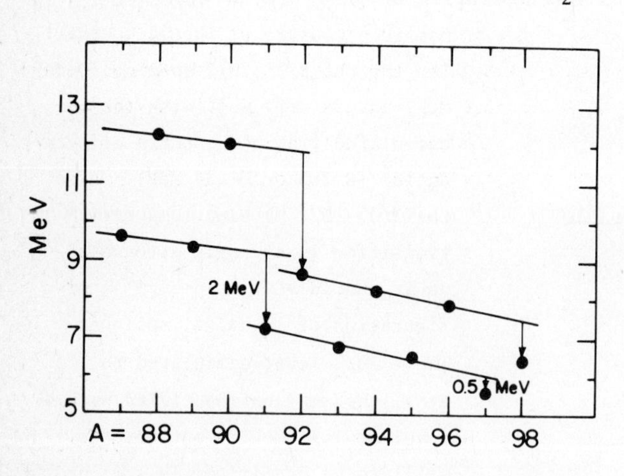

Fig. 3. Neutron separation energies of Zr isotopes.

and should thus lie on two parallel straight lines, which are separated by the pairing coefficient β. In Fig. 3 we see neutron separation energies in Zr isotopes. We see large jumps at the magic number N=50. These are followed by two parallel straight lines and then a definite although smaller drop, which may serve as another indication for the closing of the $2d_{5/2}$ sub-shell.

Although the experimental data agree very well with the mass formula (1) it is certainly not justified to reverse the argument and claim that both α and β are indeed equal to the linear combinations (2). It is clear that polarization effects of the core by the j-nucleons may contribute to the energy linear and quadratic terms in n which will be absorbed in both C and α. One may expect that such a polarization will not contribute to the pairing term and indeed in the case of $(2d_{5/2})^n$ neutrons we obtain $\beta = -1.6$ MeV from which follows $V_0-\bar{V}_2 = -1.28$. This value is in very good agreement with the center of mass of the J=2 and J=4 levels which lie in ^{92}Zr at 1.294 MeV and in ^{94}Zr at 1.280 MeV above the ground state.

In all cases considered so far, it was found that the pairing term β is rather large and attractive[12]. On the other hand, the coefficient of the quadratic term α is small, but definitely repulsive[12] (for the $2d_{5/2}$ orbit we find $\alpha = 0.2$ MeV). The fact that α is repulsive is consistent with the saturation properties of nuclear energies. If more and more identical nucleons are added to a given nucleus, the binding energy per nucleon decreases until the nucleus becomes particle unstable. This experimental fact is consistent with the existence of a strong symmetry energy in conjunction with the <u>nuclear</u> binding energy for N=Z nuclei being roughly proportional to the nucleon number A. If we attribute this repulsion to two-body interactions in lowest order, it is not easy to trace it to a definite component of the interaction. Within a single j-orbit, the multipole expansion is not unique for T=1. Thus, according to (2) repulsive α implies only that V_0 should be more attractive than $2(j+1)\bar{V}_2$.

In order to find out more about this repulsion, we have to consider nucleons in different orbits. If we add more identical nucleons to a given nucleus, the trend of binding energies does not reverse when one orbit is filled and nucleons start to fill the next orbit. Therefore, we expect that identical nucleons in different orbits repel each other on the average. There are a few cases in which such T=1 matrix elements have been determined from experiment. The average interaction energy defined by

$$\bar{V}_{T=1}(j_1j_2) = \sum_{J=|j_1-j_2|}^{j_1+j_2} (2J+1) \langle j_1j_2T=1J|V|j_1j_2T=1J\rangle \Big/ \sum_{J=|j_1-j_2|}^{j_1+j_2} (2J+1) \qquad (4)$$

turns out to be always repulsive. For example, the separation energy of a $2p_{3/2}$ neutron in ^{41}Ca is 6.3 MeV, while its separation energy in ^{49}Ca is only 5.1 MeV. In this case $\bar{V}_{T=1}(1f_{7/2}\ 2p_{3/2})$ is about 0.15 MeV. If this repulsion is due to two-body interactions it must arise from the monopole term (the monopole part of the expansion in scalar products of irreducible tensors of <u>both</u> the direct and the exchange terms). All other multipoles do not contribute to the average interaction (4). What is the origin of this monopole repulsion?

It is clear that the T=0 part of the interaction should be more attractive than the T=1 part. Still, it is not easy to see why the latter should be repulsive on the average. Various many-body calculations were unable to reproduce such a repulsion. One possible solution to this difficulty is to attribute this repulsion, or part of it, to polarization effects of the core by the valence nucleons. Since our information about this repulsion, as well as about the strong symmetry energy, comes only from binding energies this possibility may well be consistent with the data. Valence neutrons may polarize the core causing a simple (monopole) expansion of the proton distribution. When more neutrons are added, this polarization would probably become more difficult and the gain in binding energy smaller. If we had reliable effective interactions to be used in a Hartree-Fock calculation, these features could be investigated in detail. There are some indications that the matrix elements determined from binding energies do not reproduce spacings between levels in the same nucleus (e.g. the separation between T=2 and T=1, $J=0^+$ levels in ^{40}Ca or ^{40}K). This point certainly deserves closer examination.

Pure j^n configurations are not so common. If we look at N=50 nuclei, the spectra at first glance look rather irregular. A close look, however, reveals that it is only the J=0 states in even nuclei and J=9/2 states in odd ones that are strongly perturbed. It has been known for many years that these spectra can

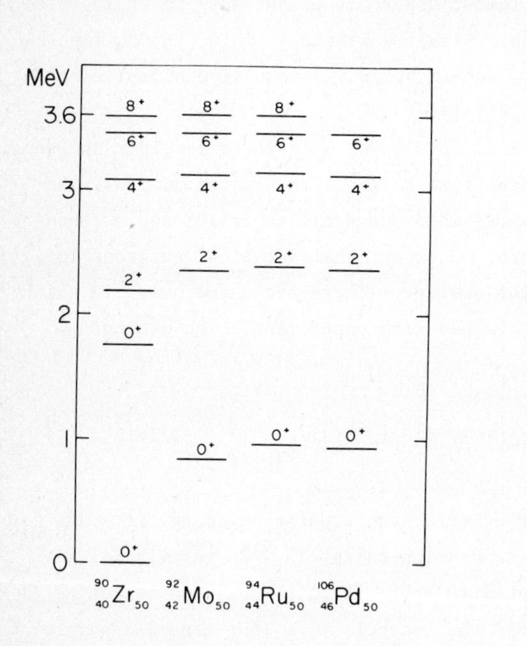

Fig. 4. Levels of even nuclei with N=50

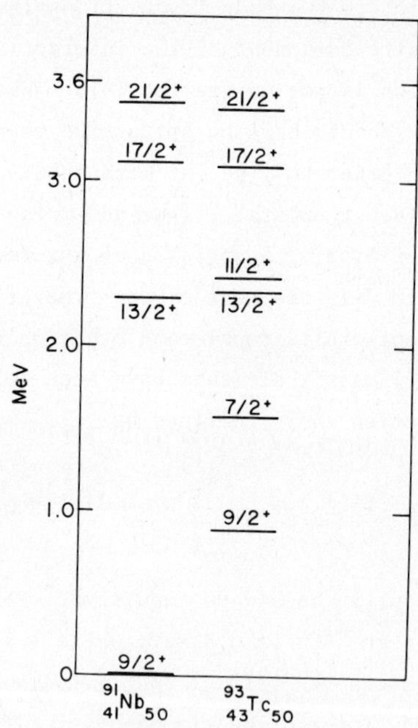

Fig. 5. Levels of odd nuclei with N=50.

be very well calculated by using both $1g_{9/2}$ and $2p_{1/2}$ proton orbits[3-5,13-16]. In this case we can see whether seniority is a good quantum number in actual nuclei also for $j > 7/2$. The matrix elements of the effective interaction obtained from these nuclei do maintain seniority as a good quantum number. This feature is clearly displayed by the fairly equal spacings between corresponding v=2 levels (Fig. 4) and v=3 levels (Fig. 5) in these nuclei.

Seniority is intimately associated with the pairing interaction operator. Yet, the latter is far from being the only interaction which is diagonal in the seniority scheme. In fact, v=2 levels with J=2,4,6,... as observed in actual nuclei are far from being degenerate. This rather strong deviation from the pairing interaction is even more pronounced in odd nuclei where the J=j-1 level with v=3 is rather close to the v=1, J=j ground state[12].

Another region in which seniority for identical nucleons can be investigated is that of $1h^n_{9/2}$ proton configurations in N=126 nuclei[17]. The three proton spectrum in ^{211}At can be accurately calculated from the two proton spectrum in ^{210}Po (Fig. 6). It is worthwhile to point out that the two particle matrix elements are much closer to those obtained from a δ-potential than in the $1g_{9/2}$ case. Also the J=7/2 level is not so close to the ground state.

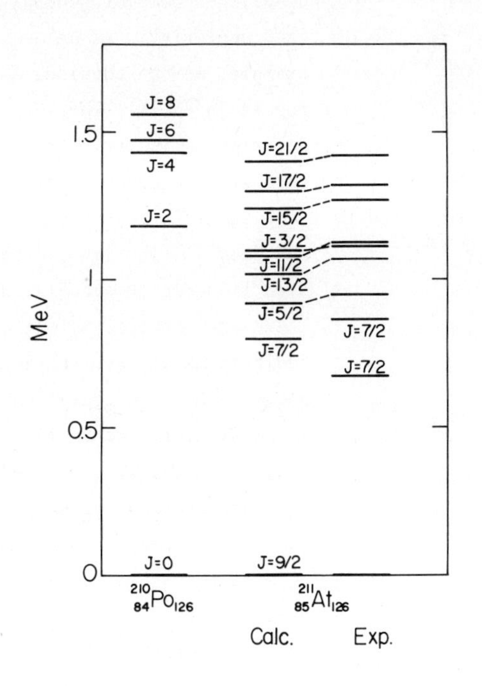

Fig. 6. Levels of $(1h_{9/2})^n$ configurations.

In other cases the complexity is even higher than for the proton configurations in N=50 nuclei. In such cases we cannot read the two particle matrix elements directly off the levels of the nucleus with two nucleons outside closed shells. A striking example is offered by neutron configurations outside ^{56}Ni. Naively we may take the ground state and first excited 2^+ state of ^{58}Ni as due to the $1p^2_{3/2}$ configuration. If we proceed to ^{60}Ni, however, we do not see any indication of subshell closure. Instead, we see the same spectrum as in ^{58}Ni and this persists even for higher nickel isotopes (Fig. 7). The spectrum looks as if it is due to a single j-orbit. The results of shell model calculations[18,19,20] show that the $2p_{3/2}$ $1f_{5/2}$ and

$2p_{1/2}$ orbits are being filled simultaneously and ground states contain very definite admixtures of these configurations given by[21,22)]

$$(S^+)^n|0> \qquad \text{where} \qquad S^+ = \sum \alpha_j S_j^+ \qquad j = 2p_{3/2},\ 1f_{5/2},\ 2p_{1/2} \tag{5}$$

and $$S_j^+ = \frac{1}{2}\sum_m (-1)^{j-m}\, a_{jm}^+ a_{j-m}^+$$

It has been shown that in such ground states, which can be characterized by generalized seniority v=0, binding energies of even nuclei follow the simple expression (1). This gives rise to a simple behaviour of pair separation energies as shown in Fig. 8.

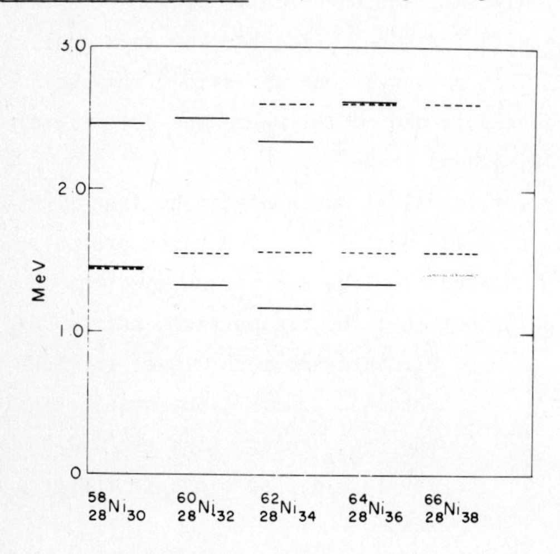

Fig. 7. Experimental (solid lines) and calculated (dashed lines) J=2 and J=4 levels in Ni isotopes.

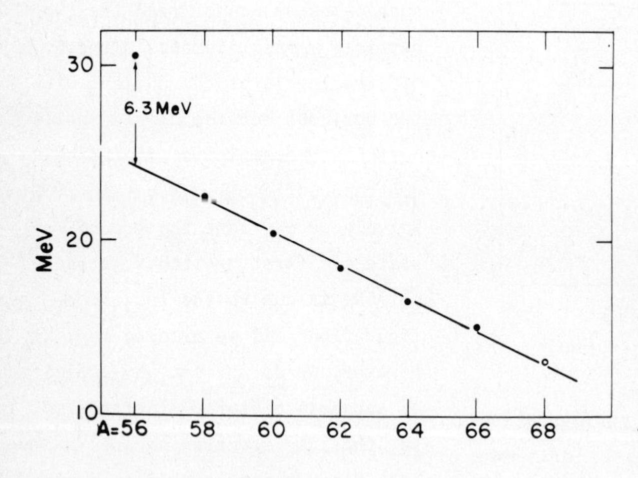

Fig. 8. Neutron pair separation energies of Ni isotopes.

If one uses some effective interactions in common use, the resulting admixtures may lead to the coupling scheme of generalized seniority. As an example, we can consider the $1f_{7/2}$ shell. The description in terms of pure $1f_{7/2}$ nucleons explains well the main features of the data[11,23,24)] as can be seen from Fig. 9 and Fig. 10. There are, however, clear indications of perturbations due to other configurations. In particular, the J=3/2 state is strongly perturbed by configurations with $2p_{3/2}$ nucleons. Good quantitative agreement has been obtained by considering both $1f_{7/2}$ and $2p_{3/2}$ orbits[25,26,27)]. It is interesting to see what is the effect on binding energies of the various admixtures. In Fig. 11 we

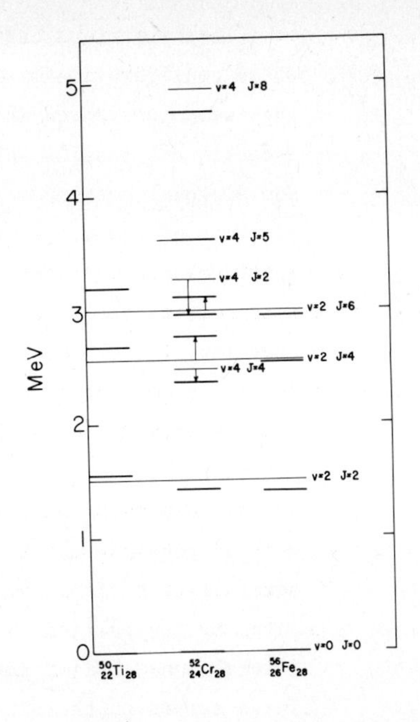

Fig. 9. Experimental and calculated (thin lines) levels in even nuclei with $1f^n_{7/2}$ proton configurations.

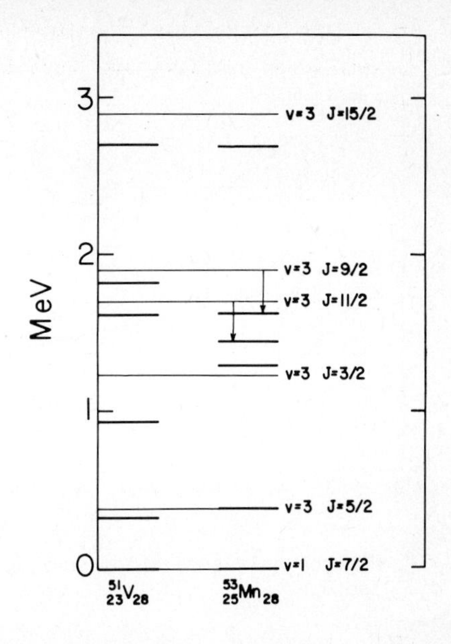

Fig. 10. Experimental and calculated (thin lines) levels in odd nuclei with $1f^n_{7/2}$ proton configurations.

see neutron pair separation energies of Ca isotopes where shell closure at N=28 is very pronounced. In Fig. 12 we see the results of several calculations. Three sets of these which give good results up to N=28 go smoothly over and fail completely to recognize the magic number N=28, in a way similar to the situation in Ni isotopes. Kuo-Brown matrix elements give the adequate shell closure at N=28 only after appreciable modifications (indicated by crosses).

This may be an interesting opportunity to see how matrix elements of the effective interaction, determined from experiment, depend upon the choice of the shell model space adopted. In Table 1 we see the effect of admixtures on the $1f_{7/2}$ matrix elements. The bigger the shell model space, the smaller the matrix elements (in absolute value) attributed to $1f^2_{7/2}$ states. It has been shown that energy levels in this region can be well reproduced by using a three-body interaction in addition to the two-body one[28]. Still, such a description would not yield the rather large admixtures of $2p_{3/2}$ states and certainly could not be extended beyond ^{48}Ca. Including nucleons in the $2p_{3/2}$ orbit greatly improves the wave functions so that it is possible to calculate transition probabilities in a much better way[29].

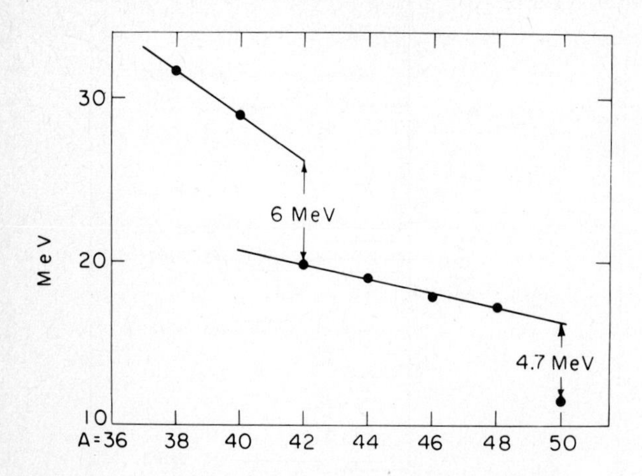

Fig. 11. Neutron pair separation energies of Ca isotopes.

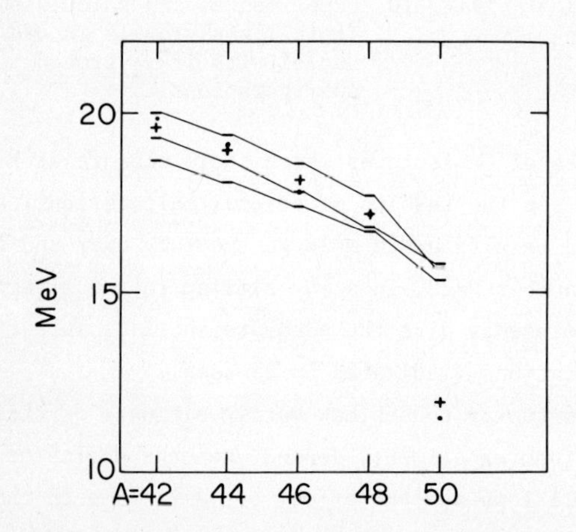

Fig. 12. Experimental (dots) and calculated neutron pair separation energies of Ca isotopes. Three sets miss the magic number N=28 while Kuo-Brown matrix elements become adequate only after appreciable modifications (crosses).

In the various cases where configuration mixings were considered in a consistent fashion, also non-diagonal matrix elements of the effective interaction were determined with some degree of certainty. They turn out to be not negligible and yet their effects are not always large. Their actual values in comparison with diagonal elements seem different from those given by the pairing interaction. Still, the large errors on the matrix elements determined from experiment make it difficult to reach a definite conclusion. The main shortcoming of the pairing interaction is that it does not reproduce the average repulsion between identical nucleons in different orbits (it acts only in two particle states with J=0). An interaction which does much better is the modified surface delta interaction[30]. Although the ratios between the various non-diagonal elements with J=0 are the same as for the pairing interaction, it does not have the shortcomings of the latter. It has been used rather successfully in several cases including the nickel isotopes[20]. In some cases where we have information from experiment

Table 1

Matrix elements $\langle j_1 j_2 J|V|j_3 j_4 J\rangle$ with $1f_{7/2}$ and $2p_{3/2}$ orbits

j_1	j_2	j_3	j_4	J	Pure $f_{7/2}$-shell	$f_{7/2}$ and some $p_{3/2}$ nucleons [a]	All $f_{7/2}$ - $p_{3/2}$ configurations [b]
7/2	7/2	7/2	7/2	0	-3.11	-2.80	-2.11
				2	-1.52	-1.29	-1.11
				4	-0.36	-0.17	-0.10
				6	+0.08	+0.34	+0.23
7/2	7/2	7/2	3/2	2		-0.50	-0.50
				4		-0.31	-0.31
7/2	3/2	7/2	3/2	2		-0.35	-0.56
				3		+0.78	+0.25
				4		-0.60	+0.28
				5		+0.60	+0.49
7/2	7/2	3/2	3/2	0		-0.78	-0.78
				2		-0.27	-0.27
7/2	3/2	3/2	3/2	2		-0.33	-0.32
3/2	3/2	3/2	3/2	0		-1.35	-1.21
				2		-0.28	-0.38

[a] P. Federman and S. Pittel, Nucl. Phys. A155 (1970) 161.

[b] J.B. McGrory, B.H. Wildenthal and E.C. Halbert, Phys. Rev. C2 (1970) 186.

on the signs of non-diagonal elements, they agree with the signs given by the surface delta interaction. In the case of the $1g_{9/2}$ and $2p_{1/2}$ protons discussed above, it was possible to determine the sign only by using information from M4 electromagnetic transitions[16)]. The sign thus determined agrees with the surface delta interaction and in this case it is opposite to the sign given by the pairing interaction.

Our knowledge about T=0 matrix elements is less reliable. The most important feature of the T=0 interaction is its being rather strong and attractive. Another important feature, which follows is that spacings between single nucleon energies become less important and thus in states with lower T values there is much more configuration mixing than in those with maximum isospin T. In particular, T=0 matrix elements are effective in mixing configurations with $j = \ell + \frac{1}{2}$ and $j' = \ell - \frac{1}{2}$ nucleons[12]. These features make the extraction of reliable values of T=0 matrix elements of the effective interaction much more difficult.

Let us first consider protons and neutrons in a single j-orbit. Had seniority provided a good scheme of eigenstates, ground states for even n would have J=0, v=0, t=1 and for odd n J=j, v=1, t=1/2. Binding energies for such states, apart from the Coulomb energy, would have been given by[6]

$$\text{B.E.}(j^n T) = \text{B.E.}(n=0) + nC + \frac{n(n-1)}{2}\alpha + [T(T+1) - \frac{3}{4}n]\beta + [\frac{n}{2}]\gamma \qquad (6)$$

where

$$\alpha = \frac{(6j+5)\bar{V}_2 + (2j+1)\bar{V}_1 - 2V_o}{4(2j+1)} \qquad \beta = \frac{(2j+3)\bar{V}_2 - 2V_o - (2j+1)\bar{V}_1}{2(2j+1)} \qquad \gamma = \frac{2(j+1)}{2j+1}(V_o - \bar{V}_2)$$

and $\bar{V}_1$ is the average interaction energy in T=0 states

$$\bar{V}_1 = \sum_{J\ \text{odd}} (2J+1)V(j^2 T=0,J) \Big/ \sum_{J\ \text{odd}} (2J+1)$$

Although the assumptions made in the derivation of (6) turn out to be incorrect, it displays in addition to quadratic and pairing terms, the symmetry energy term. This mass formula can even be applied with a moderate degree of success to actual cases[31]. The symmetry term thus obtained is strong and attractive. In the case of the $1f_{7/2}$ shell it is possible to correct for deviations from the seniority scheme. The agreement of (6) with experiment is then reasonably good. It becomes much better if a cubic term is included. Such a term could naturally arise from perturbations due to $2p_{3/2}$ nucleons which are certainly important. It is also reasonable to assume that the ^{40}Ca core energy, as well as single particle energies, undergo modifications giving rise to terms proportional to n^3 in addition to linear and quadratic terms in n.

Let us now consider how good is seniority in such configurations. If the interaction energy were diagonal in the seniority scheme, spacings between levels with the same quantum numbers of seniority v and reduced isospin t, with the same n and T, would have been independent of the latter numbers. If we compare, for example, the J=2, 4, 6 levels with v=2, t=1 in ^{42}Ca (or ^{42}Sc, T=1) and in ^{48}Sc (n=8, T=3) we can appreciate the departure from good seniority (Fig. 13). The situation is similar in the $g^n_{9/2}$ levels in ^{90}Zr and ^{88}Y (Fig. 14).

The deviations from constant level spacings can be seen, although less dramatically, by comparing level spacings in n=2, T=1 and n=4, T=0 cases. In ^{44}Ti the 0-2 separation is 1.1 MeV, as compared to 1.5 MeV in ^{42}Ca. The 0-4

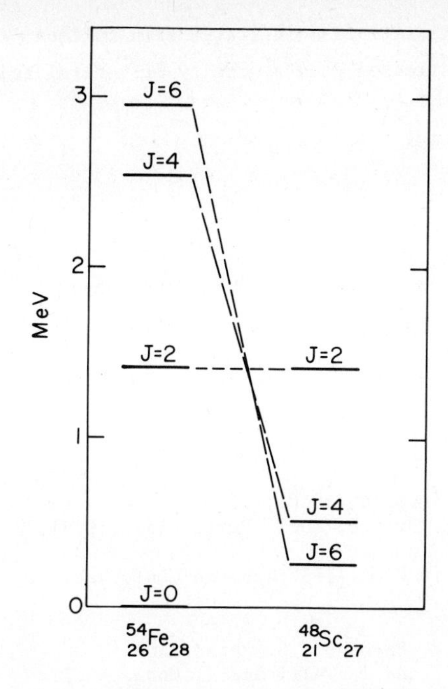

Fig. 13. J=2,4,6 level spacings for $f^2_{7/2}$ (^{42}Ca) and $f_{7/2}\ f^{-1}_{7/2}$ (^{48}Sc) configurations.

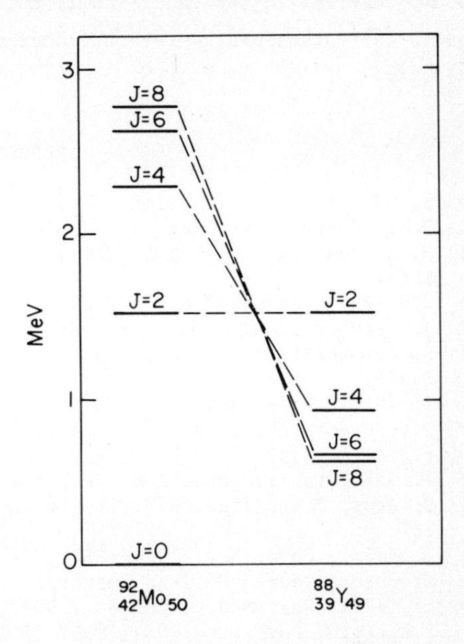

Fig. 14. J=2,4,6,8 level spacings for $g^2_{9/2}$ (^{90}Zr) and $g_{9/2}\ g^{-1}_{9/2}$ (^{88}Y) configurations.

spacing, however, does not appreciably change. This trend is in the direction leading to a rotational spectrum. Rotational-like spectra arise in several cases with T=0 and T=1/2. This feature may serve as indication that admixtures of other configurations are important for configurations with low values of isospin T.

Stronger configuration admixtures in states with lower T values emerge from many calculations. In the (1d-2s)-shell extensive calculations have been carried out assuming that nucleons occupy the $1d_{5/2}$, $2s_{1/2}$ and $1d_{3/2}$ orbits[32]. These calculations involve diagonalization of huge matrices whose matrix elements are linear combinations of 63 two-body matrix elements and three single particle energies. It is clear that even a reasonably good fit to the data cannot determine equally well all these matrix elements. Since the resulting configurations are thoroughly mixed, some simple coupling scheme is highly desirable. Such a scheme would involve less parameters so that they could be more reliably obtained from experimental data. If such a scheme works, we will have better information about the matrix elements of the effective interaction also in this case. We will also learn what features of the effective interaction give rise to rotational spectra (as are observed in the (d,s)-shell) unlike the situation in states with maximum isospin where the spectra show the characteristics of good seniority or generalized seniority.

REFERENCES

1) S. Goldstein and I. Talmi, Phys. Rev. 102 (1956), 589.
2) S.P. Pandya, Phys. Rev. 103 (1956), 956.
3) N. Auerbach and I. Talmi, Phys. Lett. 9 (1964), 153 and Nucl. Phys. 64 (1965), 458.
4) K.H. Bhatt and J.B. Ball, Nucl. Phys. 63 (1965), 286.
5) J. Vervier, Nucl. Phys. 75 (1966), 17.
6) A. de-Shalit and I. Talmi, Nuclear Shell Theory, Academic Press, New York (1963).
7) I. Talmi, Phys. Rev. 126 (1962), 2116.
8) W.V. Prestwich, B. Arad, J. Boulter and K. Fritze, Can. Jour. Phys. 46 (1968), 2321.
9) I. Talmi and I. Unna, Ann. Rev. Nucl. Science 10 (1960), 353.
10) B. Arad, J. Boulter, W.V. Prestwich and K. Fritze, Nucl. Phys. A131 (1969), 137.
11) I. Talmi, Phys. Rev. 107 (1957), 326 and in Proc. 1957 Rehovot Conf. Nuclear Structure, North-Holland Amsterdam (1958).
12) I. Talmi, Rev. Mod. Phys. 34 (1962), 704.
13) K.W. Ford, Phys. Rev. 98 (1955). 1516, B.F. Bayman, A.S. Reiner and R.K. Sheline, Phys. Rev. 115 (1959), 1627, and I. Talmi and I. Unna, Nucl. Phys. 19 (1960), 225.
14) S. Cohen, R.D. Lawson, M.H. MacFarlane and M. Soga, Phys. Lett. 10 91964), 195.
15) J.B. Ball, J.B. McGrory and J.S. Larsen, Phys. Lett. 41B (1972), 581.
16) D.H. Gloeckner and F.J.D. Serduke, Nucl. Phys. A220 (1974), 477.
17) I. Bergström, B. Fant, C.J. Herrlander, P. Thieberger, K. Wikström and C. Astner, Phys. Lett. 32B (1970), 476.
18) N. Auerbach, Nucl. Phys. 76 (1966), 321, Phys. Lett. 21 (1966), 57 and Phys. Rev. 163 (1967), 1203.
19) S. Cohen, R.D. Lawson, M.H. MacFarlane, S.P. Pandya and M. Soga, Phys. Rev. 160 (1967), 903.
20) P.W.M. Glaudemans, M.J.A. De Voigt and E.F.M. Steffens, Nucl. Phys. A198 (1972), 609.
21) I. Talmi, Nucl. Phys. A172 (1971), 1.
22) S. Shlomo and I. Talmi, Nucl. Phys. A198 (1972), 81.
23) R.D. Lawson,and J.L. Uretsky, Phys. Rev. 106 (1957), 1369.
24) J.L. McCullen, B.F. Bayman and L. Zamick, Phys. Rev. 134B (1964), 515.
25) P. Federman and S. Pittel, Nucl. Phys. A155 (1970), 161.
26) J.B. McGrory, B.H. Wildenthal and E.C. Halbert, Phys. Rev. C2 (1970), 186.
27) N. Auerbach, Phys. Lett. 24B (1967), 260.
28) I. Eisenstein and M.W. Kirson, Phys. Lett. 47B (1973), 315.
29) R.N. Horoshko, D. Cline and P.M.S. Lesser, Nucl. Phys. A149 (1970), 562.
30) I.M. Green and S.A. Moszkowski, Phys. Rev. 139B (1965), 790,= R. Arvieu and S.A. Moszkowski, Phys. Rev. 145 (1966), 830, P.W.M. Glaudemans, P.J. Brussaard and B.H. Wildenthal, Nucl. Phys. A102 (1967), 563.
31) I. Talmi and R. Thieberger, Phys. Rev. 103 (1956), 718.
32) E.C. Halbert, J.B. McGrory, B.H. Wildenthal and S.P. Pandya, Advances in Nuclear Physics, Plenum Press (New York 1971) Vol. 4, p. 315 where also many references are given.

I. TALMI: DETERMINATION OF EFFECTIVE MATRIX ELEMENTS FROM EXPERIMENTAL DATA

Zamick: Isn't it true that M1 transitions in ^{38}Cl and ^{40}K are complicated?

Talmi: Yes.

Zamick: The point you raise that the monopole average

$$\Sigma(2J + 1)<[j_1 j_2]^J|V|[j_1 j_2]^J>_{T=1}$$

is repulsive is very interesting. It is easy to show that the core polarization correction in second order for $j_1 = j_2$ makes this monopole more negative. Perhaps this then serves as a motivation for higher-order calculations, although I would not know exactly how they should be done.

Goode: In the case of ^{38}Cl/^{40}K, if one carefully calculates the discrepancies between transformed ^{38}Cl and ^{40}K, these discrepancies are considerably outside the experimental error bars. This indicates further impurities. Doesn't this mean that the simple shell-model picture is not good enough?

Talmi: In view of the drastic simplifications made in this calculation, the agreement obtained is very striking. The effects of impurities, which are certainly there, seem to be very well represented by effective two-body interactions.

Koltun: With regard to Goode's question about ^{38}Cl-^{40}K, don't you think there is any interesting question raised by the fact that no simple microscopic interaction calculation of the configuration mixing in these two nuclei gives such a very small disagreement with the Pandya transform, as is seen experimentally?

Talmi: The agreement is amazing. I do not think that microscopic calculations are sufficiently reliable to cause concern. Surely the calculated non-diagonal matrix elements could be smaller by a factor of two than the values obtained so far.

HOW DO WE DECIDE WHICH UNPERTURBED BASIS TO USE ? WHAT IS THE ROLE OF SELF-CONSISTENCY ?

P.U. Sauer

Department of Theoretical Physics, Technical University
3ooo Hannover, Germany

1. Introduction

Neutrons and protons move inside the nucleus on independent orbits. This assumption forms the basis for the nuclear shell model. The overwhelming success of the phenomenological shell model in correlating data proves the qualitative validity of the assumption. It implies for the microscopic description of the nucleus, that the many-body hamiltonian H,

$$H = K + V = H_o + H_1 \qquad (1)$$

should contain a dominant one-body piece H_o, $H_o = K + U$, the shell-model hamiltonian proper, and that the remainder H_1, $H_1 = V - U$, should yield the effective interaction W between the important shell-model configurations by some sort of perturbation theory. The single-particle hamiltonian H_o defines the active valence states $|v\rangle$, the passive hole $|\lambda\rangle$ and passive particle states $|a\rangle$ of a shell-model calculation. W is computed in the basis of H_o and given in the model space by the fully linked diagram -expansion including folded diagrams[1], schematically

$$W = H_1 + H_1 \frac{Q}{E_o - H_o} W . \qquad (2)$$

W is energy-independent. In Eq. (1) K is the operator of the kinetic energy, V the free-nucleon and U the single-particle shell-model potential. In Eq. (2) Q projects onto those many-body states of H_o excluded from the model space. The single-particle energies ε of H_o and not the experimental ones are needed for the effective interaction W. E_o, $E_o = E_o^c + E_o^v$, is the total unperturbed energy of the incoming model-space state whose W matrix elements are to be calculated, E_o^c being the core-, E_o^v the valence-space part of the unperturbed energy.

Though the many-body hamiltonian H contains a dominant one-body piece, even the low-lying states of nuclei with an extra particle (or hole) outside a doubly-closed core are not of pure single-particle nature. Thus, the single-particle hamiltonian H_o cannot be obtained by direct inspection of experimental phenomena in nuclei around magic numbers. H_o is a theoretical object. How is then H_o recognized in the many-body hamiltonian H ? How is H_o chosen in accordance with the two-body interaction V ?

First of all, H_o need not to be related to V. If one were able to sum the perturbative expansion (2) for the residual interaction W to convergence including diagrams with (-U) insertions, the shell-model results would become independent of H_o. Thus, the single-nucleon hamiltonian H_o is arbitrary within reasonable limits. Possible criteria for the selection of H_o may therefore range from rigid theory to practical convenience.

On one side, the single-particle hamiltonian may be constructed by the motivation to keep the residual interaction between nucleons outside the core as small as possible. In this spirit a best choice of H_o should either (i) maximize the overlap between the physical many-body states considered and their model-space projections or (ii) minimize the absolute size of matrix elements of the effective interaction or (iii) yield an optimal rate of convergence for the perturbative expansion (2) of W suitably rearranged to by-pass all presently encountered convergence problems.

I am not aware of any theory for the single-particle hamiltonian which proceeds in the context of the shell model according to the criteria (i) or (ii). Furthermore, complete calculations of W up to third order in the reaction matrix have only been done in one single-particle basis, the harmonic oscillator, and the convergence of the effective interaction W is not apparent in such low order. The rigid theoretical criteria are therefore of no practical value when choosing H_o.

On the other hand, H_o may be selected by sheer convenience, e.g., to make the shell-model calculations technically as simple as possible. The harmonic oscillator is from this point of view the favorite choice. Its wave functions are well studied. Summations over particle states are discrete. Though mostly used, the harmonic oscillator surely sacrifices some physical intuition about the single-particle hamiltonian H_o. First, it concentrates the valence and low-lying particle states too strongly in the nuclear volume and, second, it describes the almost kinetic particle states of high excitation, needed for the reaction matrix, rather poorly. In practice, the latter deficiency is less severe than expected. For, if the particle oscillator spectrum is shifted suitably, the shell-model reaction matrices with purely kinetic intermediate states and with oscillator intermediate states turn out quite similar[2], element by element, and calculated ground-state properties of spherical nuclei based on the two different prescriptions for particle states agree[3] very well indeed. When discussing shell-model calculations performed with different choices of the single-particle hamiltonian in Sect.3, I shall therefore leave out this difference in reaction matrices altogether.

Furthermore, it was believed for some time, that even with an harmonic oscillator hamiltonian H_o the effects of bubble and (-U) insertions on the effective interaction and on effective operators almost cancel and therefore need not be calculated. Ellis and Mavromatis[4] find this simplifying assumption to be wrong. The self-energy insertions of Fig. 1, indeed, change the oscillator shell-model results sizeably; their omission is in no way justified. The most

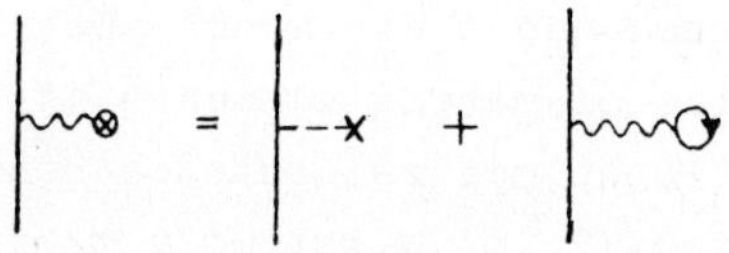

Fig. 1. Potential and bubble insertion, with convention for the sum. The wavy line indicates the reaction matrix, while the dashed line terminated by an X indicates the negative shell-model potential.

sophisticated treatment of self-energy insertions has been given by Comins and Hewitt[5] for ^{18}O. Besides including diagonal insertions implicitly up to all orders by a change of propagators, they account explicitly for diagrams with off-diagonal insertions (Fig. 2) and calculate them together with the usually considered ones up to third order in the reaction matrix. Self-energy insertions are important, but the convergence of the perturbation expansion for W, summed order by order in the reaction matrix, is not improved.

Turning now back to the practical convenience of an oscillator

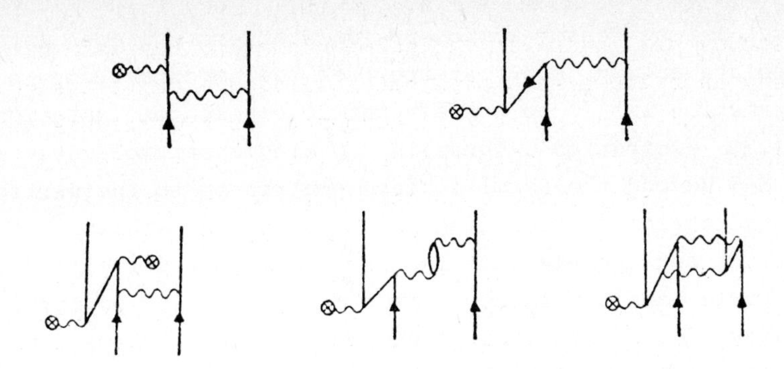

Fig. 2. Some examples of second- and third-order diagrams calculated in Ref. 5. The shown second-order diagrams contribute strongly to the effective interaction, the shown third-order diagrams are insignificant as compared to the usual ones of the same order. Sizeable wave function corrections are presumably missed in Ref. 5 by the fact that only very low passive particle states are considered.

single-particle hamiltonian, already a complete third-order calculation of the effective interaction with oscillator wave functions is quite a demanding enterprise indicating that a physically better motivated choice of the single-particle hamiltonian H_o may after all not dramatically enhance the heavy amount of labour. Especially, as long as only low orders of the perturbative expansion for W can be computed and only few selected processes be summed entirely, H_o should be picked on other grounds than technical convenience in limited calculations. Since the theoretical criteria are not helpful, physical intuition has to assist us when choosing H_o.

2. Choice of the Single-Particle Hamiltonian

When choosing the single-particle hamiltonian, is its complete definition really needed ? One might hope not. For, an ordinary shell-model calculation involves only few single-nucleon states around the Fermi surface and particle states of high excitation. The latter ones are needed for the reaction matrix, the former ones are the active valence states of the model space and the passive particle and hole states of low excitation renormalizing the bare shell-model interaction. As long as these two parts of the single-particle Hilbert space were energetically far apart, special care for a consistent choice of H_o in both was unnecessary. Using oscillator wave functions for states around the Fermi surface and non-orthogonalized plane waves for the reaction matrix reflects this sound, but unsophisticated philosophy of the early days[6]. Unfortunately, the two-nucleon interaction has a strong tensor component. It scatters nucleons into states of intermediate energy. E.g., it requires the important core-polarization correction in ^{18}O to contain particle-hole configurations of at least 12 $\hbar\Omega$ excitation[7]. Thus, in ^{18}O single-particle states up to about 15o MeV beyond the Fermi surface contribute to the particle-hole renormalization of the effective interaction, states which are also important for the reaction matrix. A complete definition of the single-particle hamiltonian H_o, consistent in all its parts, is therefore necessary. Such a consistent choice will also automatically solve the double-counting problem[2] in the two-particle ladder correction of the bare interaction.

As long as there are not many valence nucleons outside the passive core, physical intuition suggests the spherical core field as the best choice for the shell-model hamiltonian H_o. Its single-particle potential should bind all active valence states, otherwise its strength is to be increased slightly to do so. Resonances might be used as valence states, but their treatment[8] is awkward.

The Woods-Saxon potential, though phenomenological, has a sound physical basis and is therefore a legitimate single-particle potential for a shell-model calculation. Its choice constitutes a clear improvement as compared to the oscillator hamiltonian. For, the Woods-Saxon potential spreads the wave functions of states around the Fermi surface, i.e., of the valence and low-lying particle states, further out than the oscillator well allows. Of course, the Woods-Saxon potential is not tuned to the two-nucleon potential. Thus, the self-energy corrections of Fig. 1 have to be calculated as for the harmonic oscillator. Shell-model calculations with Woods-Saxon wave functions are done in Refs. 8 and 9, but have not been pushed to the sophistication of the standard harmonic oscillator scheme. I think, this is due to technical difficulties which I comment on at the end of this Section.

On aesthetic reasons and reasons of internal consistency of the theory, a microscopic definition of the single-particle hamiltonian H_o might be preferred. The problem of the shell-model hamiltonian H_o is then transferred to the microscopic theory of spherical doubly-closed nuclei and to the single-particle potential of unrenormalized or renormalized Brueckner-Hartree-Fock. Standard calculations of the core ground state aim at a satisfying description of the hole states. Their wave functions are expanded in a finite harmonic oscillator basis. The finite basis is able to describe only some low-lying particle states $|\alpha_1\rangle$ simultaneously. (The states $|\alpha_1\rangle$ form the valence

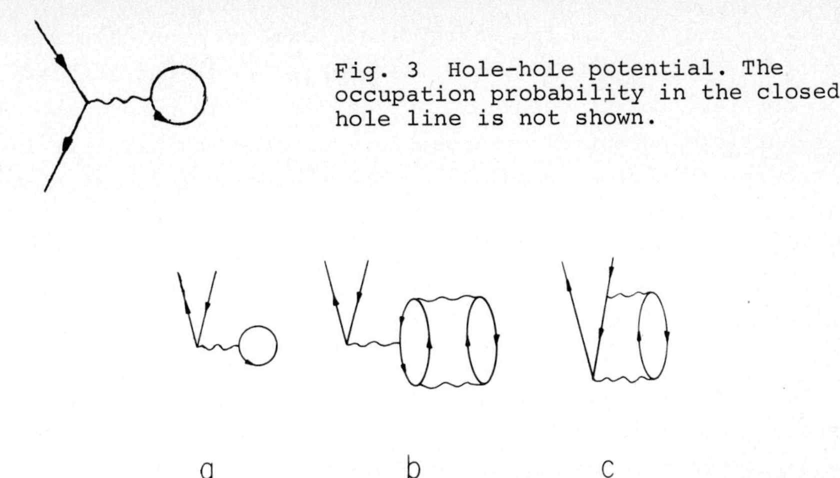

Fig. 3 Hole-hole potential. The occupation probability in the closed hole line is not shown.

Fig. 4 Particle-hole potential. The occupation probability in the closed hole line is explicitly shown in diagram b.

$|v\rangle$ and low-lying passive particle states $|a_1\rangle$ of a subsequent shell-model calculation.) Only for those states contained in the basis Brueckner-Hartree-Fock can provide a self-consistent single-nucleon potential. Its well accepted and to practical use adapted definition for the hole-hole, i.e.,

$$\langle B|U|A\rangle = \sum_C \langle BC|\tfrac{1}{2}\left[G(\varepsilon_B+\varepsilon_C) + G(\varepsilon_A+\varepsilon_C)\right]|AC\rangle P_C \ , \tag{3}$$

and the particle-hole potential is in terms of bubble insertions[10] with diagonal occupation probabilities P_A according to Figs. 3, 4a and 4b. Furthermore, a comparison of renormalized Brueckner-Hartree-Fock with the variational method of density-dependent Hartree-Fock has identified[11] the particle-hole insertion 4c, second-order in the reaction matrix, as an important saturation mechanism in finite nuclei. Its inclusion in the particle-hole potential, i.e.,

$$\langle \beta_1|U|A\rangle = \sum_C \langle \beta_1 C|G(\varepsilon_A+\varepsilon_C)|AC\rangle P_C + \\ (-\tfrac{1}{2}) \sum_{CD\gamma_1} \frac{\langle \beta_1\gamma_1|G(\varepsilon_D+\varepsilon_C)|DC\rangle P_D P_C \langle DC|G(\varepsilon_D+\varepsilon_C)|A\gamma_1\rangle}{\varepsilon_D+\varepsilon_C-\varepsilon_{\beta_1}-\varepsilon_{\gamma_1}} \tag{4}$$

is therefore recommended. Insertion 4c also removes strong

stationarity defects from the energy functional of two-body clusters[11]. The use of the particle-hole potential of Eq. (4) is not standard yet.

In contrast to the hole-hole and particle-hole potential, the particle-particle potential of Brueckner-Hartree-Fock is still in doubt. A possible choice, remodelling the finite-nucleus calculations according to current trends in nuclear matter, would be a purely kinetic particle spectrum, i.e., $<\beta_1| U |\alpha_1> = 0$. Though this choice presumably combines numerical convenience and good convergence properties for the ground state of the core nucleus, it is only reasonable for the high-lying particle states $|\alpha_h>$. It is clearly unacceptable, if one views Brueckner-Hartree-Fock as providing the input parameters for a shell-model calculation. The valence states are not purely kinetic, we expect them to be bound. The low-lying particle states are not purely kinetic either, they still see the attraction of the other nucleons and their wave functions should be modified accordingly in the nuclear volume. The transition from low-lying particle states to highly excited particle states, for which the attractive potential does not matter, should occur in nuclear matter for momenta around $2k_F$. In a finite nucleus we assume the transition to occur at the edge of the Brueckner-Hartree-Fock basis and therefore define the single-particle potential nonvanishing for all particle states within the basis. Though the bubble insertion in a particle line (Fig. 5) is not factorizable, it provides the dominant contribution

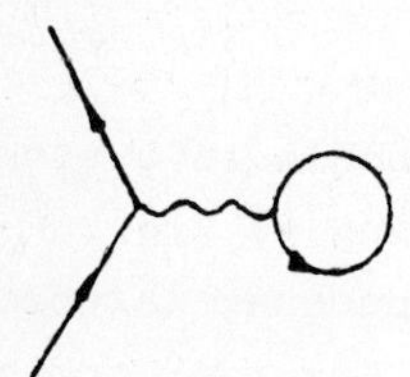

Fig. 5. Particle-particle potential. The occupation probability in the closed hole line is not shown.

to the attraction in low-lying particle states and may therefore be employed as the particle-particle potential

$$<\beta_1| U |\alpha_1> = \sum_C <\beta_1 C| \frac{1}{2} \left[G(\omega_{\beta_1} + \varepsilon_C) + G(\omega_{\alpha_1} + \varepsilon_C) \right] |\alpha_1 C> P_C , \tag{5a}$$

its off-shell energy dependence suitably averaged out,

$$\omega_{\alpha_1} = 2\bar{\varepsilon}_A - \varepsilon_{\alpha_1} . \tag{5b}$$

$\bar{\varepsilon}_A$ is the mean energy of hole states. This definition, e.g., used in the Oak-Ridge Brueckner-Hartree-Fock calculations[10], should not be extended to the high-lying particle states $|\alpha_h>$ outside the Brueckner-Hartree-Fock basis. There, the bubble insertions are to be grouped together with other higher-cluster diagrams. The single-nucleon potential involving these high-lying particle states, which are still needed for the reaction matrix, is chosen to be zero,

$$<B|U|\alpha_h> = 0, \tag{6a}$$

$$<\beta|U|\alpha_h> = 0. \tag{6b}$$

What is numerically hard, but appears feasible, is to adjust the particle states of the Brueckner-Hartree-Fock reaction matrix to the low-lying ones in a self-consistent fashion. In this respect, an ordinary shell-model calculation is faced with the same computational problem as Brueckner-Hartree-Fock. However, the partitioning of particle states in Brueckner-Hartree-Fock is the natural basis for the doubly-partitioning approach[13] to the shell model. In a doubly-partitioned Hilbert space, only the high-lying states $|\alpha_h>$ are used for the shell-model reaction matrix,

$$G(\omega) = V + V \frac{Q_{2h}}{\omega - Q_{2h}\left[K(1) + K(2)\right] Q_{2h}} G(\omega), \tag{7a}$$

$$Q_{2h} = \sum_{\alpha_h \beta_h} |\alpha_h \beta_h)(\alpha_h \beta_h|, \tag{7b}$$

and the low-lying discretized ones $|\alpha_l>$ for the particle-hole renormalization of the bare effective interaction. Eq. (7) properly orthogonalizes the purely kinetic high-lying particle states to all the other ones. Actually, the reaction matrix (7) should be modified to also contain those two-particle states, in which one is low or a valence state. This is best done as a correction to Eq. (7a).

When using the self-consistent choice (3) - (6) for H_o, all hole lines in diagrams of the shell-model interaction carry occupation probabilities P_A. This fact alone will help to reduce the ordinary higher-order corrections to the effective interaction. With the self-consistent H_o, do all diagrams of the shell-model interaction containing the self-energy insertions of Figs. 3, 4 and 5 and (-U) insertions cancel ? Not quite, but their contribution is expected to be very small. I give three characteristic examples which may stand for the general problems. Even in shell-model calculations with doubly-partitioning, Brueckner-Hartree-Fock reaction matrices of the core should be taken for the self-energy insertions and not Eq. (7).

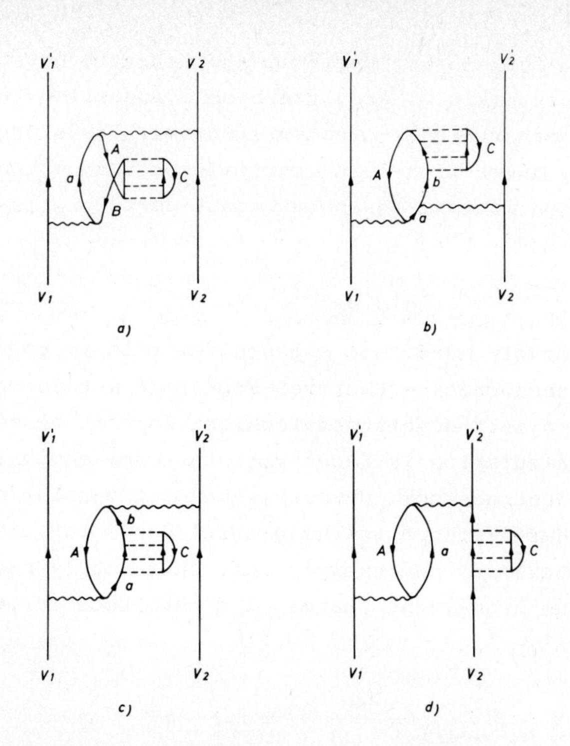

Fig. 6. Some examples of diagrams for the shell-model interaction containing self-energy insertions. The labelled particle states, $|a\rangle$ and $|b\rangle$, are low-lying ones in a doubly-partitioned approach.

(1) Since for the shell-model interaction a downward factorization scheme is used, the blown-up reaction matrix in Fig. 6a is $\langle BC|G(\varepsilon_A + \varepsilon_C)|AC\rangle$, whereas the hole-hole potential is defined according to Eq. (3). Thus, diagram 6a and the one with the corresponding (-U) insertion contribute the matrix element $\langle BC|\frac{1}{2}\left[G(\varepsilon_A + \varepsilon_C) - G(\varepsilon_B + \varepsilon_C)\right]|AC\rangle P_C$. The lack of exact cancellation is due to different factorization schemes in the shell-model and in the usual version of Brueckner-Hartree-Fock.

(2) If the valence states are non-degenerate, the blown-up reaction matrix of Fig. 6b, $\langle AC|G(\varepsilon_A + \varepsilon_C + \varepsilon_{v_1} - \varepsilon_{v_1'} + \varepsilon_{v_2} - \varepsilon_{v_2'})|bC\rangle P_C$, differs in the available energy from the first term in the potential definition of Eq. (4) by the splitting of the valence energies.

(3) Since the particle bubble does not factorize in the core nucleus, one expects contributions for the diagrams 6c and 6d and their

analogues with (-U) insertions from the mismatch of energy denominators.

Though the cancellation is not perfect, a self-consistent choice of H_o effectively reduces diagrams with self-energy insertions.

Are present Brueckner-Hartree-Fock results already suited for a subsequent shell-model calculation ? The current results are plagued by two problems: First, their agreement with experiment is still poor. Second, it is hard to assess the effect of left-out three- and higher-body cluster diagrams. If one wants to use the single-particle potential of Brueckner-Hartree-Fock for the shell model, the latter problem is inconsequential. Higher-body clusters appear in the shell-model interaction irrespective of the choice of the single-particle hamiltonian, and, as in Brueckner-Hartree-Fock, they will be calculated explicitly without modifying the single-particle potential any more. However, the disagreement with experiment is troublesome. One does not want errors in the core to obscure the shell-model results. That Brueckner-Hartree-Fock misses the total binding energy by about 4 to 5 MeV per particle throughout the periodic table does not appear serious to me. The shell-model calculation any how aims at energy differences. But it is important that the single-particle wave functions reproduce the density distributions well, such that the corresponding single-particle potential has the correct nuclear size. Among the available results[3], only the ^{16}O density is not grossly in error with experiment; this ^{16}O density is even better than the one underlying the ^{18}O shell-model calculations of Refs. 4 and 14.I expect new Brueckner-Hartree-Fock calculations at least for ^{16}O, especially using the modified particle-hole potential of Eq. (4) and producing satisfying densities[15]. Their wave functions and energies can then be taken with confidence for a subsequent shell-model calculation defining the passive hole states, valence states and passive low-lying particle states.

In the absence of successful Brueckner-Hartree-Fock calculations, especially for the ^{40}Ca and ^{208}Pb regions, the single-particle potential of density-dependent Hartree-Fock (not its effective two-body interaction) may be used as shell-model potential H_o. This single-particle potential, when derived from the Reid soft-core potential in the local-density approximation[16] will almost cancel the self-energy insertions of Figs. 3, 4 and 5, provided the reaction matrix is also derived from the Reid potential. The technical difficulties of such a semi-microscopic approach are the same as for the phenomenological Woods-Saxon potential: All particle states are continuous. The single-particle Hilbert space may be doubly-partitioned. The high-lying states can without sizeable error be taken as purely kinetic; the reaction matrix can be calculated according to Eq. (7). The continuous integrations over the low-lying particle states have to be discretized, and the wave functions of the needed low-lying particle states as well as of the hole and valence states are to be expanded in terms of harmonic oscillators in order to retain part of the simplicity of the oscillator basis.

In conclusion, we note that the present state of the computing art allows a microscopic (at least for ^{18}O), a semi-microscopic and a phenomenological choice of the single-nucleon shell-model hamiltonian H_o. Each of them is a definite improvement over the standard oscillator hamiltonian of current low-order shell-model calculations.

3. Effect of the Single-Particle Hamiltonian on the Shell-Model Results

As recalled in Sect. 1, self-energy insertions make sizeable contributions[4,5] to the effective interaction and to effective operators. We therefore have to expect non-neglegible effects on standard shell-model results from any change of the single-particle hamiltonian H_o. A proper Brueckner-Hartree-Fock potential as described in Sect. 2 has never been used yet for a shell-model calculation. Shell-model calculations in ^{18}O have been performed with Woods-Saxon wave functions[8,9] and with wave functions[4,14] obtained from Hartree-Fock for ^{16}O with Sussex matrix elements. Both calculations are not optimal: (i) In a Woods-Saxon basis the insertions of Fig. 1 should still be included, but have been left out. (ii) The ^{16}O Hartree-Fock single-particle field used in the shell-model calculations corresponds to a nuclear radius, which is by about 15% too small. As compared to the real nucleus, the hole wave functions are pulled in too much and the overlap between particle and hole states maybe unrealistically decreased. The Woods-Saxon and the Hartree-Fock single-particle potential have one common feature: The valence wave functions are spread out beyond the nuclear volume, since the valence states are states close to the potential edge. Thus, two valence nucleons are on the average further apart and exploit the two-nucleon force with reduced effectiveness. Their bare shell-model interaction, the reaction matrix, is considerably less attractive than the corresponding reaction matrix of oscillator states. The reduction is minor (less than 10%) for the $d_{\frac{5}{2}}$ state, which is still strongly bound, but up to 40% for matrix elements involving the $s_{\frac{1}{2}}$ and $d_{\frac{3}{2}}$ states. Thus, the reduction of the reaction matrix for states around the Fermi surface, i.e., of matrix elements involving valence and particle states, is the main effect of the changed single-particle hamiltonian H_o. This effect penetrates all subsequent shell-model results. When summing self-energy insertions explicitly, e.g., those of Fig. 1, this effect is pushed into higher-order diagrams of the effective interaction or of effective operators. Any change of H_o is of second or higher order in the reaction matrix for the bare interaction and of third or higher order for the core-polarization. The calculations of Ref. 5 do not test the influence of the single-particle hamiltonian on the ordinary third-order diagrams.

As a consequence of the reduced reaction matrix, the core-polarization (Ref. 17 demonstrates that the increased stability of the core in a self-consistent basis is the real physical reason for the reduction of the core-polarization correction) and the 4 particle - 2 hole renormalization of the effective interaction are also reduced, the collectivity of TDA and RPA corrections is damped. The latter facts have been demonstrated with Hartree-Fock only. I expect similar results for the Woods-Saxon single-particle hamiltonian. The qualitative nature of these results appears rather well established. The reduction of the bare interaction and its particle-hole corrections is very much a wave-function effect. There is some influence from a change in energy denominators, too, but this appears to be the less effective mechanism. However, there are unsettled quantitative problems: Using a model reaction matrix based on the Tabakin potential, Pradhan[8] carries out a careful summation of intermediate Woods-Saxon states for the core-polarization and concludes that the ratio of core polarization to bare interaction remains almost unaltered as compared to the oscillator case. Thus, there is rather little hope for a dramatic decrease in the contribution from third-order diagrams due to a change in the single-particle hamiltonian. However, the results of Refs. 4 and 14 are in quantitative disagreement with Pradhan.

What is the effect of the changed single-particle hamiltonian on real physical quantities ? All ^{18}O and ^{18}F energy levels are pushed up. Especially, the second 0^+ $T = 1$ state is affected, which gets into much better agreement with the third experimental 0^+ level, the second one of predominant two-particle nature (Fig. 7). The energy-level results are qualitatively the same for Hartree-Fock and Woods-Saxon wave functions. The effect of the changed single-particle hamiltonian on

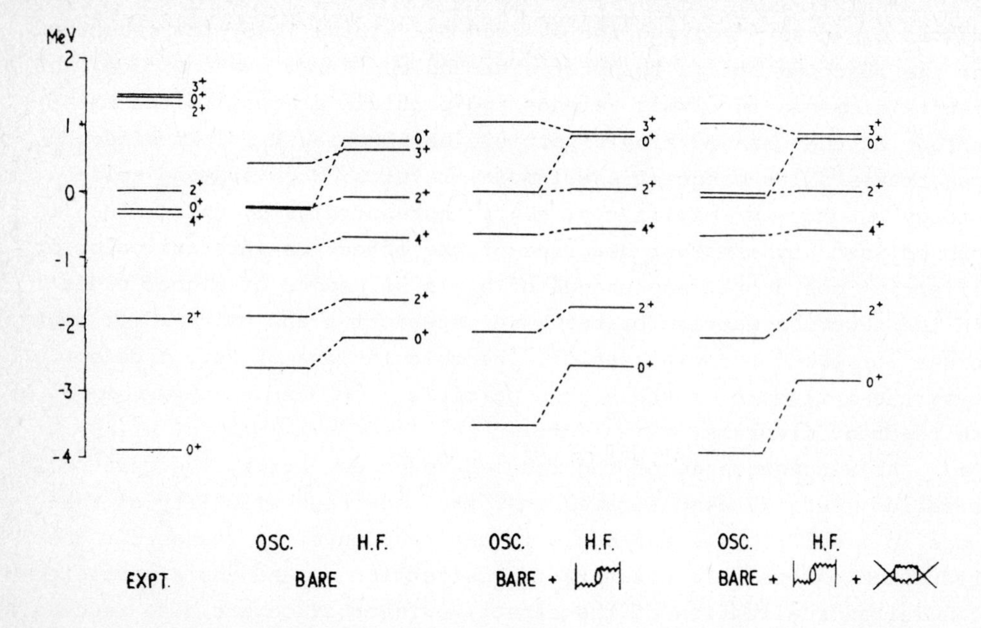

Fig. 7. Positive-parity spectra of ^{18}O. The theoretical results are from Ref. 4.Spectra obtained with oscillator and Hartree-Fock wave functions are compared.

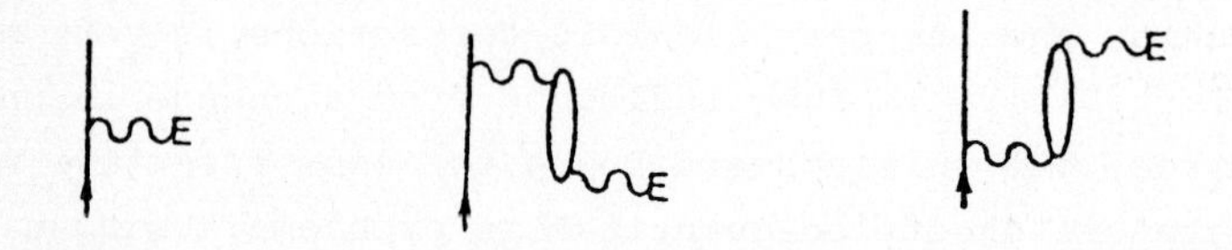

Fig. 8. The zeroth and first-order diagrams of the effective charge assuming self-consistency.

the effective charge is also sizeable. The total effective charge is reduced and individual contributions of diagrams are shifted relative to each other. The bare effective charge (diagram 1 of Fig. 8) is enhanced for protons compared to the oscillator result, its core-polarization correction (diagrams 2 and 3 of Fig. 8), however, is decreased. In the self-consistent basis, proton and neutron effective charges become very similar, but the theoretical results are too small. Experiment demands an effective charge of ≈ 0.5 in addition to zeroth order.

Shell-model calculations for a single-particle hamiltonian H_o

other than the harmonic oscillator are still very scarce. The available results indicate, however, that a physically well motivated choice of H_o is important. Such a choice is hoped to improve the reliability of shell-model calculations. The choice of the single-particle hamiltonian therefore deserves further careful theoretical attention and numerical study, though this problem is surely not the most exciting one among the many currrent problems the nuclear shell-model is faced with.

References

1. B.R. Barrett and M.W. Kirson, Advances in Nuclear Physics, ed. M. Baranger and E. Vogt, Vol. 6 (Plenum Press, New York, 1973), p. 219 and references therein.
2. B.R. Barrett and M.W. Kirson, Phys. Letters 55B, 129 (1975).
3. K.T.R. Davies, R.J. McCarthy and P.U. Sauer, Phys. Rev. C6, 1461 (1972).
4. P.J. Ellis and H.A. Mavromatis, Nucl. Phys. A175, 3o9 (1971).
5. H.N. Comins and R.G.L. Hewitt, Nucl. Phys. A228, 153 (1974). See also the contribution to this conference by Y. Starkand and M.W. Kirson.
6. G.E. Brown and T.T.S. Kuo, Nucl. Phys. A92, 481 (1967).
7. J.P. Vary, P.U. Sauer and C.W. Wong, Phys. Rev. C7, 1776 (1973).
8. H.C. Pradhan and C.M. Shakin, Phys. Letters 37B, 151 (1971); H.C. Pradhan, Ph. D. Thesis, MIT 1971, unpublished.
9. S.Kahana, H.C. Lee and C.K. Scott, Phys. Rev. 18o, 956 (1969).
1o. K.T.R. Davies and R.J. McCarthy, Phys. Rev. C4, 81 (1971) and references there.
11. K.T.R. Davies, R.J. McCarthy, J.W. Negele, and P.U. Sauer, Phys. Rev. C1o, 26o7 (1974); P.U. Sauer, Hartree-Fock and Self-Consistent Field Theories in Nuclei, ICTP Trieste 1975.
12. M. Baranger, Comments on Nuclear and Particle Physics 4, 81 (197o).
13. B.R. Barrett, Nucl. Phys. A221, 3o1 (1974).
14. P.J. Ellis and E. Osnes, Phys. Letters 41B, 97 (1972); 42B, 335 (1972); 49B, 23 (1974).
15. B. Rouben, R. Padjen and G. Saunier, Phys. Rev. C1o, 2561 (1974).
16. J.W. Negele, Phys. Rev. C1, 126o (197o).
17. D.J. Rowe, Phys. Lett. 44B, 155 (1973); H.S. Köhler, The Self-Consistent Field and the Convergence of Rearrangement and Core-Polarization Diagrams, preprint 1975.

P.U. SAUER: HOW DO WE DECIDE WHICH UNPERTURBED BASIS TO USE?
WHAT IS THE ROLE OF SELF-CONSISTENCY?

Dieperink: From your talk I get the impression that your philosophy seems to be that H_0 is a purely theoretical quantity. I agree with your attitude if you try to calculate the bulk properties of the nucleus like the total binding energy and the charge distribution. However, there exists more detailed information, from single-particle knock-out processes for example, in which one essentially measures single-particle momentum distributions and single-particle energies. Do you not feel that H_0 should be chosen in such a way that these data are also reproduced satisfactorily?, i.e., that H_0 does have some physical meaning?

Sauer: The momentum distributions and energies measured in knock-out processes do not arise from entirely pure single-particle configurations. Surely, the single-particle hamiltonian H_0 of a shell-model calculation should not be in gross disagreement with them when attempting to account for them in impulse approximation, but, within reasonable limits, H_0 is arbitrary.

Zamick: If you use oscillator parameters corresponding to a nucleus which has a radius of 10 fermi, whereas the nucleus is at equilibrium at 5 fermi then naturally you will not get self-consistency. But this is not a fault of the oscillator wave function, necessarily, but of the oscillator parameter. If you choose an oscillator parameter corresponding to 5 fermi the insertions although not zero will be much smaller.

Sauer: I agree. However, the $2s_{1/2}$ valence-state wave function of ^{18}O is even in a well of optimal size poorly described by one single oscillator. Self-energy insertions remain sizeable. They are surely artificially enhanced, once the oscillator parameter is poorly chosen. The described self-consistency corrections do not have this trivial origin.

Vary: I agree with you that in principle it is desirable to employ an H_0 which closely approximates the single-particle situation outside the core. However, in addition to the difficulties you mention there seems to be a serious problem in obtaining a local state-independent single-particle potential for nuclei beyond ^{40}Ca that satisfies simultaneously the two constraints we want most, that is, to reproduce all the single-particle energies and the density distributions.

Sauer: I am happy with a single-particle potential which reproduces the density distributions and has reasonable single-particle energies around the Fermi surface. Such a potential can be given. The one resulting from density-dependent Hartree-Fock and the Woods-Saxon potential have these desired properties.

Malta: Maybe it should be mentioned that the calculation by Ellis and Mavromatis was done with the original form of Sussex matrix elements, that is, without a hard core. The improved matrix elements yield a larger charge radius for ^{16}O.

COMPUTATION OF THE REACTION MATRIX, G*

Richard L. Becker
Oak Ridge National Laboratory
Oak Ridge, Tennessee 37830

I. Definition and Some Properties of G

Brueckner's reaction operator[1] for the interaction of two identical Fermions in a medium of the same kind of Fermions is defined by

$$G(E_s) = v + v \frac{Q}{E_s - H^o} G(E_s). \tag{1}$$

Here v is the two-body interaction, the Pauli operator Q forbids either Fermion from being scattered into a normally occupied single-particle (SP) state, and H^o is the unperturbed pair Hamiltonian

$$H^o(12) = T(1) + U(1) + T(2) + U(2),$$

where U is the SP potential which should be determined self-consistently in terms of G. In the early work the energy E_s was regarded as determined by the state (ket vector) on which G operated (to the right). Then, effectively, E_s is an operator, and in $G^+(E_s)$ E_s would have to operate to the left in order to avoid making the G-matrix non-Hermition.[2] This complication is removed and greater generality is attained by regarding E_s as a parameter held constant for all matrix elements. We thus deal with a continuous family of reaction operators, parametrized by the "starting energy",[3] E_s.

Dandow[4] has shown that the generalized-time-ordered form of perturbation theory for finite systems leads most directly to a non-Hermitian U. However, we shall assume that U is Hermitian, as it is in all shell-model and self-consistent field calculations known to us. The full interaction and Q are invariably Hermitian, so we expect that $G(E_s)$ is Hermitian. This is the case because Eq. (1) implies

*Research sponsored by the U.S. Energy Research & Development Administration under contract with the Union Carbide Corporation.

$$G(E_s) = v + v \frac{I}{Q(E_s - H^o - v)Q} v \qquad (2a)$$

$$= v + G(E_s) \frac{Q}{E_s - H^o} v \qquad (2b)$$

and from Eq. (1) $G^+(E_s)$ also satisfies (2b). Thus, if

$$v^+ = v,\ Q^+ = Q,\ U^+ = U,\ E_s = \text{real parameter} \qquad (3a)$$

then

$$G^+(E_s) = G(E_s). \qquad (3b)$$

In the Moszkowski-Scott[5] separation method $v = v_s + v_\ell$ where the separation distance, d, dividing the short- from the long-ranged part of the interaction, depends on the state (ket vector) on which v acts. In G_s^+ (calculated from v_s) d would have to act to the left in order for G_s to be Hermitian. Alternatively, d may be regarded as a parameter.

Tobocman[6] has shown that basing many-body perturbation theory on a related reaction operator

$$\overline{G}(E_s) = (v - U_2) + (v - U_2) \frac{Q}{E_s - H^o} \overline{G}(E_s)$$

or on even more general ones, has certain formal advantages. But $\overline{G}$ would be more difficult to calculate than G because U_2 depends separately on r_1 and r_2, whereas v depends on r_{12}. Thus, although $\overline{G}$ has been discussed occasionally,[7] no calculations of it have been reported. Also, the Coulomb interaction is almost always omitted from v in Eq. (1) because its long range would cause calculational difficulties, and it can be treated adequately as a perturbation.

Initially the greatest problem in computing G had to do with the strong short-range repulsive core. But this was quickly overcome by several methods. Much of the remaining difficulty arises from the Pauli operator. In degenerate perturbation theory, in which the unperturbed ("model") wave function consists of more than one configuration, there are three classes of SP states: normally occupied or "hole" states (h) with model occupation numbers $n_h = 1$; "valence" or "active" states (v) with $0 \le n_v \le 1$; and normally empty or "particle" states (p) with $n_p = 0$. There is great latitude in the choice of the active subspace in which the shell-model diagonalizations are carried out.

In the non-degenerate (closed shell) theory, in which there are no valence SP states, the Pauli operator is defined by

$$Q^{ND}(12) = Q_1(1)Q_1(2) \quad , \quad Q_1 = \sum_p^{emp} |p\rangle\langle p| . \qquad (4a)$$

We shall also define

$$P_1 = \sum_h^{occ} |h\rangle\langle h| = I_1 - Q_1 \tag{4b}$$

and

$$P^{ND}(12) = I(12) - Q^{ND}(12) = P_1(1)P_1(2) + P_1(1)Q_1(2) + Q_1(1)P_1(2). \tag{4c}$$

Q_1 and P_1 are Hermitian projection operators, i.e.

$$Q_1 = Q_1^+ = Q_1^{\;2} \quad , \quad P_1 = P_1^+ = P_1^{\;2} \tag{4d}$$

and these properties carry over to Q^{ND} and P^{ND}.

In the degenerate (open shell) theory with valence "particles" only (no valence holes) we let Q_1 and P_1 be defined as above, and

$$A_1 = \sum_{v_p}^{val.} |v_p\rangle\langle v_p| \tag{5a}$$

so that

$$P_1 + A_1 + Q_1 = I_1. \tag{5b}$$

Then Q^D is defined[8] by

$$Q^D(12) = Q_1(1)Q_1(2) + A_1(1)Q_1(2) + Q_1(1)A_1(2) \tag{6a}$$

$$= [A_1(1) + Q_1(1)][A_1(2) + Q_1(2)] - A_1(1)A_1(2) \tag{6b}$$

and

$$P^D(12) = I(12) - Q^D(12) \tag{6c}$$

$$= P_1(1)P_1(2) + P_1(1)[I(2)-P_1(2)] + [I(1)-P_1(1)]P_1(2) + A_1(1)A_1(2). \tag{6d}$$

These regions are shown in Fig. 1, from Ref. 7. The definition (6a) is most appropriate to the case in which there are few valence particles relative to the number of valence states, so that scattering into pair states of the form $|vp\rangle$ is seldom blocked by the normal occupancy of state $|v\rangle$. When there are more than two valence particles, the effective interaction differs from the G-matrix by valence-blocking corrections in addition to other corrections.

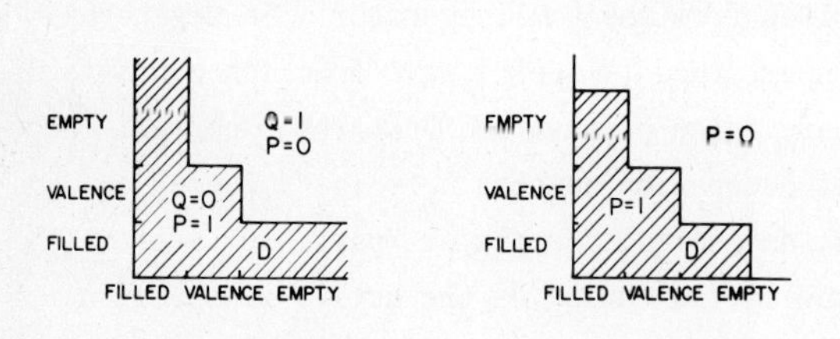

Fig. 1. Pauli projection operator P before and after truncation.

In the seldom discussed degenerate theory with valence holes only, the non-degenerate Pauli operator (4a) can be used, because only particle-particle ladders are summed by the G-matrix. However, because of (5b), the P's in (4c) must be replaced by (P+A)'s. Finally, if there are both valence particles and holes, we define

$$A_1^h = \sum_{v_h} |v_h\rangle\langle v_h| \quad , \quad A_1^p = \sum_{v_p} |v_p\rangle\langle v_p| \tag{7a}$$

so that

$$P_1 + A_1^h + A_1^p + Q_1 = I. \tag{7b}$$

Then Q^D may be defined by (6) with A_1 replaced by A_1^p and P_1 by $P_1 + A_1^h$.

The Pauli operator may be regarded as providing a dependence of G on the particle density near the interacting particles. In the non-degenerate case

$$(\vec{r}_1,\vec{r}_2|Q^{ND}|\vec{r}_1{}',\vec{r}_2{}') = [\delta(\vec{r}_1-\vec{r}_1{}') - \rho_1(\vec{r}_1,\vec{r}_1{}')][\delta(\vec{r}_2-\vec{r}_2{}') - \rho_1(\vec{r}_2,\vec{r}_2{}')] \tag{8a}$$

where ρ_1 is the SP density matrix (in position space) of the model ground state,

$$\rho_1(\vec{r}_1,\vec{r}_1{}') = \langle\Phi|\psi^+(\vec{r}_1{}')\psi(\vec{r}_1)|\Phi\rangle = \langle\vec{r}_1{}'|P_1|\vec{r}_1\rangle = \sum_h^{occ} \langle\vec{r}_1|h\rangle\langle h|\vec{r}_1{}'\rangle, \tag{8b}$$

often called the "mixed density" as distinguished from its diagonal part, $\rho(\vec{r}_1)$, the particle density at $\vec{r}_1$. An expansion about the mixed density of nuclear matter has been given.[9] Approximating the entire propagator, $Q/(E-H^o)$, by that of nuclear matter of density $\rho(\frac{1}{2}[r_1+r_2])$ is called the local density approximation (LDA).[10]

The reaction operator, as defined so far, has singularities[11] as a function of E_s. In the non-degenerate case

$$\frac{Q^{ND}}{E_s-H^o} = \sum_{p_1p_2}^{emp} \frac{|p_1p_2\rangle\langle p_1p_2|}{E_s-e_{p_1p_2}} \tag{9}$$

where $e_{p_1p_2} = e_{p_1} + e_{p_2}$, with e_p the SP energy of state $|p\rangle$. It is clear from (9) that the perturbative expansion of G has singularities for E_s equal to any $e_{p_1p_2}$. Similarly, from (2a), it is seen that a non-perturbative solution has a singularity at the eigenvalues of $Q(H_2^o + v)Q$, which lie near the $e_{p_1p_2}$. In the Brueckner-Goldstone[12] non-degenerate perturbation expansion, in its rearrangement by generalized-time-ordering,[13,3,4] and in the degenerate perturbation expansions,[8] the self-consistent E_s is always less than the lowest singularity (for non-superfluid systems). But in the SP Green's function approach to nuclear structure[14] and in the calculation of the optical potential, a transition operator is needed. The Brueckner reaction matrix must then be analytically continued to higher E_s as a Heitler reaction matrix containing a principal value operator as well as the Pauli operator. The energy dependence of G becomes linear as E_s decreases far below the lowest $e_{p_1p_2}$, which is the case for interactions far off the energy shell where $E_s = e_{k_1} + e_{k_2} - \delta E$ with δE the excitation of the medium. However, for on-shell interactions between valence particles or high-lying holes, E_s is just below the lowest singularity and there is a strong energy dependence.[15] This is seen clearly in the G-matrix elements for the s-d shell valence nucleons in ^{18}F shown in Fig. 2, taken from Barrett, Hewitt, and McCarthy.[16]

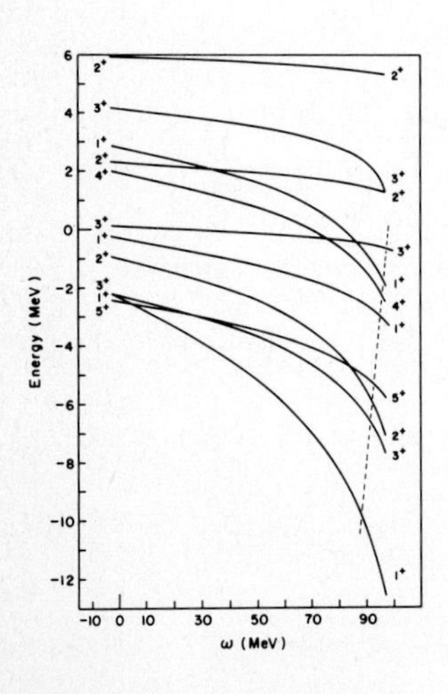

Fig. 2. T=0 spectrum in ^{18}F calculated (Ref. 16) with the G-matrix as effective interaction. The dashed line indicates the self-consistent value of the starting energy, ω,

In summary, unlike the bare interaction v, the reaction operator is dependent on the medium through Q (density dependence) and through the self-energy U (dispersive or spectral dependence), and is energy (or velocity or momentum) dependent through E_s. The latter gives rise to a "state" dependence of the effective interaction because the self-consistent E_s for a particular matrix element depends on the pair states involved in the matrix element.

II. The Bethe-Goldstone Wave Function

If v contains a hard core, Eq. (1) is singular. This difficulty is overcome by working with the equation for the Moller wave operator Ω associated with G by

$$G(E_s) = v\,\Omega(E_s), \tag{10a}$$

which satisfies

$$\Omega(E_s) = I + \frac{Q}{E_s - H^o}\, v\,\Omega(E_s). \tag{10b}$$

Acting to the right on an unperturbed pair state ϕ, it yields the correlated Bethe-Goldstone[17] pair state ψ,

$$\psi(E_s) = \Omega(E_s)\phi = \phi + \frac{Q}{E_\varepsilon - H^o}\, v\,\psi(E_s) \tag{10c}$$

which vanishes inside the hard core. A weak singularity of $v\psi$ remains, but gives no trouble as it is integrable. In relative coordinates, for the Lth partial wave

$$v_L\psi_L = \lambda_L\delta(r-c) + f_L(r) \tag{11}$$

where $f_L(r)$ is finite.[17]

It is convenient also to define the defect operator

$$\chi(E_s) = I - \Omega(E_s) = \frac{Q}{H^o - E_s}\, v\,\Omega(E_s) \tag{12a}$$

and the defect wave function

$$\zeta(E_s) = \chi(E_s)\phi = \phi-\psi(E_s) = \frac{Q}{H^o-E_s} v\,\psi(E_s). \tag{12b}$$

In infinite nuclear matter the "small parameter" κ of the compact cluster expansion[4] is

$$\kappa = \rho W \tag{13a}$$

where ρ is the nuclear density and W is the "wound integral"[18]

$$W = \int |\zeta(\vec{r})|^2 \, d^3r, \tag{13b}$$

which is a very characteristic quantity for the interaction, v.[19] The transformation from G to Ω is very useful even if v does not have a hard core.

In infinite matter where the hole state spectrum is continuous and the unperturbed pair states are taken to be plane waves, it follows from (10b) that the BG wave function for two normally occupied states and for $E_s = e_{h_1h_2}$ has no phase shift. It "heals"[18] to the unperturbed wave function because the final states permitted by Q are of higher unperturbed energy than E_s. In fact the defect function in the Lth partial wave decays as

$$\zeta_L \sim \frac{\text{const}}{r} \cos(k_F r + \eta_L)$$

where k_F is the Fermi wave number.[17] In a harmonic potential, U, the BG wave function oscillates about ϕ and "heals" to it before ϕ becomes negligible beyond the nuclear surface where it tunnels into the potential. This rapid decay is the most important property which any approximate defect function must have.

In their calculations for infinite nuclear matter, Brueckner and Gammel[20] solved the integral equation (10c) numerically after approximating the Green's function for a fixed average momentum of the pair, $\vec{K}$,

$$\langle \vec{r},\vec{K} \left| \frac{Q}{H^o-E} \right| \vec{r}',\vec{K}\rangle = \int d^3k \; Q(\vec{K}+\vec{k},\vec{K}-\vec{k};k_F) \; \frac{e^{i\vec{k}\cdot(\vec{r}-\vec{r}')}}{e_{\vec{K}+\vec{k}} + e_{\vec{K}-\vec{k}} - E}$$

first by "angle-averaging" over the angle between $\vec{K}$ and $\vec{k}$,[21] which restores spherical symmetry and uncouples different partial waves, and second by truncating at some k_{max}. The angle-averaged, nuclear-matter Pauli operator is[21]

$$Q^{AANM}(k,K;k_F) = \begin{cases} 0 \text{ if } k^2+K^2 < k_F^2 \text{ and } 1 \text{ if } k > k_F+K \\ (k^2+K^2-k_F^2)/2kK, \text{ otherwise} \end{cases} \tag{14a}$$

where

$$\vec{k} = \tfrac{1}{2}(\vec{p}_1-\vec{p}_2) \quad \text{and} \quad \vec{K} = \tfrac{1}{2}(\vec{p}_1+\vec{p}_2). \tag{14b}$$

It has been found to be a quite accurate approximation.[22] In their excellent review of methods (through 1967) for calculating G in nuclear matter Dahll, Ostgaard, and Brandow[23] found Brueckner's method could be very accurate, and found ways to improve it.

III. Representations of G in Terms of P Rather than Q

The operator Q is of infinite dimensionality for both particles, whereas each term of P is of infinite dimensionality for at most one particle. Two ways of expressing G or Ω in terms of P rather than Q are known. The first is to multiply Eq. (10c) by H^o-E_s, which leads to the Bethe-Goldstone integro-differential equation[17]

$$(H^o+v-E_s)\psi = (H^o-E_s)\phi + P\, v\psi. \tag{15}$$

Several ways of solving this equation, when P is truncated, will be described below.

A second, more complicated formulation,[7] can be derived from a familiar identity[24] for a matrix partitioned by the projection operators P and Q:

$$(M^{-1})_{QQ} = [M_{QQ} - M_{QP}(M_{PP})^{-1} M_{PQ}]^{-1} = [M-MP(M_{PP})^{-1}PM]^{-1}. \tag{16}$$

With $M = (E-h)^{-1}$, where h is H^o or H^o+v, the inverse of (16) is

$$\frac{Q}{Q(E-h)Q} = \mathcal{P}\frac{I}{E-h} - \mathcal{P}\frac{I}{E-h} P[P \frac{\mathcal{P}}{E-h} P]^{-1} P\, \mathcal{P}\frac{I}{E-h} \tag{17}$$

where once again $\mathcal{P}$ stands for the Cauchy principal value. The equation for G in terms of the full Green's function, Eq. (2a), becomes

$$G(E) = G^I(E) - \chi^{I\dagger}(E)\, P\, A(E) P\, \chi^I(E) \tag{18}$$

where we have let $G^I(E)$ denote the reaction matrix for two interacting particles in the potential U but isolated from the medium ($Q \to I$), which satisfies

$$\begin{aligned} G^I(E_s) &= v + v\,\mathcal{P}\frac{I}{E-H^o-v}\,v = v + v\,\mathcal{P}\frac{I}{E-H^o}\,G^I(E_s) \\ &= v[I-\chi^I(E_s)] = [I-\chi^{I\dagger}(E_s)]v, \end{aligned} \tag{19}$$

and where

$$A(E_s) = \left[\left(\mathcal{P}\frac{I}{E_s-H^o-v}\right)_{PP}\right]^{-1}. \tag{20}$$

The inversion in Eq. (20) can be done easily because the space P is of finite dimension. However, the evaluation of $\left(\mathcal{P}\frac{I}{E_s-H^o-v}\right)_{PP}$ can be done only approximately, in terms of a truncated set of eigenfunctions of the two-particle Schroedinger equation.

IV. The Integral Equation Relating Two Reaction Operators

As it is not possible to solve for G or ψ exactly, various approximation methods have been developed. These involve simplifying the interaction or the propagator. We should like to know in principle how the exact G is related to an approximate one so we can estimate correction terms. Fortunately, different reaction matrices are related exactly by identities. If the spectrum is continuous, these identities are integral equations. The rigorous version of a comprehensive identity of Moszkowski and Scott,[5] which allows all quantities to vary, can be derived as follows:[2,3]

$$\Omega_A(E_A) - I - \frac{Q_A}{E_A-H_A^o} G_A(E_A) = 0 \tag{21a}$$

$$\Omega_B^\dagger(E_B) - I - G_B^\dagger(E_B)\left[\frac{Q_B}{E_B-H_B^o}\right]^\dagger = 0. \tag{21b}$$

Multiplying (21a) on the left by $G_B^\dagger(E_B)$, subtracting it from (21b) multiplied on the right by $G_A(E_A)$, and using (10a), one obtains

$$G_A(E_A) = G_B^\dagger(E_B) + \Omega_B^\dagger(E_B)(v_A-v_B^\dagger)\Omega_A(E_A) + G_B^\dagger(E_B)\left\{\frac{Q_A}{E_A-H_A^o} - \frac{Q_B}{E_B-H_B^{o\dagger}}\right\}G_A(E_A). \tag{22}$$

Incidentally, a special case of this in which only E_s varies yields[25,13,26]

$$\frac{dG(E)}{dE} = -G(E)\left(\frac{Q}{E-H^o}\right)^2 G(E) = -\chi^\dagger(E)\chi(E). \tag{23}$$

One sees that the diagonal matrix elements of G are non-positive. The propagator-correction term in (22) sometimes is split[5] into a Pauli and a spectral (dispersion) term:

$$\frac{Q}{E-H^o} - \frac{Q_A}{E-H_A} = \frac{Q-Q_A}{E-H^o} + Q_A\left[\frac{I}{E-H^o} - \frac{I}{E-H_A}\right], \text{ if } [Q_A,H^o] = 0 \tag{24a}$$

$$= \frac{Q-Q_A}{E-H_A} + Q\left[\frac{I}{E-H^o} - \frac{I}{E-H_A}\right], \text{ if } [Q,H_A] = 0. \tag{24b}$$

V. Two Simple Approximations Which Provide Insight

A. The Moszkowski-Scott Separation of the Interaction. Different parts of the interaction produce quite different effects. The strong repulsive core must be treated to all orders, whereas a weak interaction need be kept only to low orders. Regions of rapid variation induce high Fourier components in the defect function, whereas slowly varying parts induce only low components. Eden and Emery,[27] Gomes, Walecka, and Weisskopf,[18] and others considered separation of the hard core, the tensor force, etc. In the Moszkowski-Scott[5] separation method, with

$$v = v_s + v_\ell, \tag{25}$$

the short-ranged part, v_s, includes, along with the repulsive core (which may be soft), the strong, rapidly varying attraction just beyond the core. The remaining long-ranged part, v_ℓ, is weak and slowly varying. A reaction matrix, G_s, obtained from v_s is defined. Since v_s produces the short-range correlation in the BG wave function, which involves primarily admixtures of high-lying unperturbed states, it is a good approximation to replace Q by I in the equation for G_s, so that

$$G_s^I(E) = v_s + v_s \,\mathcal{P}\frac{I}{E-H^o}\, G_s^I(E). \tag{26}$$

The especially clever feature of the method is that the separation distance, d, is chosen in principal such that each diagonal element of G_s, proportional to tan δ, is zero for the self-consistent value of the starting energy. The BG wave function, ψ_s, then heals to ϕ at the separation distance (see Fig. 3, from Ref. 5).

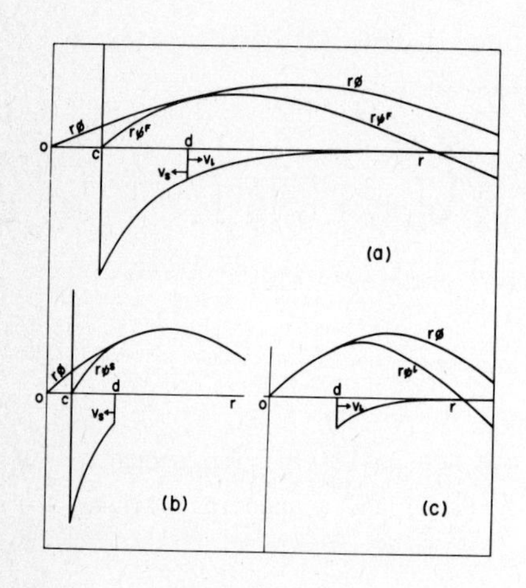

Fig. 3. Illustration of the MS separation method (Ref. 5).

Since v_ℓ is too weak to produce much wave distortion, ψ_s may be used as a good approximation to the correct ψ. The identity (22) yields

$$G(E) = G_s^I(E,d) + \Omega_s^I(E,d)^\dagger v_\ell(d)\Omega(E) + G_s^I(E,d)\,\mathcal{P}\frac{P}{H^o-E}\, G(E) \tag{27a}$$

with diagonal elements (for the self-consistent values E_α and d_α, where α labels the pair state)

$$\langle\alpha|G(E_\alpha)|\alpha\rangle = \langle\alpha|v_\ell(d_\alpha) + v_\ell(d_\alpha)\frac{Q}{E_\alpha-H^o}v_\ell(d_\alpha) + 2v_\ell(d_\alpha)\frac{Q}{E_\alpha-H^o}G_s^I(E_\alpha,d_\alpha) + \ldots|\alpha\rangle. \tag{27b}$$

For simplicity in the evaluation of G_s^I the free particle Hamiltonian, T, was used, and corresponding dispersion corrections were included in the formulation:

$$G_s^F(E) = v_s + v_s \,\mathcal{P}\frac{I}{E-T}\, G_s^F(E) \tag{28a}$$

and

$$G(E) = G_s^F(E) + \Omega_s^F(E)^\dagger\, v_\ell(E)\Omega_s(E)$$
$$+ G_s^F(E)\, \mathcal{P} \frac{P}{H^o - E} G(E) + G_s^F(E)\, \mathcal{P} \left[\frac{I}{E-H^o} - \frac{I}{E-T}\right] G(E). \qquad (28b)$$

In triplet even states Scott and Moszkowski[5] found a large contribution in (27b) from the quadratic term in the long-range tensor force,

$$v_{T\ell} = v_T(r)\ \theta(r-d)\ S_{12}, \qquad (29)$$

where S_{12} is the tensor operator. In their calculations with the separation method for finite nuclei, Kuo and Brown[28] used the free-particle spectrum and the angle-averaged, nuclear-matter, Pauli operator, $Q^{AANM}(k_F)$, with a fixed Fermi wave number, $k_F = 1.36\ \mathrm{fm}^{-1}$, appropriate to the saturation density. Because Q^{AANM} commutes with T

$$v_{T\ell} \frac{Q}{E_s - H^o} v_{T\ell} \rightarrow \int d^3k\ d^3K\ v_{T\ell}|\vec{k}\vec{K}\rangle \frac{Q(k,K,k_F)}{E_s - \frac{\hbar^2}{2m}(k^2+K^2)} \langle\vec{k}\vec{K}|v_{T\ell}. \qquad (29a)$$

They found that this could be fairly accurately approximated in a closure approximation

$$v_{T\ell} \frac{Q^{AANM}(1.36)}{E_s - T} v_{T\ell} \approx v_{T\ell} \frac{I}{E_s - \langle T\rangle_{eff}} v_{T\ell} = (-8+2S_{12}) \frac{[v_T(r)]^2\theta(r-d)}{E_s - \langle T\rangle_{eff}}. \qquad (29b)$$

In a later paper Kuo[29] evaluated

$$\langle\psi_s^F|V_{T\ell} \frac{Q^{AANM}}{E_s - T} V_{T\ell}|\psi_s^F\rangle,$$

again finding closure to be accurate. However, Dahll, _et al._[23] showed that in nuclear matter the infinite series of terms in the tensor force converges slowly, because there is no separation distance beyond which the coupled distortions in the $^3S_1$ and $^3D_1$ partial waves both vanish, and Köhler[5] discovered that the MS expansion does not treat the large dispersion effect adequately. In the "modified separation method" (MMS)[3] the equation for G_s contains Q, and Eq. (28b) is modified accordingly.

B. The Reference Spectrum. We have seen that the exact BG wave function heals to the unperturbed function because the Pauli operator keeps the intermediate state spectrum above the self-consistent values of E_s, i.e. above twice the Fermi energy, e_F. If Q is replaced by I, the intermediate spectrum extends further downward, overlapping the self-consistent E_s, and the "healing" property is lost. But if E_s is chosen off-shell, below the hole-pair spectrum, the healing property is regained. Alternatively, the healing property is regained if the hole-pair spectrum is replaced by an upward-shifted auxiliary spectrum, lying above

$2\,e_F$ (Ref. 30 and page 111 of Ref. 5). In the "reference spectrum" (RS) method of Bethe, Brandow, and Petschek,[3] such an auxiliary spectrum is introduced. Then Ω^{RS}, which satisfies

$$\Omega^{RS}(E) = I + \frac{I}{E-H^{RS}}\, v\, \Omega^{RS}(E), \tag{30a}$$

can have the main qualitative behavior of the correct Ω, while being much easier to calculate because it lacks Q. Then

$$G(E) = G^{RS}(E) + G^{RS}(E)\left[\frac{I}{H^{RS}-E} - \frac{Q}{H^{o}-E}\right] v\, \Omega(E). \tag{30b}$$

The RS method works for all partial waves, unlike the separation method. It shows very simply that the repulsive core contribution grows rapidly as E_s goes off shell. For example, in nuclear matter with an interaction containing a hard core of radius c, for the Lth partial wave

$$\langle\phi_{\vec{k}}|v_L|\psi^{RS}_{\vec{k}_o}\rangle = \frac{\hbar^2}{m^*}\,\{(\gamma^2+k_o^2)\int_o^c \mathcal{J}_L(kr)\,\mathcal{J}_L(k_o r)\,dr$$
$$+\,\mathcal{J}_L(kc)\frac{d}{dr}(\mathcal{J}_L-\mathcal{H}_L)\Big|_c\} + \int_c^\infty [\mathcal{J}_L(kr) - \mathcal{H}_L(\gamma r;k)]\,v_L(r)\,u_L^{RS}(r)\,dr \tag{31}$$

where $\mathcal{J}_L$ and $\mathcal{H}_L$ are Riccati-Bessel and Riccati-Hankel functions, and u_L^{RS} is the RS radial wave function. The matrix element contains contributions from inside the core, the core edge, and outside the core. The inner core contribution is proportional to $\frac{\hbar^2}{m^*}\gamma^2$, which is the negative of the relative-state starting energy plus a constant.

Bethe, _et al_. carefully worked out the parameters of a reference spectrum in nuclear matter of the form

$$e^R_{p_1} = A^R + p_1{}^2/2m^* \tag{32}$$

which would approximate closely the off-energy shell Brueckner-Hartree-Fock spectrum for $2k_F < k < 4k_F$ where the Fourier transform of the defect function is large. A similar reference spectrum is illustrated in Fig. 4, taken from Ref. 31. In the "compact cluster" expansion of Brandow,[4] in which U contains only self-energies which are placed on the energy shell by generalized time ordering, the SP potential of "particles" is nearly zero, so that $Q\,H^{gto}\,Q \cong QTQ$. Then G^{RS} becomes just the reaction matrix for free, isolated particles, similar to G^F_s, Eq. (28a),

$$G^F(E) = v + v\,\mathcal{P}\frac{I}{E-T}\,G^F(E). \tag{33}$$

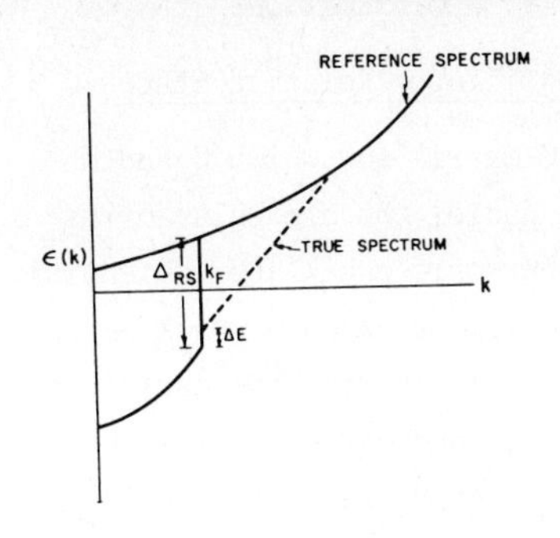

Fig. 4. The BBP (Ref. 3) reference spectrum in nuclear matter. The dashed line is an interpolation. From Ref. 31.

Sprung, et al.[32] have made many calculations of nuclear matter with the RS method.

The earliest calculations of G^{RS} for finite nuclei were those of Day, of Brandow, and of C. W. Wong.[33] Kuo and Brown[28] evaluated G^{RS} for relative p-states (where v is not attractive enough for the separation method to be used) in ^{18}O. Later Wong[34] calculated Pauli and spectral corrections, which can be expressed as

$$\Delta G(E) = G^{RS}(E)\left[\frac{I}{H^{RS}-E} - \frac{Q}{H^{o}-E}\right]\left[G^{RS}(E) + \Delta G(E)\right]$$
$$= -\,[\Delta_1\chi(E)]^{+}[G^{RS}(E) + \Delta G(E)], \tag{34}$$

$$\Delta\chi(E) = \frac{-I}{H^{RS}-E}\,G^{RS}(E) + \frac{Q}{H^{o}-E}\,G(E)$$
$$= \Delta_1\chi(E) + \frac{Q}{H^{o}-E}\,\Delta G(E). \tag{35}$$

Here the leading correction to the defect operator, $\Delta_1\chi$, can be separated into Pauli and spectral parts as

$$\Delta_1\chi(E) = \left[\frac{-I}{H^{RS}-E} + \frac{Q}{H^{o}-E}\right]G^{RS}(E) = \Delta_1^{P}\chi(E) + \Delta_1^{S}\chi(E) \tag{36}$$

with

$$\Delta_1^{P}\chi(E) = -P\,\frac{I}{H^{RS}-E}\,G^{RS}(E) = -P\,\chi^{RS}(E) \tag{37}$$

and

$$\Delta_1^{S}\chi(E) = \frac{Q}{H^{o}-E}\,G^{RS}(E) - Q\,\chi^{RS}(E). \tag{38}$$

Wong used the free-particle reference spectrum and a local, angle-averaged, nuclear-matter Pauli operator

$$Q(k,R) = Q^{AANM}(k,K_{eff}(R);k_F(R)) \tag{39a}$$

with

$$k_F^3(R) = \frac{3\pi^2}{2}\,\rho(R)\quad,\quad K_{eff}^2(R) \approx 0.3\,k_F^2(R). \tag{39b}$$

This Q commutes with the relative kinetic energy, and its variation with R gives a distinct improvement in the nuclear surface. Wong also introduced another approximate Pauli operator (denoted by Q^W below), defined in the relative and center-of-mass (rel-cm) oscillator representation. It is an approximation to the single-oscillator-configuration, two-particle Q. He found that his local Q gave results similar to the more accurate shell model one, thereby justifying his somewhat refined version, Eq. (39), of the local density approximation.[10]

VI. Accurate Calculations with an Approximate Q in Oscillator Relative States

For finite nuclei one would like to calculate a G-matrix for which Q and U correspond to the self-consistent orbitals. In light nuclei the orbitals correspond closely (except in the surface) to those of the harmonic oscillator potential. Moreover, the harmonic potential contains the only spatial dependence which separates in both 2-particle and rel-cm coordinates. For these reasons the oscillator basis is a preferred basis in which to calculate a finite-nuclear G. It is special in comparison with more realistic potentials such as the Woods-Saxon in having no continuous spectrum. A two-particle harmonic-oscillator state $|n_1\ell_1m_1, n_2\ell_2m_2>$ is expressible[35] as a linear combination of rel-cm states $|n\ell m, NLM>$ with

$$\rho \equiv \bar{n}_1+\bar{n}_2 = \bar{n}+\bar{N} \quad , \quad \bar{n} \equiv 2n+\ell. \tag{40}$$

The states of given ρ lie on a line of angle -45° in a plot such as Fig. 1. Moshinsky and Brody[36] have tabulated the coefficients of this transformation with the angular momenta coupled to total orbital angular momentum, λ. For the reduction of G to relative states we need a propagator which is both a good approximation to the self-consistent (SC) propagator and diagonal in N, L, and ℓ.

There are two common choices of approximate Hamiltonian. One is the shifted oscillator (SO),[37,38] which can be generalized to include an effective mass,[39]

$$H^{SO} = \frac{-\hbar^2\nabla^2}{2m^*} - C + \frac{Kr^2}{2}. \tag{41}$$

The other approximate Hamiltonian, H^{QTQ}, is defined such that[40,41,4]

$$Q^{SOC}\frac{I}{H^{QTQ}-E}Q^{SOC} = Q^{SOC}\frac{I}{Q^{SOC}TQ^{SOC}-E}Q^{SOC} \tag{42}$$

where SOC stands for "single oscillator configuration". It is based on two assumptions: that U = 0 for virtual particles, a good approximation for the compact cluster expansion;[4] and that $Q_1^{SC} = Q_1^{SOC}$. Unlike the cruder approximation[28,34]

$$Q^{SOC}\frac{I}{T-E}Q^{SOC}, \tag{42a}$$

Eq. (42) preserves the orthogonality of the particle and hole states (similar to the orthogonalized-plane-wave approximation in solid state theory); and unlike Köhler's approximation,[42]

$$Q^{SOC}\frac{I}{\frac{1}{2}H^{OSC}-E}Q^{SOC}, \tag{42b}$$

Eq. (42) allows for the non-diagonality of T in the oscillator basis. However, the additional approximation of keeping only diagonal elements of T_{cm},

$$\langle NL|Q[QTQ-E]^{-1}Q|N'L'\rangle = \delta_{NN'}\delta_{LL'}\langle NL|Q[QT_{rel}Q + \tfrac{1}{2}e^{osc}_{NL}-E]^{-1}Q|NL\rangle, \qquad (42c)$$

is still made in the calculations.

Two approximations to P^{SOC} have been widely used. Eden and Emery[27] proposed

$$P^{EE}(\rho) = \begin{cases} 1, & \rho \le \rho_{max} \\ 0, & \rho > \rho_{max}. \end{cases} \qquad (43)$$

It is diagonal both in 2-body and in rel-cm oscillator states. Wong's[34] approximation is defined in the rel-cm oscillator basis, for closed L-shell configurations, by

$$(n\ell NL|P^{W}|n'\ell'N'L') = \frac{\delta_{\ell\ell'}\delta_{LL'}}{(2L+1)(2\ell+1)}\sum_{\lambda}(2\lambda+1)\sum_{n_1\ell_1 n_2\ell_2} M^{n\ell NL}_{n_1\ell_1 n_2\ell_2}(\lambda)\, M^{n'\ell N'L}_{n_1\ell_1 n_2\ell_2}(\lambda) \qquad (44)$$

where M is a Moshinsky coefficient and the sum is over pair states $|n_1\ell_1, n_2\ell_2\rangle$ for which (see Fig. 1) P^{SOC} is unity. The averaging over λ is just an angle-averaging in the classical (vector model) limit. P^{W} is more accurate than P^{EE} and is preferable if one is not going to calculate "residual Pauli corrections"[38] involving P-$P^{rel\text{-}cm}$. Moreover, P^{W} can be easily generalized to j-j coupling, to non-oscillator radial functions, and to fractional occupancy.[34] However, P^{W} should not be used if residual Pauli corrections are to be made, because P^{W} is not defined in the 2-body oscillator representation; whereas P_{EE} is defined there, and by Eq. (40) takes the same simple form, Eq. (43). A related difficulty with P^{W} is that it is not a projection, i.e. is not idempotent, $(P^{W})^2 \neq P^{W}$, because of the dropping of the off-diagonal elements.

In (44) $N'+n' = N+n$. Köhler and McCarthy[42,43] have made the additional truncation in which $N' = N$:

$$(n\ell NL|P^{KM}|n'\ell'N'L') = \delta_{nn'}\delta_{NN'}\delta_{\ell\ell'}P^{W}(NLn\ell). \qquad (45)$$

Kallio and Day[44] also have required full diagonality, but have kept a dependence on λ by omitting the average over λ:

$$(n\ell NL|P^{KD}(\lambda)|n'\ell'N'L') = \delta_{nn'}\delta_{\ell\ell'}\delta_{NN'}\delta_{LL'}\sum_{n_1\ell_1 n_2\ell_2}\left[M^{n\ell NL}_{n_1\ell_1 n_2\ell_2}(\lambda)\right]^2. \qquad (46)$$

They show that even small differences in the Pauli operator significantly affect the asymptotic behavior of the defect function and hence such quantities as U insertions and the rms radius.

Next we turn to the methods for calculating G with a propagator diagonal in N and L. We let $g = \langle NL|G|NL\rangle$. All the methods involve truncating the projection operator P (as in Fig. 1) or Q to a finite number of states. As the oscillator pair-energy parameter ρ increases, the fraction of the line ρ=const for which

$P = 1$ decreases rapidly. Moreover, Wong[34] has shown that $P^W(\ell L,\rho)$ falls off, as ρ increases, even more rapidly than this geometrical argument would suggest. Sauer[45] found that g-matrix elements calculated with the maximum relative radial quantum number equal to 5 agreed with the matrix elements for $n_{max} = 15$ to within 1%.

We shall describe four nearly exact methods for solving for g with a truncated P or Q: two for solving the BG equation, one involving g^I and the reaction matrix identity, Eq. (22), and one involving expansion in eigenfunctions of the Schroedinger equation for an isolated pair.

The BG equation (15) may be regarded as an inhomogeneous equation in which the inhomogeneity is a linear combination of oscillator orbitals. Eden and Emery[27] suggested calculating Green's functions for each of these inhomogeneities and taking that linear combination which satisfies the boundary conditions. MacKellar and Becker[37,38] further developed this Green's function method, including the first exact treatment of the tensor force through coupled partial waves. Figure 5 shows the $^3S_1$-$^3D_1$ defect function for several values of the starting energy.[46]

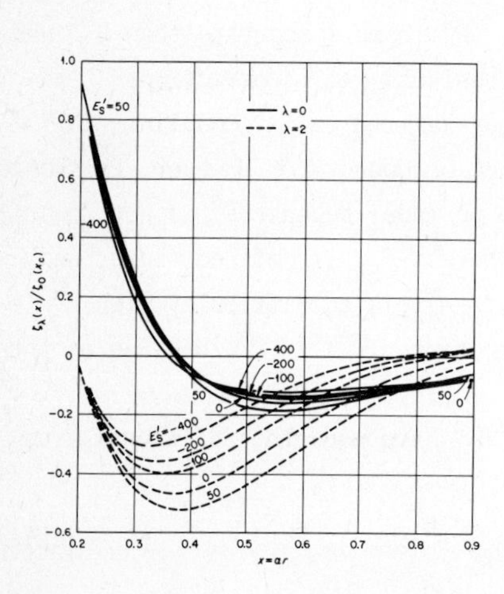

Fig. 5. Triplet relative defect functions for N-L-ℓ-0, n=5 for the Hamada-Johnston interaction for several values of the shifted starting energy $E_s' \equiv E_s + 2C$, $\alpha = (m\omega/2\hbar)^{1/2} = 0.4\ fm^{-1}$, and $\rho_{max} = 5$. From Ref. 46.

Kallio and Day[44] solved the BG equation by iterating the inhomogeneity, $P^{KD}(\lambda)v\psi$. They also applied this inhomogeneity-iteration method to nuclear matter. Siemens[47] has done extensive calculations of nuclear-matter matrix elements with the iteration method. Its only drawback is that convergence becomes slower as k_F increases and is not efficient for calculations at high density.

Köhler and McCarthy[42,43] first calculate the reference matrix

$$g^I(E) = v + v \frac{I_{rel}}{E - \langle H^o_{cm} \rangle_{NL} - H^o_{rel}} g^I(E) \quad (47a)$$

and then solve either in perturbation theory[42] or more accurately by matrix inversion,[43] in a truncated basis of relative oscillator states,

$$\left[I_{rel} - g^I(E) \frac{P^{KM}_{trunc}}{\langle H^o_{cm} \rangle_{NL} + H^o_{rel} - E} \right] g(E) = g^I(E). \quad (47b)$$

Here P^{KM}_{trunc} is the approximation of (45) in which the sum over $(n_1\ell_1, n_2\ell_2)$ in (44) is truncated. By energy conservation (40) this implies a truncation of n and ℓ. In Ref. 42 they have used, instead of the QTQ prescription, $\frac{1}{2} H^{osc}$ (42b); and in the other papers[43] an oscillator spectrum. Both of these are diagonal in relative oscillator states, so the propagator is

$$\sum_{n\ell} |n\ell)\, P^W_{trunc}(NLn\ell)\,[\langle H^o_{cm}\rangle_{NL} + \langle n\ell|H^o_{rel}|n\ell\rangle - E]^{-1}(n\ell|. \tag{48}$$

The reference matrix has singularities in the desired range of starting energies, so one must be careful to calculate $g^I(E)$ for values of E well removed from these singularities in order to avoid loss of numerical accuracy in the matrix inversion. The method is mathematically equivalent to the method of Green's functions described above; however, the work is arranged differently.

Sauer[45] has applied the matrix inversion method to the QTQ problem,

$$G(E) = v + v\,\frac{Q^{SOC}}{Q^{SOC}(E-T)Q^{SOC}}\,G(E) \tag{49}$$

where it involves additional approximations, because T_{rel} is not quite diagonal in relative oscillator states, and some arbitrariness, because after truncation Q is no longer a projection operator. In terms of the reference matrix,

$$g^I(E) = v + v\,\frac{I_{rel}}{E-\frac{1}{2}e_{NL}-T_{rel}}\,g^I(E), \tag{50}$$

Sauer's approximation is to solve in a truncated space of $(n\ell)$ states

$$\sum_{\bar{\ell}=\ell,2j-\ell} \sum_{\bar{n}'}^{n_{max}} \{\delta^{n\ell}_{\bar{n}'\bar{\ell}} - \sum_{\bar{n}}^{n_{max}} (n\ell|g^I|\bar{n}\bar{\ell})(\bar{n}|\Delta_{\bar{\ell}}|\bar{n}')\}(\bar{n}'\bar{\ell}|g|n'\ell') = (n\ell|g^I|n'\ell') \tag{51}$$

where the Pauli-correction kernel is

$$(\bar{n}|\Delta_{\bar{\ell}}|\bar{n}') = Q^W(NL\bar{n}\bar{\ell})\,\langle\bar{n}|A_{\bar{\ell}}^{-1}|\bar{n}'\rangle\,Q^W(NL\bar{n}'\bar{\ell}) - \langle\bar{n}|B_{\bar{\ell}}^{-1}|\bar{n}'\rangle$$

with

$$(n|A_\ell|n') = E - \tfrac{1}{2}e_{NL} - Q^W(NLn\ell)(n\ell|T_{rel}|n'\ell)Q^W(NLn'\ell) \tag{52a}$$

$$(n|B_\ell|n') = E - \tfrac{1}{2}e_{NL} - (n\ell|T_{rel}|n'\ell). \tag{52b}$$

Notice that as $\bar{n}\to\infty$, $Q^W(NL\bar{n}\bar{\ell})\to 1$ and $(n|A_\ell|n')\to(n|B_\ell|n')$ fairly rapidly.

Butler, et al.[48] have suggested that an expansion of a BG wave function in terms of eigenfunctions of the Schroedinger equation[49]

$$[H^{osc}_{cm} + H^{osc}_{rel} + v]\psi_{NL,i} = (e^{osc}_{NL} + E^{rel}_i)\psi_{NL,i} \tag{53}$$

should converge rapidly. Barrett, Hewitt, and McCarthy[16] have implemented this idea by expanding the reference BG wave function in ψ_i's and then using the matrix inversion method. The reference BG equation corresponding to (53) is

$$[H^{osc}_{cm} + H^{osc}_{rel} + v - E]\psi^{I}_{NLn\ell}(E) = (e^{osc}_{NLn\ell}-E)\phi_{NL,n\ell} = (e^{osc}_{n\ell}-E^{rel})\phi_{NL,n\ell}. \quad (54)$$

Then

$$\psi^{I}_{NLn\ell}(E) = (e^{osc}_{n\ell}-E^{rel}) \sum_i \frac{\psi_{NL,i}\,(\psi_{NL,i}|\phi_{NL,n\ell})}{E^{rel}_i-E^{rel}} \quad (55)$$

and

$$(n'\ell'|G^{I}_{NL}(E)|n\ell) = (e^{osc}_{n\ell}-E^{rel}) \sum_i \frac{(E^{rel}_i-e^{osc}_{n'\ell'})}{(E^{rel}_i-E^{rel})} (\phi_{NL,n'\ell'}|\psi_{NL,i})(\psi_{NL,i}|\phi_{NL,n\ell})$$

$$= (e^{osc}_{n\ell}-E^{rel}) \left[\delta_{n',n}\delta_{\ell',\ell} - (e^{osc}_{n'\ell'}-E^{rel}) \sum_i \frac{(\phi_{NL,n'\ell'}|\psi_{NL,i})(\psi_{NL,i}|\psi_{NL,n\ell})}{E^{rel}_i-E^{rel}}\right]. \quad (56)$$

In practice one truncates the sum over eigenfunctions i. Equation (56) has the advantage that the dependence on the starting energy is explicit. It can provide a very accurate energy derivative.

VII. Accurate Calculations in the Two-Particle Oscillator Basis

Finally, we discuss the case in which a propagator defined in terms of individual particle states is used. The only known method is to use Eq. (22) again in a truncated basis, i.e. the matrix inversion method. The equation to be solved is

$$G(E) = G^{RCM}(E) + G^{RCM}(E)\left[\frac{Q}{E-H^{o}} - \frac{Q^{RCM}}{E-H^{RCM}}\right]G(E) \quad (57)$$

where RCM labels the approximate quantities defined in terms of rel-cm states. In (57) these quantities are assumed to be re-expressed in terms of two-particle oscillator states. We recall that I and Q^{EE} can be expressed in two-particle states, but Q^{W} cannot.

The first use of (57) was with $Q^{RCM} = Q^{EE}$, $H^{RCM} = H^{SO}$, $Q = Q^{SOC}$ and G(E) on the right-hand side approximated by G^{RCM}.[38] Later the full matrix inversion was used.[50] The case $Q = Q^{SOC}$ has also been treated by calculating G^{RCM} with $Q^{RCM} = I$ and $H^{RCM} = H^{osc}$ by the eigenfunction expansion method.[16] Equation (56) transforms to the two-particle oscillator basis $\alpha,\beta,\ldots$ as

$$G^{I}_{\beta\alpha}(E) = (e^{osc}_{\alpha}-E)\left[\delta_{\beta\alpha} - (e^{osc}_{\beta}-E) \sum_i^{i_{max}} \frac{b_{i\beta}\, b_{i\alpha}}{E_i-E}\right] \quad (58a)$$

where

$$b_{i\alpha} = (\psi_i|\phi_\alpha). \quad (58b)$$

The H^{o} in (57) can be allowed to contain level-shifts,[27,38,16]

$$H^{o} = H^{SO} + \sum_{\alpha} (e_{\alpha} - e_{\alpha}^{SO}) |\phi_{\alpha}><\phi_{\alpha}|, \tag{59}$$

so as to have self-consistent energies of the low-lying "particle" states.

Self-consistency of the Pauli operator with the orbitals of a self-consistent field calculation can be obtained by expanding the SC orbitals in a truncated basis of oscillator states and then solving (57) with $Q = Q^{SC}$, or, if G^{SOC} has already been obtained from (57), by solving

$$G(E) = G^{SOC}(E) + G^{SOC}(E) \left[\frac{Q^{SC}}{E-T-U^{SC}} - \frac{Q^{SOC}}{E-H^{SO}} \right] G(E). \tag{60}$$

MacKellar[51] initiated the first residual Pauli corrections from Eq. (60) with $U^{SC} = U^{SO}$. This refinement of G made enough difference in the saturation properties of ^{16}O to warrant its inclusion in other calculations.[52,53] It is expected to become more important the heavier the nucleus. Spectral corrections, $U^{SC} \neq U^{SO}$, were also included in Ref. 53. Equation (58) could be applied easily, with ϕ_{α} in (58b) becoming a self-consistent pair state expanded in oscillator pair states.

We have now reached the stage where G is essentially exact, limited only by the truncation of the oscillator basis and the uncertainty in the best definition of the potential U for virtual particles.

References

1. K. A. Brueckner, Phys. Rev. 97, 1353 (1955).
2. R. L. Becker, Phys. Rev. 127, 1328 (1962) App. C.
3. H. A. Bethe, B. H. Brandow, and A. G. Petschek, Phys. Rev. 129, 225 (1963).
4. B. H. Brandow, Ann. Phys. (N.Y.) 57, 214 (1970); Phys. Rev. 152, 863 (1966).
5. S. A. Moszkowski and B. L. Scott, Ann. Phys. 11, 65 (1960); Scott and Moszkowski, ibid. 14, 107 (1961); Nucl. Phys. 29, 665 (1962); S. Köhler, Ann. Phys. 16, 375 (1961).
6. W. Tobocman, Phys. Rev. 107, 203 (1957).
7. S. F. Tsai and T. T. S. Kuo, Phys. Lett. 39B, 427 (1972).
8. See M. H. MacFarlane, in Proc. Int. School of Physics, "Enrico Fermi", Course 40, Varenna, 1967 (Academic Press, New York, 1968): B. H. Brandow, Revs. Mod. Phys. 39, 771 (1967).
9. J. W. Negele and D. Vautherin, Phys. Rev. C5, 1472 (1972); ibid. C11, 1031 (1975).
10. K. A. Brueckner, J. L. Gammel, and H. Weitzner, Phys. Rev. 110, 431 (1958).
11. See V. J. Emery, Nucl. Phys. 12, 69 (1959).
12. K. A. Brueckner, Phys. Rev. 100, 36 (1955); J. Goldstone, Proc. Roy. Soc. (London) A293, 267 (1957).
13. D. J. Thouless, Phys. Rev. 112, 906 (1958).
14. R. W. Jones, F. Mohling, and R. L. Becker, Nucl. Phys. A220, 45 (1974).
15. R. Mercier, E. U. Baranger, and R. J. McCarthy, Nucl. Phys. A130, 322 (1969); R. L. Becker, et al., Oak Ridge Nat. Lab. Report ORNL-4395 (1968) p. 116.
16. B. R. Barrett, R. G. L. Hewitt, and R. J. McCarthy, Phys. Rev. C3, 1137 (1971).
17. H. A. Bethe and J. Goldstone, Proc. Roy. Soc. A238, 551 (1957).
18. L. C. Gomes, J. D. Walecka, and V. F. Weisskopf, Ann. Phys. (N.Y.) 3, 241 (1958).
19. R. K. Bhaduri and M. A. Preston, Can. J. Phys. 42, 696 (1964); C. W. Wong, Nucl. Phys. 56, 213 (1964).
20. K. A. Brueckner and J. L. Gammel, Phys. Rev. 109, 1023 (1958).
21. K. A. Brueckner and W. Wada, Phys. Rev. 103, 1008 (1956).
22. E. J. Irwin, Thesis, Cornell Univ., 1963; G. E. Brown, G. T. Schappert, and C. W. Wong, Nucl. Phys. 56, 191 (1964).
23. G. Dahll, E. Ostgaard, and B. Brandow, Nucl. Phys. A124, 481 (1969).
24. See, e.g. G. G. Hall, "Matrices and Tensors" (MacMillan, New York, 1963) p. 32.
25. M. L. Goldberger, Phys. Rev. 84, 929 (1951).
26. R. J. McCarthy and K. T. R. Davies, Phys. Rev. C1, 1644 (1970).
27. R. J. Eden and V. J. Emery, Proc. Roy. Soc. (London) A248, 266 (1958); Eden, Emery, and S. Sampanthar, ibid. A253, 177, 186 (1959).
28. T. T. S. Kuo and G. E. Brown, Nucl. Phys. 85, 40 (1966).
29. T. T. S. Kuo, Nucl. Phys. 103, 71 (1967).
30. R. E. Peierls, in Lectures in Theor. Phys., Vol. 1 (Boulder, 1958) ed. by W. Brittin and L. Dunham, (Interscience, New York, 1959).
31. G. E. Brown, Unified Theory of Nuclear Models and Forces, 3rd ed. (North-Holland, Amsterdam, 1971).
32. See D. W. L. Sprung and P. K. Banerjee, Nucl. Phys. A168, 273 (1971); P. K. Banerjee and Sprung, Can. J. Phys. 49, 1899 (1971).
33. B. Day, Phys. Rev. 136, B1594 (1964); B. H. Brandow, Thesis, Cornell Univ., 1964; C. W. Wong, Thesis, Harvard Univ., 1965.
34. C. W. Wong, Nucl. Phys. A91, 399 (1967).
35. I. Talmi, Helv. Phys. Acta 25, 185 (1952).
36. M. Moshinsky, Nucl. Phys. 13, 104 (1959); T. A. Brody and M. Moshinsky, Tables of Transformation Brackets for Nuclear Shell Model Calculations (Monografias del Instituto di Fisica, Mexico, 1960); M. Baranger and K. T. R. Davies, Nucl. Phys. 79, 403 (1966); D. H. Feng and T. Tamura, "Calculations of Harmonic Oscillator Brackets", preprint, Univ. of Texas, May 1975.
37. A. D. MacKellar and R. L. Becker, Phys. Lett. 18, 308 (1965); A. D. MacKellar, Thesis, Texas A & M, Oak Ridge Nat. Lab. Report ORNL-TM-1374 (1966); Becker

and MacKellar, Phys. Lett. 21, 201 (1966).
38. R. L. Becker, A. D. MacKellar, and B. M. Morris, Phys. Rev. 174, 1264 (1968).
39. R. L. Becker, unpublished; employed by Becker, Phys. Rev. Lett. 24, 400 (1970) and by Becker, Morris, and Patterson (to be published).
40. M. Baranger, in Proc. Int. School of Physics, "Enrico Fermi", Course 40, Varenna, 1967 (Academic Press, New York, 1968).
41. M. M. Stingl and M. W. Kirson, Nucl. Phys. A137, 289 (1969).
42. H. S. Köhler and R. J. McCarthy, Nucl. Phys. 86, 611 (1966); McCarthy and Köhler, ibid. A99, 65 (1967).
43. H. S. Köhler and R. J. McCarthy, Nucl. Phys. A106, 313 (1968); McCarthy, ibid. A130, 305 (1969).
44. A. Kallio and B. D. Day, Nucl. Phys. A124, 177 (1969).
45. P. U. Sauer, Nucl. Phys. A150, 467 (1970).
46. R. L. Becker and A. D. MacKellar, "Test of Day's Approximation of the Defect Function", to be published.
47. P. J. Siemens, Nucl. Phys. A141, 225 (1970).
48. S. J. Butler, R. G. L. Hewitt, B. H. J. McKellar, I. R. Nicholls, and J. S. Truelove, Phys. Rev. 186, 963 (1969); J. S. Truelove and I. R. Nicholls, Australian J. Phys. 23, 231 (1970).
49. B. P. Nigam, Phys. Rev. 133, B1381 (1964); Y. E. Kim, Phys. Lett. 19, 583 (1965).
50. R. L. Becker, K. T. R. Davies, and M. R. Patterson, Phys. Rev. C9, 1221 (1974).
51. R. K. Tripathi, A. Faessler, and A. D. MacKellar, Phys. Rev. C8, 129 (1973); Faessler, MacKellar, and Tripathi, Nucl. Phys. A215, 525 (1973).
52. R. K. Tripathi, A. Faessler, and H. Müther, Phys. Rev. C10, 2080 (1974).
53. R. L. Becker and N. M. Larson, ORNL-5025 (1974) p. 12; R. L. Becker, Bull. Am. Phys. Soc. 20, 554 (1975).

R.L. BECKER: COMPUTATION OF THE REACTION MATRIX G

Kümmel: I want to make three remarks, the first one being almost a psychological one.
Our method to solve the BG equation (H. Kümmel and J.G. Zabolitzky, Phys. Rev. C6 (1972) 1606 and J.G. Zabolitzky, Nucl. Phys. A228 (1974) 272) also is numerically exact and very fast, including self-consistent Pauli operators and exact c.m. treatment. It uses a mixed representation (coordinate space for relative and oscillator states for c.m. pair motion). I always wondered why few people pay any attention to this work.

Becker: I'm sorry, I was not aware of your method. Does it fall into one of the categories I discussed?

Kümmel: It is somewhat similar to the Kallio-Day method.
Investigating the c.m. motion we found that it is a bad approximation to leave out the non-diagonal terms. Then it is better not to subtract T_{CM}, and after performing all calculations to subtract the average c.m. kinetic energy.

Becker: I believe you are speaking of the center-of-mass of the entire nucleus. I was referring to the center-of-mass of a pair of interacting nucleons.

Kümmel: After the three-body calculations being performed by us, one can say that the question which single-particle potential for particles should be used, is settled - at least on the level of approximation we are dealing with here. It turns out that one may mock up the three-body effect by a constant single-particle potential of - 8 MeV for particles.

Talmi: Could you comment on the numerical agreement, in some cases, between the various approximation methods of calculating the G-matrix?

Becker: Meaningful comparison of the different methods requires calculations for the same interaction, starting energy, oscillator parameter, Pauli operator, and single-particle potential. The older methods, which involved a first approximation and a truncated series of correction terms, differed noticeably. Bob McCarthy has compared matrix elements calculated by the newer methods and has found them to be nearly identical. The truncation of the two-particle Pauli operator is still a source of error. This truncation error is thought to be quite small when about eight or more major shells are included in the matrix which is diagonalized. Most existing calculations of the effective interaction have been done with older, more approximate G-matrix elements. Use of the "exact" ones should eliminate one source of unreliability in future calculations.

Towner: When using the two-particle harmonic oscillator as basis functions, can the G-matrix be calculated in r_1, r_2 coordinate space, rather than transforming to relative and center-of-mass coordinates?

Becker: For an interaction $\nu(r_{12})$ with strong variations at short distances the Slater multipole expansion into a sum of terms in r_1, r_2 and θ_{12} is inconvenient and fails entirely if ν has a hard core. An alternative, the expansion of ν into a sum of separable terms, $f_j(r_1)f_i(r_2)$, is also inconvenient because it requires a very great number of terms. Thus, the Talmi transformation to relative coordinates seems almost essential.

CORE POLARIZATION

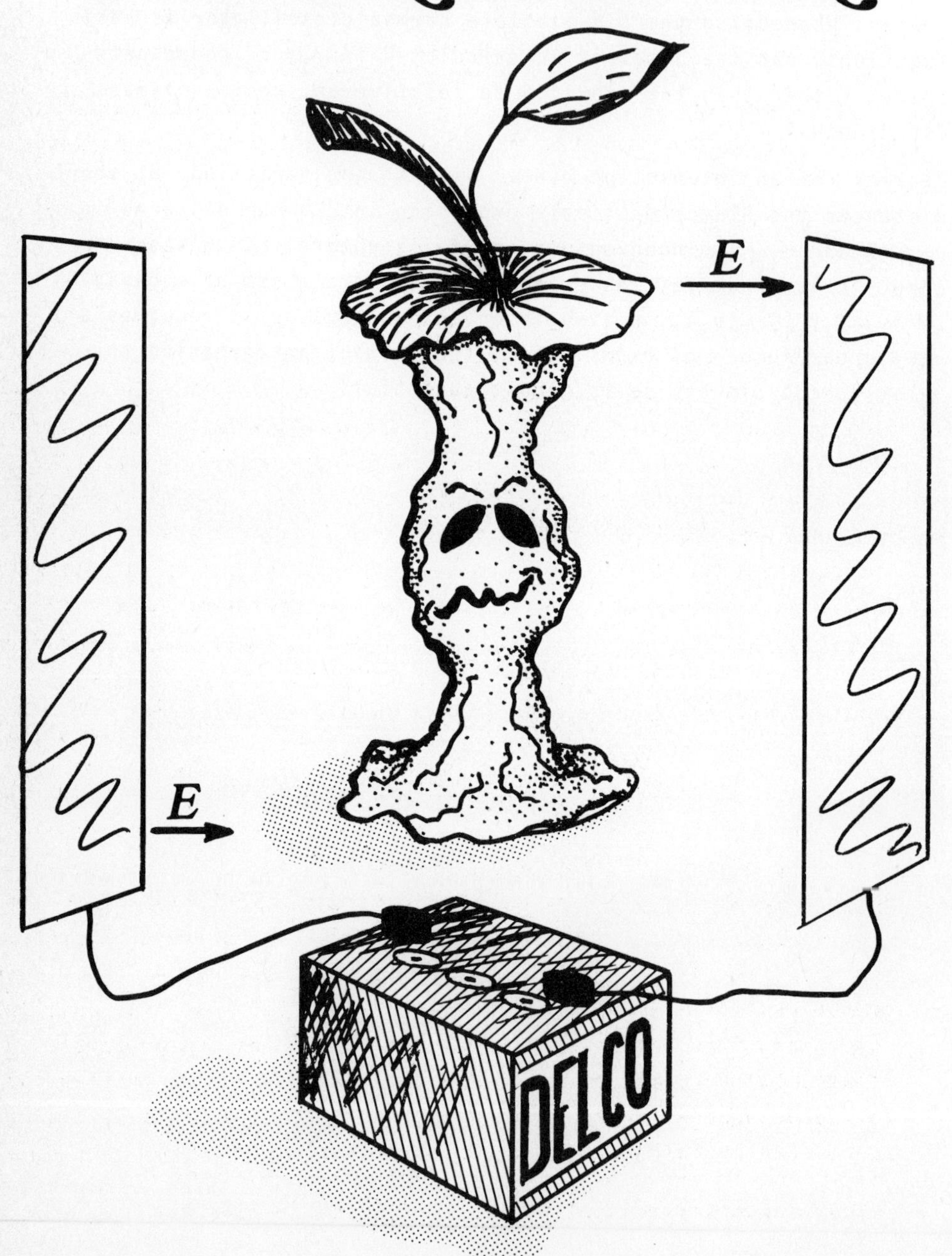

PERTURBATION CALCULATION IN A DOUBLE-PARTITIONED HILBERT SPACE *

Bruce R. Barrett
Department of Physics
University of Arizona
Tucson, Arizona 85721 USA

I. INTRODUCTION

In previous talks we have learned about the microscopic theory for determining the effective interaction $\mathcal{V}$ for any number of valence nucleons outside a closed-shell core. The question remains as to the most efficient and meaningful way to compute $\mathcal{V}$. The usual procedure has been to compute $\mathcal{V}$ as a perturbation expansion in the Brueckner reaction matrix G. This expansion is usually of the Bloch-Horowitz-Brandow (BHB) form [1,2] described in the General Theory session by Dr. Brandow. In spite of convergence problems, which will be discussed in a later paper by Prof. Weidenmüller, one still wants to determine the most accurate possible perturbation-theory value for $\mathcal{V}$, since at the very least the value determined in low-order perturbation theory will serve as the input to a more sophisticated and accurate procedure for obtaining $\mathcal{V}$. These more sohpisticated procedures will be described in Session VII; what I want to discuss now is a method for performing the most accurate possible computation of $\mathcal{V}$ in low-order perturbation theory. This is the method of successive partitioning, of which I will consider in detail only the case of double partitioning.

II. THEORY OF DOUBLE-PARTITIONING

The principal idea of successive partitioning as first discussed by Brandow [3] and later outlined by Brown [4] is to separate the intermediate-state spectrum used in the calculation of $\mathcal{V}$ into several parts, applying a different but more accurate approximation for computing $\mathcal{V}$ in each of these successive energy regions. In the double-partition approach the intermediate-state spectrum used in computing $\mathcal{V}$ is separated into two parts, a high-energy part to be related to the computation of G and a low-energy part to be related to the determination of $\mathcal{V}$ in perturbation theory. This is shown schematically in Figure 1. In this way one hopes to improve the

* Work supported in part by the NSF (Grant GP-39030X1)

accuracy of the calculation of $\mathcal{V}$ by moving intermediate excitations out of a calculation in which they were poorly accounted for (such as two-particle ladders in G at low excitation energies) and into a complementary calculation where their effect will be well approximated (such as leaving the low-lying two-particle ladders out of G and explicitly including them in the perturbation expansion for $\mathcal{V}$).

Basically the double-partition procedure is to determine an effective interaction $\mathcal{V}_1$, for a fairly large model space, which contains the smaller shell-model space of interest, and then to use the resulting $\mathcal{V}_1$ in place of V, in determining $\mathcal{V}_2$ in the smaller model space. Formally, the effective interaction $\mathcal{V}_1$ for the outer model space is defined by the relation

$$\mathcal{V}_1 \Psi_{D_1} = V \Psi$$

so that

$$\mathcal{V}_1(E) = V + V \frac{Q_1}{E - H_0} \mathcal{V}_1(E) \qquad (1)$$

Then using $\mathcal{V}_1$ in the role of V gives

$$\mathcal{V}_2 \Psi_{D_2} = \mathcal{V}_1 \Psi_{D_1}$$

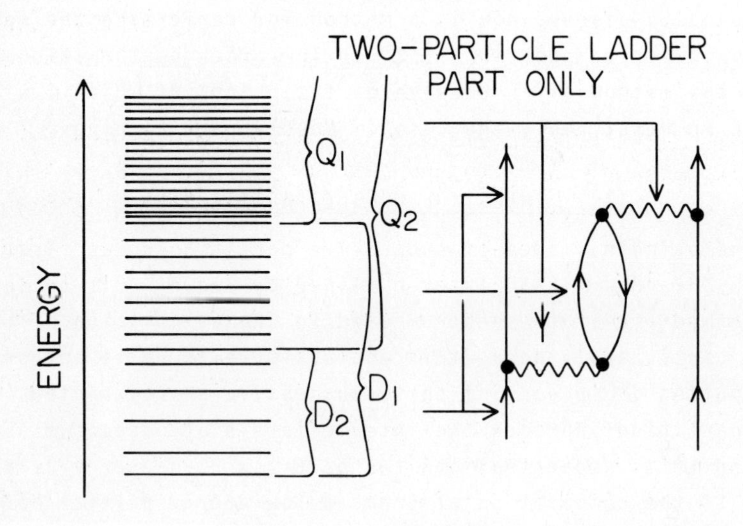

Fig. 1 Schematic representation of how the Hilbert space can be doubly partitioned for calculating the core-polarization process.

for effective interaction $\mathcal{V}_2$ in the smaller model space D_2. This leads to the following equation for $\mathcal{V}_2$ in terms of $\mathcal{V}_1$,

$$\mathcal{V}_2(E) = \mathcal{V}_1(E) + \mathcal{V}_1(E)\,\frac{Q_2 - Q_1}{E - H_0}\,\mathcal{V}_2(E) \tag{2}$$

The Pauli projection operator Q_1 excludes all states within the larger model space D_1, whereas Q_2 excludes all states within the smaller model space D_2 (i.e. Q_1 and Q_2 are equal to 1 for the energy regions marked in Figure 1 and are zero otherwise). Hence the only states for which the difference $Q_2 - Q_1$ is non-zero are those within the outer model space but outside the inner model space.

Thus, $\mathcal{V}_1$ as defined in eq. (1) includes only those states that are outside the outer model space. These are relatively high-energy states, and hence represent an interaction over a short distance. Because of the low density of nuclei and the relative smallness of the contribution of three-body clusters in nuclear matter calculations, we shall approximate the intermediate-state sum specified by Q_1 as consisting solely of two-particle states. Because of this, $\mathcal{V}_1$ becomes a G matrix, although not the one that would normally be used in the inner model space. Instead, it is the proper G for use in calculations in the outer model space. Therefore, the approximation described above can also be looked at as approximating $\mathcal{V}_1$ by the first term of its BHB expansion in terms of G. Although this approximation will be used exclusively here, it is clear that one might usefully include more terms of the expansion, for example, the core-polarization term.

No matter what approximation one uses for $\mathcal{V}_1$, the next step is to plug it into eq. (2). The presence of $Q_2 - Q_1$ in the homogeneous part of the equation restricts the intermediate-state sum to those states between the two model-space boundaries, including the two-particle ladder terms, such as those shown in Figures 2 and 3. One way of looking at the reason for the introduction of these new diagrams is to observe that the intermediate states of eq. (2) complement those of the defining equation for $\mathcal{V}_1$ and thus states left out of the latter must be included in the former.

For the method of double-partitioning to be useful and meaningful, it must be possible to separate the excitation space into the "large" and "small" model spaces described above. However, this separation is not simply dependent on the mathematical structure of the theory, but rather on the detailed behavior of the nuclear force. One way of looking at what is accomplished by

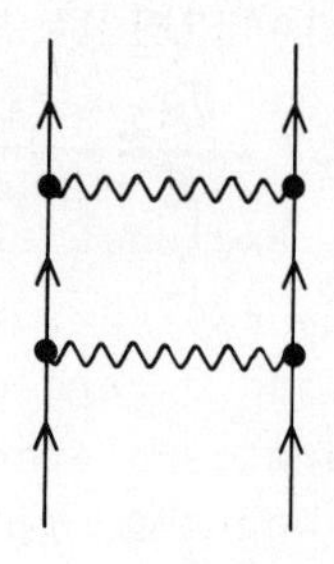

Fig. 2 Second-order ladder diagram

double partitioning is to note that $\mathcal{V}_1$ is calculated at relatively high energies and $\mathcal{V}_2$ is calculated at relatively low energies. This allows one to use an approximation good at high energies (in this case, using G for $\mathcal{V}_1$) to treat high-energy excitations, and a different one (in this case, the first three terms of the BHB perturbation expansion), which is hopefully good for low-energy excitations. Each approximation will be good over a certain energy range. If the nuclear force is such that each range can be made to contain all the important excitations for which it is effective, the highest accuracy will be attained. Figure 4 shows two possibilities for an outcome for this comparison, each with its optimum outer model-space boundary shown. The abscissa of each graph represents the energy of excitation in arbitrary units. The ordinate is meant to schematically represent the relative contribution to $\mathcal{V}_2$ by excitations of each energy (also in arbitrary units).

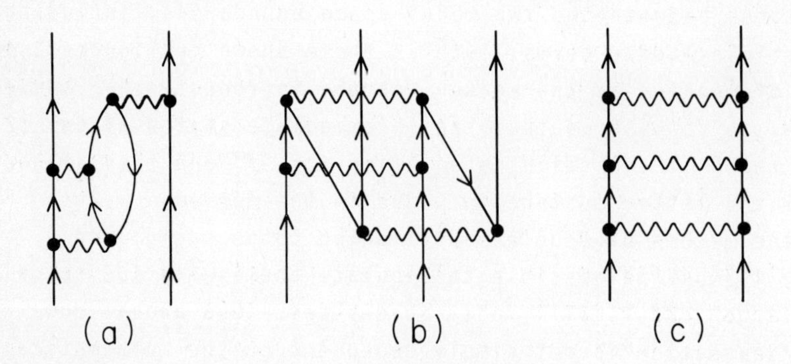

Fig. 3 Third-order diagrams containing two particle ladders.

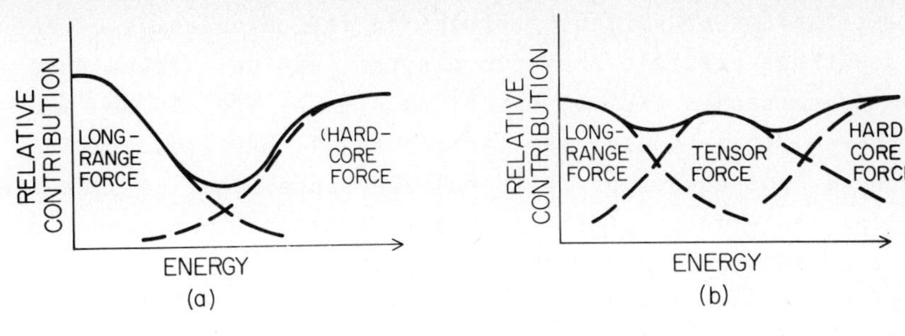

Fig. 4 Two hypothetical plots of the relative contribution of v_2 as a function of excitation energy.

If the case shown in Figure 4a represents reality in the system under study, then the double-partitioning technique should be expected to be a considerable improvement. That is, if there is a considerable gap between the energies of the dominant short-range excitations and the energies of the dominant multi-particle excitations, so that an outer model-space boundary can be placed in the middle of this gap, each important excitation will be accounted for by a valid approximation. If, on the other hand, there is a sizeable overlap in the energies of these two categories of excitations, as is shown in Figure 4b, then each approximation will be trying to calculate effects it was not meant to handle, and will fail.

The way selected here to find the best partition position is to repeat a doubly-partitioned calculation several times, each time with an enlarged model space (a "higher" partition position). As the partition moves through the region of unimportant excitation energy, if any, the rate of change in the effective interaction with respect to the changing energy of the partition will go through a minimum. This is because the excitations transferred from the one approximation to the other will be the unimportant ones. Further, the depth of the minimum in the rate of change of strength-shifting should correspond to the depth of the minimum in excitation importance. The usefulness of the double-partitioning technique is proportional to the depths of these minima.

One of the obstacles to an effective use of double-partitioning seems to be the tensor force. Its range is intermediate in the frame-work of this study, and so could fill in the minima mentioned above. In fact, Vary, Sauer and Wong [5] have investigated the importance of excitations of various energies in the calculation of the second-order three-particle-one-hole diagram, and have found that the tensor force causes excitations of as high as $10\hbar\Omega$ to be important. Of course, their G used excitations of all energies (their calculation was not double partitioned) and hence might be suspected of scrambling the effects of the two approximations, but other indications are that this is not the main problem.

It might be mentioned in passing that there is nothing to keep one from triply partitioning the Hilbert space as an attack on the problem of the tensor force with its contribution at the intermediate energies. The idea would be to partition off each of the three categories of excitation; short range (hard core), intermediate range (tensor force), and long range (multi-particle), into regions of energy where each would be dealt with by the appropriate approximation. A good approximation for the tensor force's effects would be needed for this, of course, and to date no satisfactory one is known. If this requirement is met, however, the criterion for gauging relative success would still be the depths of the minima mentioned above.

The procedure for triple-partitioning would be a simple extension of that for double partitioning. One would expand $\mathcal{V}_1$ in the appropriate G, probably keeping just the first term as at present, then solve eq. (2) with some approximation suitable for the tensor force, using a $Q_2 - Q_1$ that would select the intermediate region. Then one should calculate the BHB expansion for $\mathcal{V}_3$ in terms of $\mathcal{V}_2$.

III. DOUBLE-PARTITION CALCULATIONS FOR MASS-18 NUCLEI

I now want to present the results of Barrett [6)] and Herbert [7)] regarding double-partition calculations for mass-18 nuclei. In their calculations the inner model space is defined as the sd shell and the outer model space consisted of those states not more than a given energy above the sd shell. The outer model space for Barrett was the sd and pf shells, which are a good approximation to the set of intermediate states not more than $2\hbar\Omega$ above the inner model space. The outer model space for Herbert was the space of all states not more than $4\hbar\Omega$ above the sd shell.

They used the method of Barrett, Hewitt and McCarthy (BHM) [8)] to compute the G matrix elements, since the BHM method treats the two-particle Pauli projection operator Q exactly. A two-particle

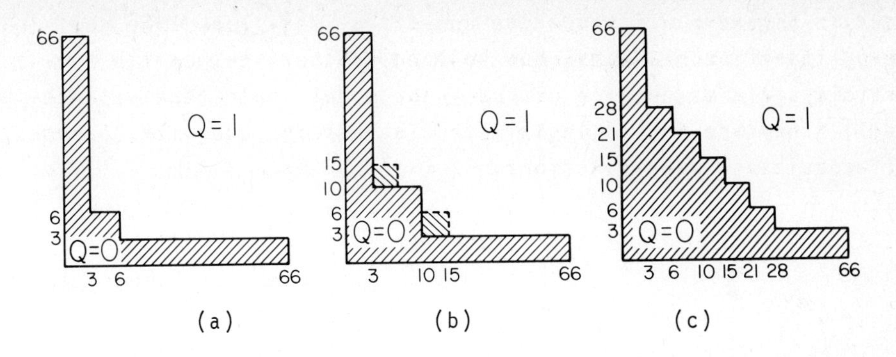

Fig. 5 Division of the two-particle Hilbert space for G for (a) the I6 partition, (b) the I10 partition and (c) the I15 partition.

harmonic oscillator basis, the Hamada-Johnston hard-core potential and $\hbar\Omega$ = 14 MeV were employed in determining G. Figure 5 shows the three different Pauli operators Q used in their calculations. The shaded area represents the two-particle states excluded by Q in computing G. The first figure (5a) shows the Q operator for a standard calculation for A = 18 nuclei; this calculation will be referred to as I6, since there are six single-particle states up to and including the sd shell. The second figure (5b) gives the Q operator for a double-partition calculation, in which the pf shell is also excluded from G. This will be called the I10 calculation. The small areas cross-hatched in the opposite direction represent $2\hbar\Omega$ excitations, which were <u>not</u> excluded from G in Barrett's calculations. Finally, the third figure (5c) illustrates the Q operator for Herbert's double partition calculation, in which all excitations $4\hbar\Omega$ above the sd shell are excluded from G. This will be denoted by I15.

These G matrix elements were then used to determine the second- and third-order terms contributing to $\mathcal{V}_2$. These higher-order terms were computed using the formulas of Barrett and Kirson (BK) [9], and their notation will be used here. The calculations of Barrett [6] also introduced the new ladder diagrams in Figures 2 and 3. The introduction of $4\hbar\Omega$ excitations by Herbert [7] produced two new effects. One is that a number of diagrams --- BK numbers 6(1), 6(2), 11(1) and 11(2)--- can now be excited and must be included in the calculation. Second, there are two diagrams which were not listed in BK which should be included. These are shown in Figure 6. It should be noted that they look like self-energy-insertion diagrams, but so long as the intermediate two-particle state is outside of the inner model space (the valence shell), they are indeed legitimate diagrams for use with (assumed) self-consistent wave functions.

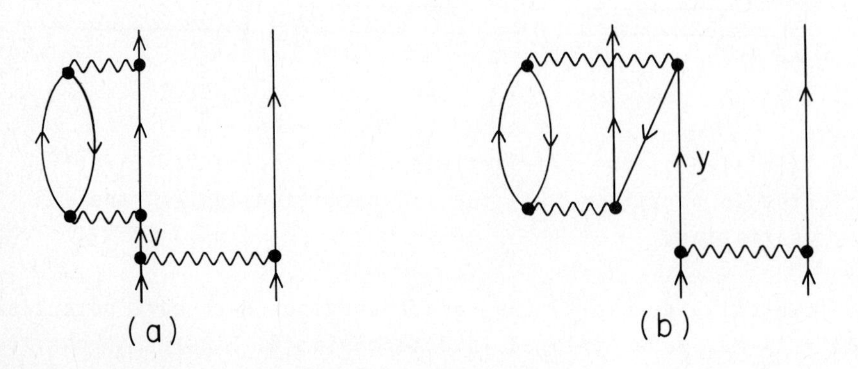

Fig. 6 Two new third-order diagrams included the I15 but not in the I10 calculation.

The effective interaction $\mathcal{V}_2$ through third-order in G was then diagonalized, using the experimental single-particle energies from ^{17}O, i.e. $\varepsilon_{0d_{5/2}} = 0.0$, $\varepsilon_{1s_{1/2}} = 0.871$ and $\varepsilon_{0d_{3/2}} = 5.083$ (all in MeV), for the unperturbed sd-shell energies, to obtain the J = 0, T = 1 spectrum.

This whole procedure was carried out for three values of the starting energy of G, namely ω = -3,63 and 82 MeV. Using a different starting energy is equivalent to making a different choice for the intermediate-state spectrum [8)], since varying ω is related to varying the gap between the occupied and unoccupied single-particle states.

Figure 7 shows the variation in the I15 J = 0,T = 1 eigenvalue spectrum as the different orders are included for each value of ω. Figure 8 shows the variation in the eigenvalue spectrum as a function of the partition position. On the whole the change in the spectrum (and also in the individual matrix elements of $\mathcal{V}_2$) decreases in magnitude as one goes from I6 to I10 to I15. Similarly the convergence of the perturbation series for $\mathcal{V}_2$, as one goes from first to second to third order in G, in general improves from I6 to I10 to I15. It should be noted that the experimental energy

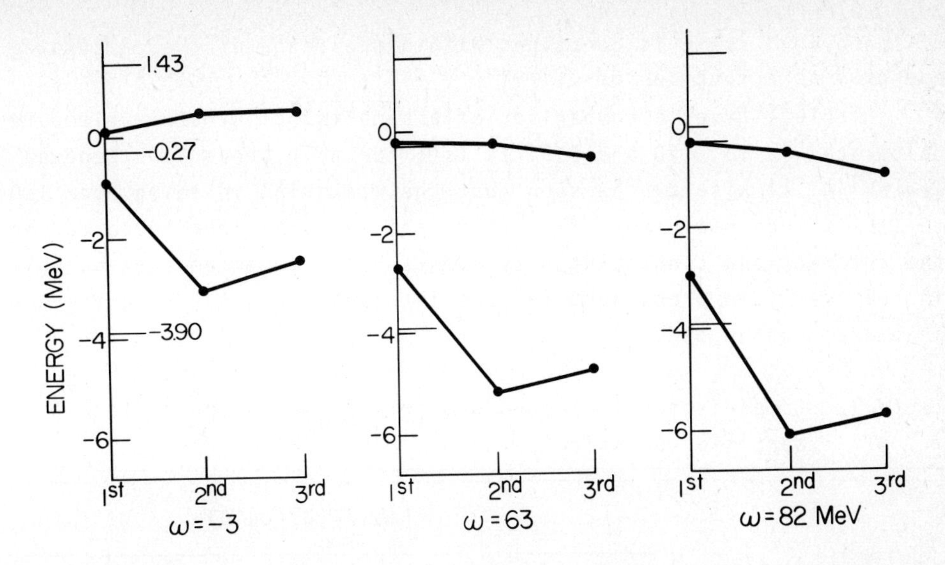

Fig. 7 Variation in the J = 0, T = 1 eigenvalue spectrum of $\mathcal{V}_2$ for mass-18 for the I15 calculations as successive orders of G are summed through third order. Results are given for three values of ω: -3, 63 and 82 MeV. Experimental values are marked on the vertical axis.

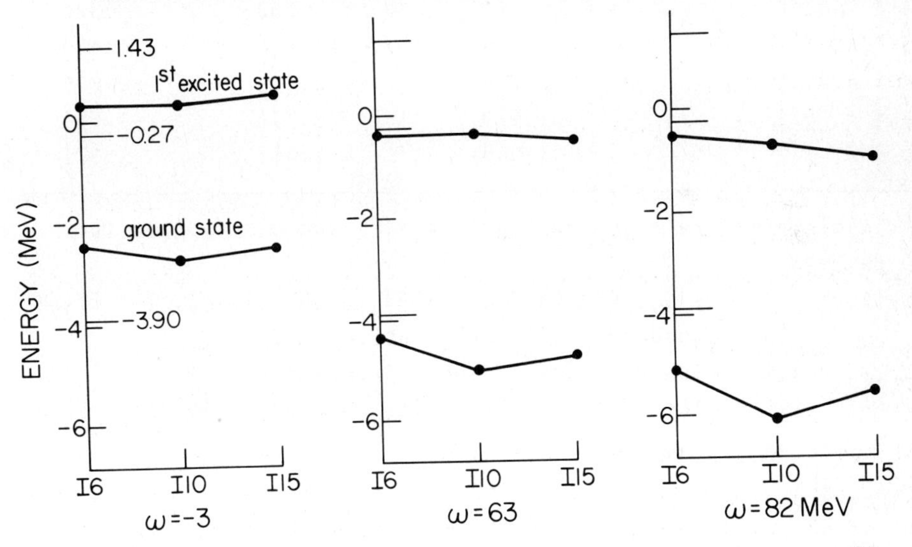

Fig. 8 Variation in the J = 0, T = 1 eigenvalue spectrum of $\mathcal{V}_2$ for mass-18 as a function of the partition position.

of the ground state is contained within the range of the calculated energies as a function of ω.

Table 1 gives a comparison of a sampling of G-matrix elements calculated in I6, I10 and I15. It is clear that there is a general weakening. It also can be seen that the weakening in going from I10 to I15 is much diminished from that seen in going from I6 to I10 in the interactions connecting only even-angular-momentum states. This is not nearly so pronounced for the interactions connecting the odd-angular-momentum states.

Table 1. Comparison of selected G-matrix elements for different partition positions

	<(ab)JT|G|(cd)JT>		
a b c d, $J^{\pi}T$	I6	I10	I15
1 5 4 4, $1^{+}0$	1.706	1.449	1.135
2 4 2 4, $1^{-}0$	-3.242	-1.736	-1.174
4 4 4 4, $1^{+}0$	-1.844	-1.425	-.708
2 4 4 7, $1^{-}0$	1.921	.763	.372
1 5 4 4, $0^{+}1$	1.514	1.444	1.420
2 4 2 4, $1^{-}1$	-2.700	-2.338	-2.234
4 4 4 4, $0^{+}1$	-1.920	-1.744	-1.655
2 4 4 7, $1^{-}1$	1.848	1.467	1.371

Matrix elements are given for various J, T and parity (π), for ω = 82 MeV.

Notation: $1 \equiv 0s_{1/2}$, $2 \equiv 0p_{3/2}$, $4 \equiv 0d_{5/2}$, $5 \equiv 1s_{1/2}$, $7 \equiv 0f_{7/2}$

I6 ≡ standard A = 18 calculation of G

I10 ≡ also excludes pf shell from G (Ref. 6)

I15 ≡ also excludes all excitations up to $4\hbar\Omega$ from G

IV. ANALYSIS AND CONCLUSIONS

The convergence of the expansion for $\mathcal{V}_2$ (eq. (2)) does seem to be improved in going from I6 to I10 to I15. The smaller G-matrix elements undoubtedly play a role in this. It is also very likely that moving more of the tensor force excitations out of the outer model-space effective interaction and into the multi-particle excitations improved the calculation from the standard-point of the tensor force. The tensor force would be especially unlikely to be well-approximated by a G with a large gap between low-lying shells, and hence the convergence might be expected to improve the most for the case of the small gap, which it does.

Not only is the convergence improved in the I15 calculation but also the individual terms of the expansion diminish in absolute magnitude as the partition is moved from I6 to I10 to I15. The change from I6 to I10 is negative, while the change from I10 to I15 is positive but of a smaller amount.

The improvement in the apparent convergence for the I10 partition appears to be significantly greater than the additional improvement to be had in going to I15. Hence one might conclude that the partition, when at I15, may be in the vicinity if one of the minima pictured in Figure 4, because of this decreased improvement.

One can also see the effect noted by Vary _et al_. [5)] in the core-polarization diagram (BK number 2). They discovered that the sum over excitations in this diagram does not converge rapidly when one uses Kuo G-matrix elements. By including higher excitations, as one does in double-partitioning, one obtains major changes in the value of this term. It is important to note, however, that Vary _et al_. used G-matrix elements corresponding approximately to those used here with ω = -3 MeV.

Table 2 compares some of the doubly-partitioned core-polarization results of Barrett [6)] and Herbert [7)] with those obtained by Vary _et al_. [5)]. The numbers in Table 2 suggest that certainly at ω = -3 MeV the same behavior is seen in both studies. In the ω = 82 MeV calculation, the situation as regards the convergence of the intermediate state sum seems to be improved, since the results are not changing so drastically (relative to their total values) as for ω = -3 MeV. However, the improvement is not very great.

To summarize, the double-partition calculations of Barrett and Herbert suggest the following situation. The results of Vary _et al_. are corroborated in the sense that in the case of a large gap in the excitation spectrum the intermediate-state sum converges slowly. Since Vary _et al_. indicate that this difficulty is due to the tensor force, it appears that Figure 4b is a more accurate picture of the situation than in Figure 4a. However, since the change between I6 and I10 is greater than that from I10 to I15, a minimum is apparently being approached in the range $2\hbar\Omega$ to $4\hbar\Omega$, in the sense

Table 2 Comparison of Selected Contributions to the Core-Polarization Term for J = 0, T = 1 as the Partition Position is Varied with the Results of Vary et al. [5].

Energy Extent of the Intermediate - State Sum	Present Calculation (c,a) contribution in MeV			
	ω = -3 MeV		ω = 82 MeV	
	(4,4)	(5,5)	(4,4)	(5,5)
I6($2\hbar\Omega$) [a]	-.53	.18	-2.51	-.50
I10($2\hbar\Omega$) [a]	-.61	.15	-2.32	-.64
I15($4\hbar\Omega$) [b]	-.43	.35	-1.86	-.38
	Vary et al.			
	(4,4)	(5,5)		
$2\hbar\Omega$	-.71	.14		
$4\hbar\Omega$	-.58	.28		
$22\hbar\Omega$	-.08	.84		

Notation: <(c,c)JT| core-polarization |(a,a)J,T>
J = 0, T = 1, 4 $\equiv$ $0d_{5/2}$, 5 $\equiv$ $1s_{1/2}$

a) Ref. 6
b) Ref. 7

of Figure 4. One might therefore conclude that this minimum is the first one shown in Figure 4b. Since this is not the minimum where a partition could be placed in order to include the tensor force effects within the inner model space, one would have to do one of two things in order to make an effective doubly-partitioned calculation. The first thing would be to calculate $\mathcal{V}_1$ by some method suitable for approximating the effects of the tensor force, instead of just using G for $\mathcal{V}_1$. The other possibility would be to triply partition the Hilbert space in the manner discussed in sect. 2. Either way, however, a nonperturbational technique would probably be necessary in order to obtain an accurate approximation for $\mathcal{V}$, because of the convergence difficulties to be discussed tomorrow by Prof. Weidenmüller.

Finally, more work should be done to investigate the role of the tensor force in determining nuclear structure and to try to include its effects accurating in nuclear structure calculations. Work along this line is presently in progress.

ACKNOWLEDGEMENT

I would like to thank Dr. Floyd Herbert for many helpful discussions and for his contributions, based on his Ph.D. thesis, to the preparation of this talk.

REFERENCES

1. C. Bloch and J. Horowitz, Nucl. Phys. 8 (1958) 91
2. B.H. Brandow, Rev. Mod. Phys. 39 (1967) 771
3. B.H. Brandow, Lectures in Theoretical Physics, Vol. XIB, ed. K.T. Mahanthappa and W.E. Brittin (Gordon and Beach, New York, 1969) p. 55
4. G.E. Brown, Rev. Mod. Phys. 43 (1971) 1
5. J.P. Vary, P.U. Sauer and C.W. Wong, Phys. Rev. C7 (1973) 1776
6. B.R. Barrett, Nucl. Phys. A221 (1974) 299
7. F.L. Herbert, University of Arizona, Ph.D. Thesis, April 1975, to be submitted for publication
8. B.R. Barrett, R.G.L. Hewitt and R.J. McCarthy, Phys. Rev. C3 (1971) 1137
9. B.R. Barrett and M.W. Kirson, Nucl. Phys. A148 (1970) 145; A196 (1972) 638

B.R. BARRETT: PERTURBATION CALCULATION IN A DOUBLE-PARTITIONED HILBERT SPACE

Negele: I don't see the motivation for triple or higher partitioning. If the contribution as a function of E has no pronounced separation into regions where different components of the force clearly dominate, then partitioning seems artificial and doesn't seem to me to accomplish anything in principle. Perhaps the point is one of technical convenience; i.e. double partitioning allows one to build in cancellations in the model space which a crummy Pauli operator might not. Is there a corresponding technical advantage for triple partitioning?

Barrett: If the nuclear force clearly divides itself into three parts - long-range central, tensor and short-range strong repulsion- which cause different contributions to the effective interaction in different energy regions, then one would need to triple partition in order to "accurately" determine the effective interaction in each of these energy regions.

Ellis: Shouldn't one worry about Hartree-Fock effects, one already knows that the second-order diagrams are large?

Barrett: Yes, one should work in a self-consistent basis. Because of computer limitations, we chose to work in a harmonic oscillator basis. Since we did this for all three calculations, we at least had a consistent model for studying the double-partition approach. Malcolm Harvey has now provided us with BHM G-matrix elements in a Hartree-Fock basis, and we plan to use them in future calculations.

Zamick: The results you got in going from first to second to third order looked convergent to me. Or is this an optical illusion?

Barrett: No, they look convergent to me also.

Goode: How does the cancellation among the terms in the third-order number-conserving sets vary as you go from I6 to I10 to I15 ?

Barrett: The cancellation remains about the same. although it is somewhat better for I6 and I10 than for I15, and for the second number-conserving set than for the first.

Sauer: Do you see a mechanism for cutting down the effect of the tensor force in third order? If not, should you not double partition in your calculations at 10 $\hbar\Omega$?

Barrett: I know of no such mechanism. Consequently, one should include the effect of the tensor force in third order either by increasing the partition energy or by triple partitioning.

THE AVERAGE EFFECTIVE INTERACTION

Philip Goode*
Department of Physics
Rutgers University, New Brunswick, New Jersey 08903

A critical test for any shell model theory for nuclear structure is the determination of the effective interaction between two nucleons inside a nucleus starting from their interaction in free space. The usual many body problem for the nuclear case is

$$(H_o + V)\,|\psi_i\rangle = E_i|\psi_i\rangle\ , \text{ where } H_o|\psi_i\rangle = \varepsilon_{oi}|\psi_i\rangle$$

and where V is the two body interaction in free space and the ε_{oi}'s give the one body spectrum and the E_i's are the exact energies. $|\psi\rangle$ represents some complete basis which, for example, could be an oscillator with an infinite number of basis states. To limit this problem to a few degrees of freedom we solve instead a model problem

$$(H_o + v)\,|\phi_j\rangle = E_j|\phi_j\rangle\ ,$$

where the E_j's represent some subset of all the E_i's in the general problem and where $P|\psi_i\rangle \equiv |\phi_i\rangle$ and $Q|\psi_i\rangle \equiv$ all the rest of the basis states. Formally [1,2],

$$v = V + V\,\frac{Q}{E-H_o}\,v\ ,$$

where v is the model effective interaction. To exploit this formalism in the shell model problem where V is the two nucleon interaction, we need the series for v to be rapidly convergent. Otherwise the formalism would not yield a tractable shell theory.

The series as it stands is not rapidly convergent, the basic hope, then, is to rearrange the series via partial summation into a more useful form. One such rearrangement is the summation of two nucleon ladder terms. This partial summation is reflected in a rearrangement of the series so that

$$v = G + G\,\frac{Q'}{E-H_o}\,v\ ,$$

where G is the Bruckner G and two particle ladders are now excluded

---

*Work supported in part by the National Science Foundation

from Q'. This re-arranged series has been the object of considerable investigation in which order by order calculations have been attempted as well as partial summations to suggest further re-arrangements. None of these calculations has yielded entirely satisfactory results.

Selection of diagrams for partial summation is difficult. The largest diagrams in any order for the two nucleon case come from the single interaction of two dressed nucleon lines. These diagrams constitute the so-called number conserving sets. For example, in third order in G one number conserving set is illustrated in fig. 1. Clearly, the individual members of this set are large because each configuration contributes quadratically to the dressing of the single nucleon

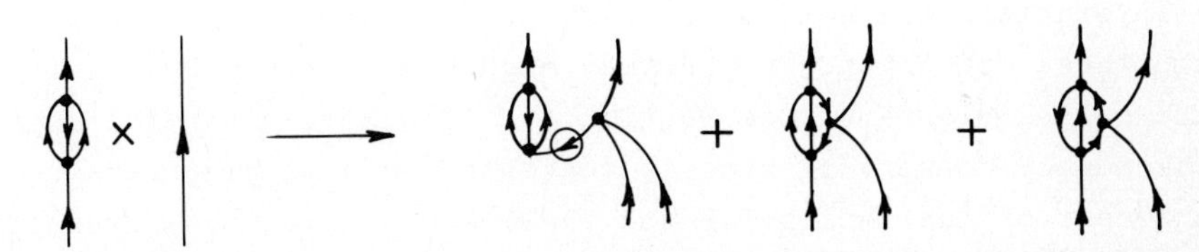

Fig. 1. A number conserving set in third order. The multiplication sign represents that the constituents of the set are formed by the single interaction between the dressed and undressed lines.

line (i.e. $\frac{G^2}{\Delta E}$). However, Brandow [3] has argued that the members of a number conserving set will strongly self-cancel because of the equivalent number of particle-particle and particle-hole interactions between the two dressed single-nucleon lines. The number conserving sets have been ignored in the partial summations to date.

Barrett and Kirson [4] have calculated a shell model effective interaction through third order in G, for the $J = 0, T = 1$ configurations in mass-18. They showed that these number conserving sets strongly self-cancel in third order in G. In addition, they showed that there are other sizeable terms in third order. Ellis, Osnes, and Jackson [5] have shown that these number conserving sets do not cancel so well if $J \neq 0$ in third order. Thus, it would seem hopeless to suggest any useful rearrangement of the series based on the third order results. Furthermore, for fourth order in G there are about a thousand diagrams. Therefore, even if the term-by-term calculation of fourth order diagrams were very interesting, the full calculation would seem to be intractable.

This talk is largely concerned with a calculation by myself and

Dan Koltun [6]. We calculated the average shell model effective interaction through fourth order in G (the theoretical method for this calculation is discussed here in Section 2). The fourth order averages are as large as the corresponding second or third order averages (see Section 3). This fact is almost entirely due to a failure of the number conserving sets to self-cancel. Furthermore, this failure can be directly attributed to a nuclear size effect (see Section 4).

We conclude that the large fourth order averages are not connected to the intruder state problem (see Section 5). We suggest that the convergence properties of the series for the shell model effective interaction are not improved by using a Hartree-Fock basis although these convergence properties might be improved by elimination of the folded diagrams (see Section 5). Finally, a partial summation of the number conserving sets is suggested as an appropriate rearrangement of the series for the shell model effective interaction.

Section 2. Theory

In this section it is shown that the average of an open shell effective interaction can be related to a closed shell problem. We want to calculate the average of

$$\langle A\alpha B\beta \ldots N\nu | v^{(n)}(m) | A\alpha B\beta \ldots N\nu \rangle ,$$

where, for example, in $A\alpha$, A represents the good single nucleon quantum number n, ℓ, s, j, and t, and α represents the corresponding z-components m_j and m_t. Thus, we are interested in the average of the linked m-body shell model effective interaction of nth order in an N-body valence space ($N > m$). This calculation proceeds from a trace theorem.

The trace is defined as

$$\mathrm{Tr}\, v^{(n)}(m,A,B\ldots N) = \sum_{\alpha,\beta,\ldots\nu} \langle A\alpha B\beta \ldots N\nu | v^{(n)}(m) | A\alpha B\beta \ldots N\nu \rangle ,$$

where, for example, if $A = B$ then $\alpha > \beta$. The average is given by

$$\langle v^{(n)}(m,A,B,\ldots N)\rangle = \mathrm{Tr}\, v^{(n)}(m,A,B,\ldots N)/\mathrm{Tr}\, 1 ,$$

where

$$\mathrm{Tr}\, 1 = \sum_{\alpha,\beta,\ldots\nu} \langle A\alpha B\beta \ldots N\nu | A\alpha B\beta \ldots N\nu \rangle .$$

The trace theorem can be stated as

$$\text{Tr } v^{(n)}(m,A,B,\ldots N) = (-1)^{\phi}\ G\ (A,B\ldots N)\ ,$$

where the trace of fig. 2a is the Goldstone diagram, G, depicted in fig. 2b. ϕ is the number of upward going lines created by closing the valence lines in fig. 2a. The diagram of fig. 2b is an improper Goldstone diagram but is evaluated in the same way as a proper closed shell diagram.

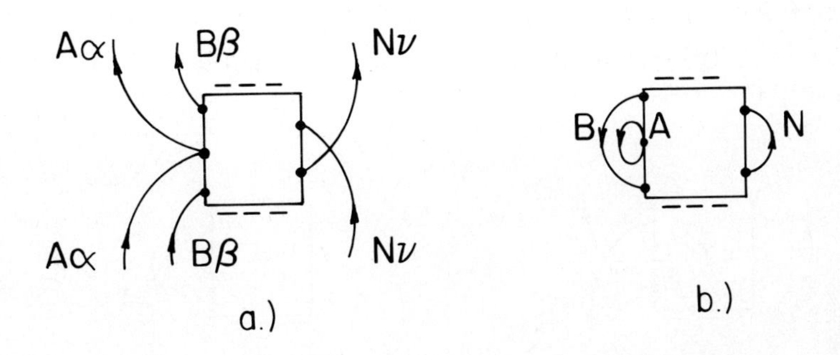

Fig. 2. (a) represents (2.1) and (b) represents its trace. The Goldstone diagram is improper because it has a down-going valence line(s) and a valence insertion(s). Only three of the N lines are shown.

In taking the trace, the order of interaction vertices remains intact, only the incoming and outgoing valence lines are modified.

To prove the theorem, consider fig. 3a which is the same as fig. 2a, ignoring all valence lines except B lines. The expression for this diagram is

$$D_x(B\beta) = \langle B\beta|U_x(A)|B\beta\rangle\ ,$$

where the one body operator,

$$U_x(B) = \sum_{\beta} \langle A\alpha\ldots N\nu|v^{(n)}(m)|A\alpha\ldots N\nu\rangle\ C^{+}(B\beta)C(B\beta)\ .$$

We wish to establish that trace of this diagram is represented in fig. 3b. The expression for fig. 3b which corresponds to a filled B shell is

$$C_x(B) = \langle \text{filled B}|U_x(B)|\text{filled B}\rangle\ .$$

By direct substitution of $U_x(B)$ into the previous expression we see that

$$C_x(B) = \sum_{\beta} D_x(B\beta) \ .$$

In a similar fashion we next consider fig. 3c, which ignores all

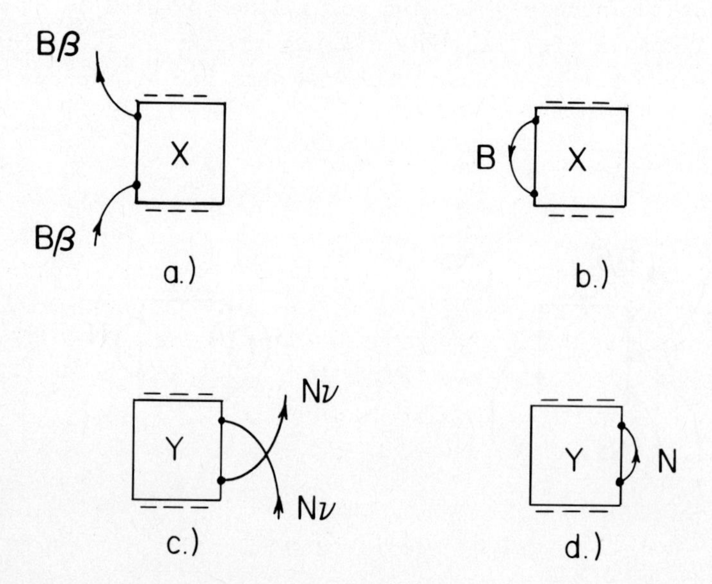

Fig. 3. The (a), (c) and (b), (d) terms are the same as those of fig. 2 (a) and (b), respectively.

but N valence lines. The expression for this diagram is

$$D_y(N\ ) = \langle N\nu|U_y(N)|N\nu\rangle \ ,$$

where,

$$U_y(N) = -\sum_{\nu} \langle A\alpha B\beta..N\nu|v^{(n)}(m)|A\alpha B\beta..N\nu\rangle C(N\nu)C^{+}(N\nu) \ .$$

The order CC^+ is chosen to correspond to the creation of the outgoing N line at a lower vertex. Straightforwardly,

$$D_y(N\nu) = \langle A\alpha,B\beta..N\nu|v^{(n)}(m)|A\alpha,B\beta,..N\nu\rangle \ ,$$

plus an unlinked term which is subtracted away. The expression for fig. 3d is,

$$C_y(N) = \langle 0|U_y(N)|0\rangle = -\sum_{\nu} D_y(N\nu) \ .$$

Obviously, the above arguments can be extended to the remaining valence lines to obtain the Goldstone diagram of fig. 2b with the phase $(-1)^{\phi}$, which yields the trace.

Section 3. The Calculation and Results

We wish to calculate the two body average through fourth order in G. The two body trace in a two particle valence space is

$$\mathrm{Tr}\, v^{(n)}(m,A,B) = \sum_{\alpha,\beta} \langle A\alpha B\beta | v^{(n)}(2) | A\alpha B\beta \rangle$$

$$\equiv \sum [JT] \langle AB;JT | v^{(n)}(2) | AB;JT \rangle \; .$$

The trace calculation proceeds directly from the Goldstone diagrams which appear in figs. 4a-c. All of the Goldstone diagrams used in the second and third order calculations appear in figs. 4a and 4b, respectively. The basic types of fourth order diagrams appear in fig. 4c. The various stretches of the vertices of these fourth order time independent diagrams are not shown, however, the stretching procedure can be deduced by inspection of fig. 4b, the fourth order terms with stretched vertices are included in the calculation. All the possible folded diagrams are included in the calculation as well.

To calculate the trace of a particular order, we consider each Goldstone diagram in that order. For each diagram in the order of interest all possible particle and hole labellings are considered. Then each labelled diagram can have the valence nucleon labels A and B replace any two of the nucleon labels. Thus, each of the labelled diagrams gives way to at least one of the "improper" type of Goldstone diagrams. In these latter diagrams the remaining particle (hole) indices are summed. Each of the diagrams can then be opened (breaking the A,B lines) to give a proper valence diagram. For example, consider the diagram shown in fig. 5a. One contribution to the trace would come if a = A and c = B. The expression for the trace for this example would be

$$-\frac{1}{4} \sum_{\substack{JT \\ b,d,e,f,gh}} [JT](Ab;JT|G|Bd;JT)(Bd;JT|G|ef;JT)(ef;JT|v|gh;JT)$$

$$\times\ (gh;JT|v|Ab;JT)/D$$

where the energy denominator, D, is yet to be specified. The (-1) comes from the closing of the A line and the 1/4 is for equivalent pairs and each G matrix element is unnormalized and antisymmetrized. By opening the A and B lines in fig. 5a, we obtain fig. 5b. The trace of fig. 5b is given by the expression above.

Consistent with many previous calculations for v, we use

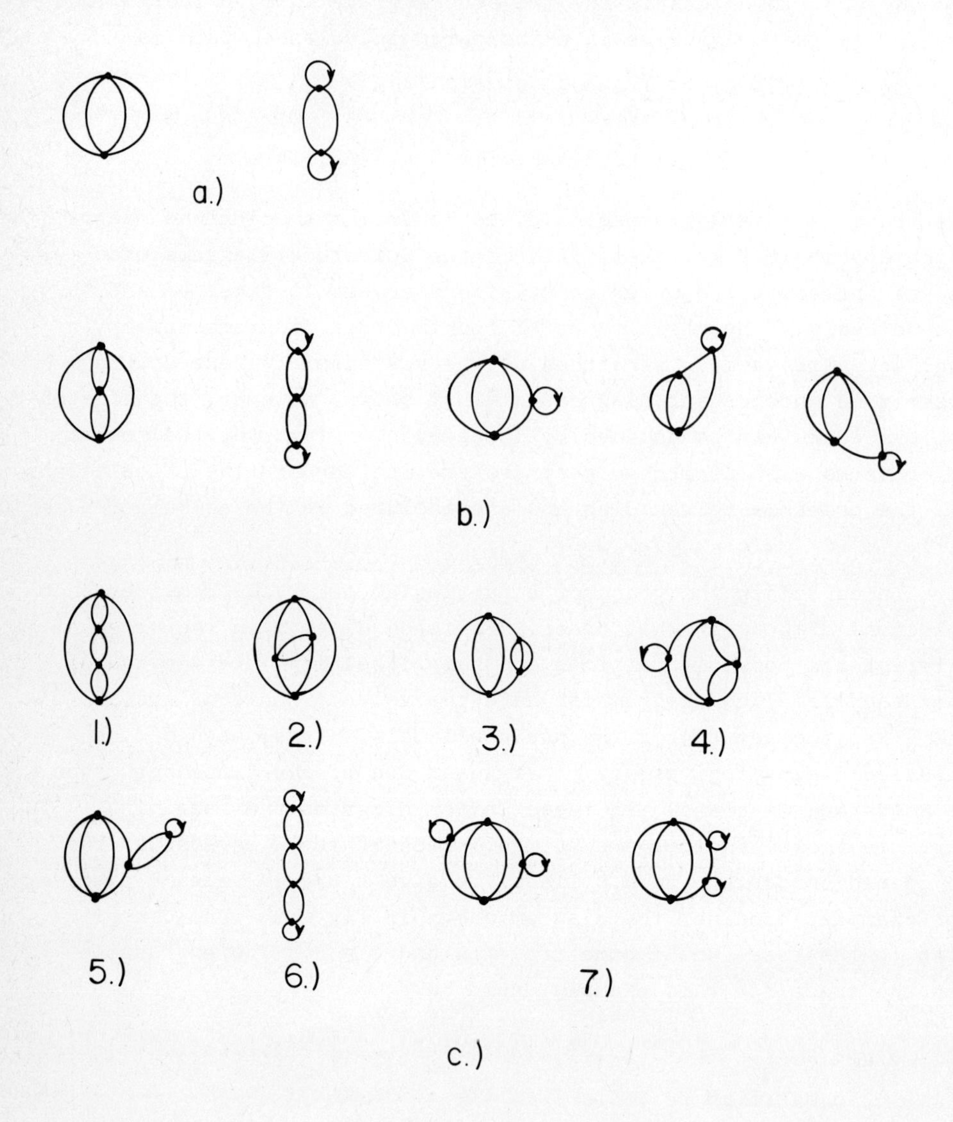

Fig. 4. Figure 4(a) and (b) represent all the second and third order terms in Goldstone form. (c) represents all the unstretched fourth order terms in Goldstone form. The stretching procedure is illustrated in (b) and can be easily done for (c).

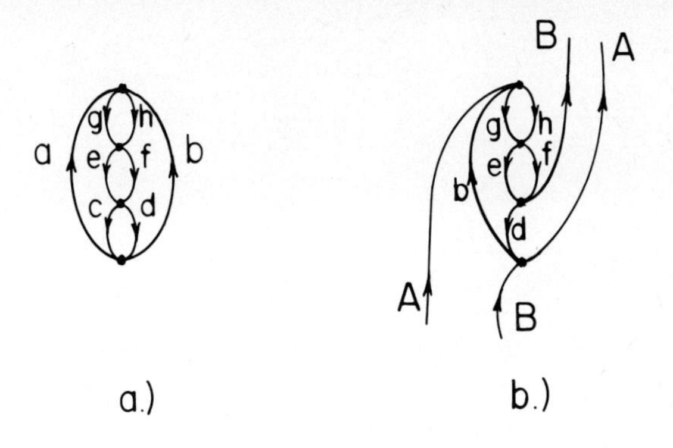

Fig. 5. (a) represents a particular fourth order labelled Goldstone diagram. In (a) the closed valence lines are arbitrarily fixed by a = A and c = B. (b) is obtained from (a) by breaking the A and B lines.

degenerate harmonic oscillator basis states including excited configurations which are $2\hbar\omega$ MeV away from the valence space. Thus, for fig. 5, $D = (-2\hbar\omega)^3$. We will discuss the results for the mass-6 system only, where $2\hbar\omega = 21.75$ MeV. The mass-6 calculations were performed using Barrett, Hewitt, McCarthy G-matrix elements [7,8]. $2\hbar\omega$ ladders are excluded from this G and therefore must be included explicitly in the calculation of the average. Self-consistency is assumed in the calculation of the average. The G-matrix elements were calculated with $\Omega = 0$ MeV (a large gap between the occupied and unoccupied levels).

The p^2 averages, $\langle v^{(n)}(2,A,B)\rangle$ for mass-6 appear in Table 1. The fourth order average is quite large and attractive. In fact, the fourth order contributions to the average are as large, if not larger, than those of second or third order.

In fourth order, there are many terms for each configuration which are close to the total size for that configuration. Still, the results are quite systematic (e.g. all are negative). In fact, the systematic nature of these results persist even when (i) extremes of the starting energy are chosen for the calculation of G, (ii) the calculations are performed in mass-18, (iii) the calculations are performed with different interactions, and (iv) when center-of-mass effects are removed. Thus, the series for v, as presently ordered is quite far from rapid convergence. Still, the systematic nature of these numbers has a simple origin suggesting a possible rearrangement of the series.

Table 1

The results (in MeV) of the $<v^{(4)}(2,A,B)>$, mass-6 calculation presented in terms of the diagrams in fig. 4(c). $2 \equiv 1p_{3/2}$ and $3 \equiv 1p_{1/2}$.

	22	23	33
1	-0.0662	-0.1354	-0.5914
2	0.1017	-0.1396	0.1363
3	0.0038	0.1724	0.2290
4	-0.0553	0.1568	-0.4950
5	0.0859	0.1312	-0.0267
6	-0.0035	0.0049	-0.0106
7	-0.0562	-0.0178	-0.1401
1f	0.0078	0.0331	0.0478
2f	0.0119	0.0152	0.0468
3f	-0.1168	-0.2215	-0.1837
4f	-0.0807	-0.3039	0.0114
5f	-0.0832	-0.1275	0.0031
	-0.2508	-0.4321	-0.9731
2nd order	-0.159	0.405	-0.451
3rd order	0.107	0.269	-0.611
4th order	-0.251	-0.432	-0.973

Section 4. Why the Expansion is Poorly Behaved

We have seen that the fourth order averages tend to be as large as the second or third order averages. In considerable measure, the large fourth order results are due to the fact that the number conserving sets do not internally cancel. The degree of internal cancellation of the individual members of the sets depends on the rate of change in G across the shell closure - a nuclear size effect. This size effect assures us that the number conserving sets will be important for the third, as well as the fourth order averages in light nuclei.

To see this in detail, we compare the individual configuration weighted contributions to the average of the number conserving set in fig. 6a. This set appears in detail in fig. 7. In fig. 7, (i) and (iii) represent the case in which the intermediate vertex is particle-particle in nature, whereas it is particle-hole in nature for (ii). In all three cases presented, this latter term dominates the other two. This dominance is due to the smaller interaction

Fig. 6. (a) represents a second third order number conserving set and (b) represents a fourth order number conserving set.

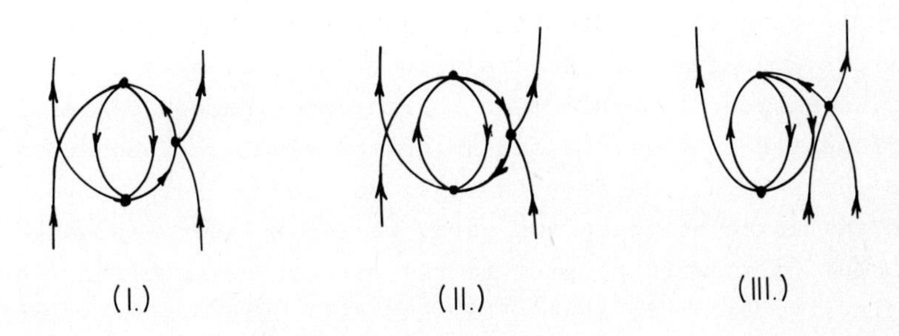

Fig. 7. (i)-(iii) represent the valence (open) form of the number conserving set of fig. 6(a).

volume available to the interaction between a core and a valence nucleon as compared to that between two valence nucleons. Since this is a size effect, we expect it to be more important for smaller A nuclei. In Table 2, the results for the configuration weighted average are presented for masses-6, -18, and -42 (for -18 and -42, we used the Kuo-Brown G) [9]. As expected, the internal cancellation of the set improves with increasing A. For mass-6, the sum of the members of the set is about twice as large as that of the terms represented in (i) and (iii) of fig. 7, whereas, in mass-18 and mass-42 the sum of the members of the set is less than any individual member. This size argument does not work so well for the other third order number conserving set because one of the "hole" lines is actually a valence (folded) line.

Table 2

The configuration weighted averages (in MeV) of the number conserving set of fig. 7 are compared for masses-6, -18, and -42.

	Mass-6	Mass-18	Mass-42
(i)	0.071	0.071	0.050
(ii)	-0.386	-0.228	-0.157
(iii)	0.111	0.092	0.066
Total	-0.204	-0.065	-0.041

Based on this size effect argument, we anticipate that the number conserving sets represent a significant piece of the large fourth order averages. To see this in detail, we make a crude argument. Consider the third and fourth order number conserving sets represented in figs. 6a and 6b, respectively. These sets are drawn in their Goldstone form in figs. 8a and 8b, respectively. To estimate the value of these two sets, we first ignore exchange (non-bubble) terms with respect to the corresponding direct (bubble) terms. In fig. 8a, this leaves (i) and (ii) which would exactly cancel in Brandow's argument. However, as we have seen, these terms do not cancel because the interaction of a valence particle with a core particle is stronger than the interaction between two valence particles. Again, these third order averages are important for small A nuclei.

This being the case we will ignore particle-particle interactions with respect to particle-hole interactions. Thus, we approximate the average of fig. 6a by fig. 9a. To compare fourth order averages to those of third order we consider the average of fig. 6b, we make the same approximations and ignore particle-hole interactions with respect to hole-hole interactions. We, therefore, approximate the terms in fig. 8b by those in fig. 9b.

The expressions for the terms represented in figs. 9a and 9b are (suppressing isospin)

$$\sum_J X_J = \frac{1}{(-2h\omega)^2} \sum_{\substack{h_1,h_2,\\ p,J,I}} [J] \langle Ap;J|G|h_1h_2;J\rangle^2 \left[\frac{I}{h_1}\right] \langle h_1A;I|G|h_1A;I\rangle$$

$$Y = \frac{1}{(-2h\omega)^3} \sum_{\substack{h_1,h_2,h_3,h_4\\ p,J,I}} [J] \langle Ap;J|G|h_1h_2;J\rangle \langle Ap;J|G|h_3h_4;J\rangle$$

$$\times \langle h_1h_2;J|G|h_3h_4;J\rangle \left[\frac{I}{h_1}\right] \langle h_1A;I|G|h_1A;I\rangle \quad .$$

For mass-6, the holes are confined to the $1s_{1/2}$ shell, so that

$$Y = \sum_J X_J \langle h_1h_2;J|G|h_3h_4;J\rangle/(-2h\omega) \quad .$$

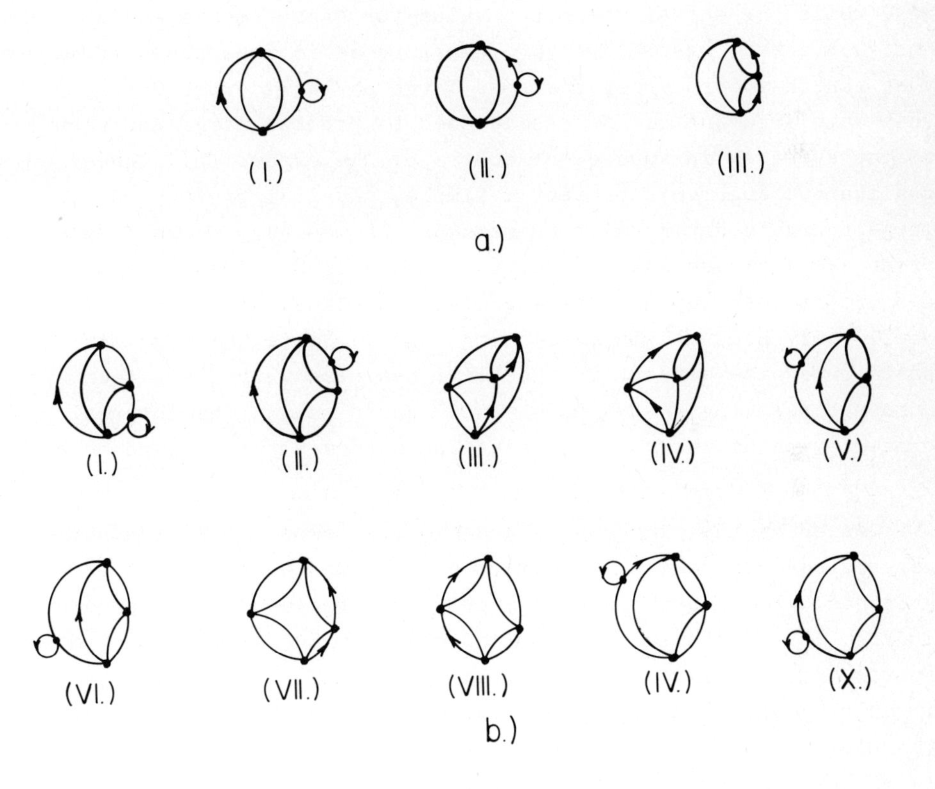

Fig. 8. (a) represents the Goldstone diagrams (traces) for fig. 6(a) and (b) represents the Goldstone diagrams for fig. 6(b). The labelled lines correspond to the closed incoming and outgoing valence lines.

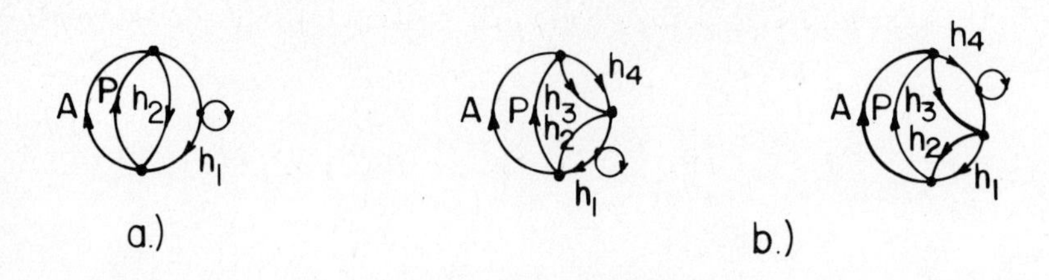

Fig. 9. (a) and (b) represents the approximations to the terms of fig. 8(a) and (b), respectively.

The unnormalized, antisymmetrized two body matrix element is quite large, so that $Y \approx \sum_J X_J$.

In mass-6 for $AB = 1p_{1/2}1p_{1/2}$, $\sum_J X_J = -0.670$ MeV and Y = 0.498 MeV, while the actual numerical value for these sets are -0.653 MeV and -0.562 MeV, respectively. Therefore it is reasonable to expect that fourth order averages are as large as third order averages. This same argument can be easily used to predict large contributions to the averages in subsequent orders of the expansion. These qualitative arguments present a simple, physical picture of the large fourth order results in mass-6. In mass-18, where fourth order averages are also large, the change in G across the shell closure is less rapid. Therefore, the qualitative argument is really only accurate in predicting signs. Since the argument is based on a size effect, if the Fermi sea were very deep, then the number conserving sets should cancel quite well. The internal cancellation of the number conserving sets may also depend on extending the Q-space beyond $2\hbar\omega$ in energy.

Based on the preceding discussion, we expect that the number conserving sets make a sizeable contribution to the fourth order averages. The fourth order number conserving sets are represented in fig. 10. In Table 3, their calculated value is compared to the fourth order total. The agreement between the two sets of numbers is quite good. The largest terms in fourth order are the TDA (particle-hole chain) terms. If we include in the subset the TDA term plus the folds of TDA terms from second and third order, then the agreement between the subset and the entire set is even better. This same type of agreement holds in mass-18 as well except that the TDA terms are somewhat more important. The subset constitutes less than 10% of all the fourth order terms. The remaining 90% of the terms are not so sensitive to the change in G across the shell closure. These remaining terms tend to be small and self-cancelling.

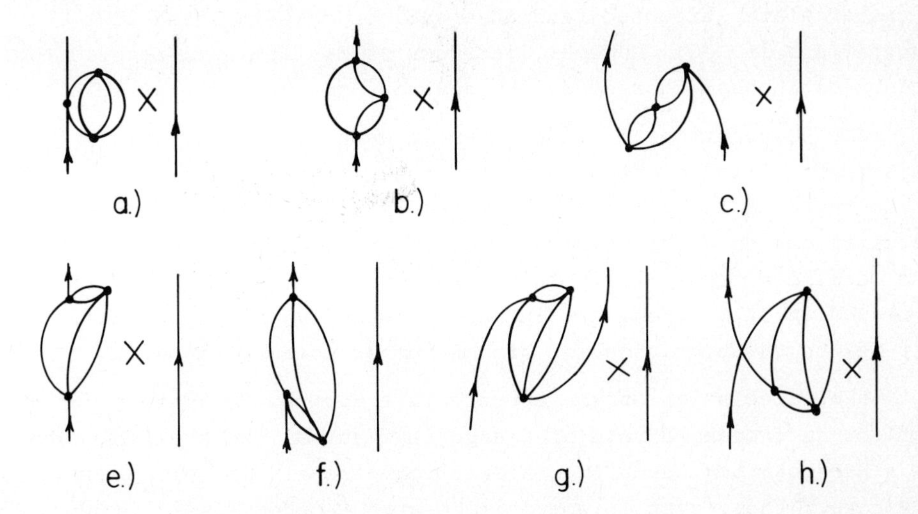

Fig. 10. (a)-(h) represent the valence form of the fourth order number conserving sets.

Table 3

The first three columns contain partial contributions (in MeV) to $\langle v^{(4)}(2,A,B)\rangle$. (1) is the contribution of the number conserving sets, (2) is the contribution of the TDA terms. The third column is the sum of the first two columns which is to be compared to the fourth order total in the fourth column.

	(1)NCS	(2)TDA	Subset ((1)+(2))	Total Set
22	-0.272	0.010	-0.262	-0.251
23	-0.488	0.024	-0.464	-0.432
33	-0.999	0.019	-0.980	-0.973

Thus, the change of G across a shell closure leads us to suggest that a partial summation of the number conserving sets and TDA terms would yield a useful rearrangement of the series appropriate for the P and Q spaces we have used.

Section 5. Further Discussion of the Results

In this section we will discuss the role of intruder states, Hartree-Fock bases, and folded diagrams in terms of the results and conclusions of the previous sections.

In the expansion for the average effective interaction, fourth order results are as large as those from lower orders. This is not due to the presence of intruder states [10]. The mass-6 and the mass-18 expansions are equally poorly behaved. Yet, there are intruder states expected (and observed) in mass-18 based on the excitations in ^{16}O. However, intruder states are not expected (or observed) in mass-6 based on the energies of excited states in ^{4}He.

The calculations discussed herein were performed in an oscillator basis. It is often suggested that it would be more appropriate to do these calculations in a Hartree-Fock basis. In fact, such calculations do serve to reduce second order effects compared to the oscillator case [11]. This result is obtained essentially by a reduction of G. However, in all probability this approach would not fix up the poor behavior of the number conserving sets.

The calculation of the average in a Hartree-Fock basis would reflect an increased rate of change in G across the shell closure. In a Hartree-Fock basis the interaction strength is pulled into the core. Thus, even in competition with reduced G's it is unlikely that the internal cancellation of the number conserving sets would be improved.

We have performed the calculations for the average shell model effective interaction within formal framework of Brandow which includes folded diagrams. The folded diagrams arise in the series for the effective interaction from the expansion of the exact energy from the denominator. In the Bloch-Horowitz expansion, the exact energy is left in the denominator and unlinked terms are included. The calculated averages can be readily put into a Bloch-Horowitz form. To do this, we make the reasonable approximation that the exact energy can be replaced by the unperturbed energy ($E \rightarrow E_0$). Since there will be no unlinked diagrams due to the truncation of $2\hbar\omega$ energy intermediate states, we need only to ignore folded diagrams to obtain the Bloch-Horowitz result. The folded diagram averages are compared to the results of the linked, folded expansion in Table 4.

Table 4

The mass-6 fourth order averages (in MeV) $\langle v^{(4)}(2,A,B)\rangle$, are compared with the corresponding sum of folded diagrams.

	22	23	33
Folded	-0.261	-0.605	-0.075
Total	-0.251	-0.432	-0.973

Section 6. Summary

A method for calculating the average of the valence effective interaction has been developed. This method involves the calculation of open shell averages directly from diagrams with a closed shell form.

The method was exploited to calculate the two body average through fourth order in G. The fourth order averages are large. This is true for different mass regions (mass-18) and different interactions.

The fourth order numerical results are dominated by the number conserving sets. In turn, the number conserving sets are sensitive to the rate of change in G across the shell closure. The importance of this effect persists beyond fourth order. Thus, the linked cluster expansion, $v = G + G \frac{Q'}{E-H_o} v$, is poorly behaved in low order for light nuclei due to a nuclear size effect.

It does not seem likely that the internal cancellation of the number conserving sets would be improved in a Hartree-Fock basis.

The expansion, $v = G + G \frac{Q'}{E-H_o} v$, is not rapidly convergent. To improve its convergence properties, a summation of the number conserving sets and TDA terms is suggested.

References

1. C. Bloch and J. Horowitz, Nucl. Phys. 8 (1958) 91.
2. B. H. Brandow, Rev. Mod. Phys. 39 (1967) 771.
3. B. H. Brandow, Boulder Lecture Notes, Lectures in Theoretical Physics 11 (Gordon and Breach, New York, 1968).
4. B. R. Barrett and M. W. Kirson, Nucl. Phys. A148 (1970) 145.
5. P. J. Ellis, A. D. Jackson, and E. Osnes, Nucl. Phys. A196 (1972) 161.
6. P. Goode and D. S. Koltun, Nucl. Phys. A243(1975)44.
7. B. R. Barrett, R. G. L. Hewitt, and R. J. McCarthy, Phys. Rev. C3 (1971) 1137.
8. B. R. Barrett, private communication.
9. T. T. S. Kuo and G. E. Brown, Nucl. Phys. 85 (1966) 40.
10. T. H. Schucan and H. A. Weidenmüller, Ann. Phys. (N.Y.) 73 (1972) 108; 76 (1973) 483.
11. P. J. Ellis and E. Osnes, Phys. Lett. 41B (1972) 97.

P. GOODE: BEHAVIOR OF THE G-MATRIX EXPANSION FOR THE AVERAGE EFFECTIVE INTERACTION

Brandow: I am not surprised that the degree of cancellation within the number-conserving sets is rather poor. This is a reflection of the effective momentum dependence of the G-matrix elements, which nuclear matter experience shows to be quite strong. Nevertheless, there is some tendency towards cancellation. What is rather surprising is that this non-cancellation turns out to be the dominant contribution to the 4th order sum. My question is, can you be more specific about your proposal to partially sum the number-conserving sets? Just how is one supposed to carry this out?

Goode: The number conserving sets do not self cancel because of the size effect discussed in the talk. The terms to be partially summed are those which come from the single interaction of two dressed single-nucleon lines. I do not know any easy way to do this partial summation.

Scheerbaum: The work of Vary, Sauer and Wong indicates that intermediate states of quite high excitation energy play an important role in the core-polarization diagram. Can you comment on the possible importance of such high-lying intermediate states in the diagrams you calculate?

Goode: It is hard to guess in the absence of a calculation, but at least the denominators go as $(-2\hbar\omega)^{2 \text{ or } 3}$. It would be an interesting calculation. Maybe Vary has something to say about this.

Vary: Let me summarize what we might reasonably expect the role of the tensor force to be as one goes to third order and beyond. There are basically three factors which enter, two of which help and one of which hinders convergence of particle-state sums.The fact that one has the square of an energy denominator helps, and the fact that in some diagrams a passive particle line may coincide with the vertex which pushes the interaction further off shell and weakens it. Both these should assist convergence over the situation found in the core-polarization diagram. On the other hand, the nice kinematic constraint found in the 3p-1h diagram, which is actually responsible for convergence, would be lacking in many third-order diagrams. Thus, it is an open and interesting question what the convergence over particle-state sums will be like in orders higher than second.

Vary: Blomquist, Bergstrom and a host of collaborators have been performing extensive studies on the neutron deficient isotopes of ^{208}Pb along the lines you have described. Their results, using empirically determined matrix elements from ^{207}Pb and ^{206}Pb provide a remarkably accurate description of the high spin spectra and magnetic properties of the ^{205}Pb, ^{204}Pb and ^{203}Pb nuclei. The rms level deviations between theory and experiment are of order 10 keV. From these results one can deduce useful implications for the direction of our microscopic efforts. For example, they make useful conclusions regarding the upper limit of the effective three-body force in the lead region.

Goode: I am glad that you mentioned this beautiful work which I did not have time to describe. Their results are very impressive.

Barrett: What conclusion can you draw regarding convergence for individual terms based on your averages? After all the core-polarization is negative for low J values and repulsive for high J values.

Goode: Since the averages are large the effects on some individual states must be sizeable. Based on experience from a third order JT calculation, we might expect the effects to be concentrated in states with light statistical weighting.

Barrett: Michael Kirson and I found that the vertex renormalization term in third-order was very large. Is it large in your third-order average? And why is it small in your fourth-order averages?

Goode: In the third-order averages, the vertex corrections are small but not negligible. This is reflective of the fact that the vertex corrections are large in third-order for $J = 0$, as you mention, but tend to be small for $J \neq 0$. In the fourth-order averages, vertex corrections are smaller still and thus are associated with the aforementioned size effect.

Koltun: I would like to comment on Barrett's question about behavior of individual J-T matrix elements from the calculated averages. We can only say that the fourth-order terms cannot all be smaller than the averages! Since we get large, not small results, this is no problem. Had we found very small fourth-order averages, we would not be certain of the result for each JT.

ALGEBRAIC STRUCTURE OF EFFECTIVE INTERACTIONS AND OPERATORS. CONVERGENCE PROPERTIES OF THE PERTURBATION EXPANSIONS.

Hans A. Weidenmüller
Max-Planck-Institut für Kernphysik,
Heidelberg, Germany

Abstract. Algebraic expressions for effective interactions and operators are used to show that the calculation of both eigenvalues and transition matrix elements is possible without additional normalization problems only if these operators are Hermitean. Infinitely many such Hermitean representations exist, and some special choices are pointed out. Independently of the representation chosen, the convergence of the perturbation expansions hinges on the positions of branch points connected with intruder states. Practical aspects of this problem and the rate of convergence of the series in the absence of close-lying branch points are discussed.

1. Introduction. Convergence properties of the perturbation expansion for the effective interaction in nuclei have been studied in some detail in recent years. Another problem that has received less attention concerns the calculation of matrix elements of effective operators, and the convergence properties of the associated expansions. Both problems can be studied jointly by displaying the algebraic structure of the underlying expressions. This is done in the first part of this paper. Some of the material is new and the result of a collaboration of J. Richert, T.H. Schucan, M.H. Simbel and the author [1] . A study of the algebraic structure like the one presented here cannot, of course, replace the perturbation expansion. It helps rather to elucidate the structure of the theory, and to show which difficulties, if any, beset the perturbation expansion. The second part deals with convergence properties of the perturbation expansions. The existence of branch points defining the radius of convergence is recalled, and practical implications of this result, as well as some other aspects of convergence, are discussed.

I. The Algebraic Structure.

2. The Problem. The matrix elements of the residual interaction used in shell-model calculations can be determined from experimental data [2] . The understanding of these matrix elements in

terms of the fundamental nucleon-nucleon force presents a considerable challenge to nuclear theorists. In view of the regularities displayed by the empirical matrix elements over a wide range of mass numbers [3], the calculation of the "residual interaction" or "effective interaction" seems to offer a problem of a more general nature than would be the case for the calculation of an individual nuclear level scheme in terms of the nucleon-nucleon force. One of the obstacles encountered in this programme is the strongly repulsive character of the nucleon-nucleon interaction at short distances [4]. It necessitates the introduction of the Bethe-Brueckner G-matrix [5]. In the following, we disregard this complication and write the nuclear Hamiltonian in the form $H = H_o + V$ where H_o is the shell-model Hamiltonian, and V the fundamental nucleon-nucleon interaction (if it is chosen to be sufficiently soft) or the G-matrix, each minus the average potential contained in H_o. [In doing so, we temporarily also disregard the problem of double-counting which arises through the presence of two-particle G-matrix ladders. See section 7 below.] For purposes of discussing the perturbation expansion, we multiply V by a dimensionless parameter x and choose x so that it assumes the value unity for the actual strength of the interaction. We accordingly write

$$H = H_o + xV \tag{1}$$

The levels of H_o form groups of nearly degenerate states, typical of a shell model. Let the lowest group contain M levels, and let the M-dimensional space spanned by the associated eigenfunctions be called S_o, the projector onto this space P_o with $P_o^2 = P_o = P_o^+$. The effective interaction $W(x)$ is defined by the requirement that the effective operator

$$H_{eff}(x) = P_o H_o P_o + W(x) \tag{2}$$

which is defined on the model space, should have M eigenvalues which coincide with M eigenvalues of $H(x)$. Effective β- and γ-transition operators etc. $A_{eff}(x)$ are correspondingly defined by the requirement that their matrix elements with eigenfunctions of $H_{eff}(x)$ be equal to the transition matrix elements of the full problem. It is our aim to construct closed expressions for $H_{eff}(x)$ and $A_{eff}(x)$ which display some essential properties of these operators. In all of part I of this paper, we keep x fixed and real.

3. The effective Interaction.

Let $\lambda_i(x)$, $i = 1, \ldots, M$ be the M eigenvalues of $H(x)$ we wish to reproduce, $|\psi_i(x)\rangle$, $i = 1, \ldots, M$ the associated orthonormal eigenvectors. The operator

$$P(x) = \sum_{i=1}^{M} |\psi_i(x)\rangle\langle\psi_i(x)| \tag{3}$$

with $P^+(x) = P(x) = P^2(x)$ projects onto a space $S(x)$ of M dimensions. We are not interested in the full $H(x)$ but only in that part of $H(x)$ which contains the eigenvalues λ_i, $i = 1, \ldots, M$. It is given by

$$P(x)\,H(x)\,P(x) = \sum_{i=1}^{M} |\psi_i(x)\rangle\,\lambda_i(x)\,\langle\psi_i(x)| . \tag{4}$$

The operator (4) is defined on $S(x)$. Since both $S(x)$ and S_o have dimension M , an operator $H_{eff}(x)$ with the desired properties is simply obtained by mapping $S(x)$ onto S_o , and transforming (4) correspondingly. Let $U^{-1}(x)$ be a one-to-one mapping of $S(x)$ onto S_o , so that

$$S_o = \mathcal{U}^{-1}(x)\,S(x) \quad \text{and} \quad S(x) = \mathcal{U}(x)\,S_o . \tag{5}$$

The eqs. (5) imply

$$\mathcal{U}(x)\,\mathcal{U}^{-1}(x) = P(x), \quad \mathcal{U}^{-1}(x)\,\mathcal{U}(x) = P_o \tag{6}$$

The operator $H_{eff}(x)$ is simply given by

$$H_{eff}(x) = \mathcal{U}^{-1}(x)\,P(x)\,H(x)\,P(x)\,\mathcal{U}(x) . \tag{7}$$

Using eq. (4), we can cast this into the form

$$H_{eff}(x) = \sum_{i=1}^{M} \mathcal{U}^{-1}(x)|\Psi_i(x)\rangle \, \lambda_i(x) \, \langle \Psi_i(x)| \mathcal{U}(x) . \tag{8}$$

This shows that $H_{eff}(x)$ has the M eigenvalues λ_i, i=1, ..., M, the associated right-hand eigenfunctions ϕ_i^R and left-hand eigenfunctions ϕ_k^L with

$$|\phi_i^R\rangle = \mathcal{U}^{-1}(x)|\Psi_i\rangle \quad , \quad |\phi_k^L\rangle = \mathcal{U}^{+}(x)|\Psi_k(x)\rangle . \tag{9}$$

Note that $|\phi_i^R\rangle \neq |\phi_k^L\rangle$ unless U is unitary which we have not required. It follows that

$$\begin{aligned} \langle \phi_i^L | \phi_k^R \rangle &= \langle \Psi_i | \Psi_k \rangle = \delta_{ik} \\ &= \langle \phi_i^R | \mathcal{U}^+ \mathcal{U} | \phi_k^R \rangle = \langle \phi_i^L | \mathcal{U}^{-1} (\mathcal{U}^{-1})^+ | \phi_k^L \rangle , \end{aligned} \tag{10}$$

while

$$\langle \phi_i^R | \phi_k^R \rangle = \langle \Psi_i | (\mathcal{U}^{-1})^+ \mathcal{U}^{-1} | \Psi_k \rangle ; \quad \langle \phi_i^L | \phi_k^L \rangle = \langle \Psi_i | \mathcal{U} \mathcal{U}^+ | \Psi_k \rangle . \tag{11}$$

Neither the $|\phi_i^R\rangle$ nor the $|\phi_k^L\rangle$ are mutually orthogonal unless $U^+ = U^{-1}$.

4. Effective Operators.

Several different possibilities exist of defining effective operators. We consider one of them by defining

$$\langle \phi_i^L | A_{eff}(x) | \phi_k^R \rangle = \langle \Psi_i(x) | A | \Psi_k(x) \rangle \tag{12}$$

where A is the full operator. Eq. (12) is supposed to hold for all i, k = 1, ..., M. A simple substitution yields

$$A_{eff}(x) = \mathcal{U}^{-1}(x) \, P(x) \, A \, P(x) \, \mathcal{U}(x) . \tag{13}$$

We note the complete analogy between eqs. (7) and (13). Another possibility to define the effective operator consists in writing

$$\langle \phi_i^R | \bar{A}_{eff}(x) | \phi_k^R \rangle = \langle \Psi_i(x) | A | \Psi_k(x) \rangle \tag{14}$$

which yields

$$\bar{A}_{eff} = \mathcal{U}^+(x)\, P(x)\, A\, P(x)\, \mathcal{U}(x) . \tag{15}$$

Note that $\bar{A}_{eff}(x)$ is Hermitean even if $H_{eff}(x)$ is not. A closer inspection of eqs. (13) and (15) reveals the following normalization problem. The functions ϕ_i^L and ϕ_k^R are, as eigenfunctions of $H_{eff}(x)$, each determined up to a normalization constant. The condition $\langle\phi_i^L|\phi_k^R\rangle = \delta_{ik}$ is not sufficient to determine this constant since $\langle\phi_k^L|\phi_k^R\rangle$ remains unchanged if we multiply each $|\phi_k^R\rangle$ by $\alpha_k \neq 0$ and each $|\phi_i^L\rangle$ by $(\alpha_k^*)^{-1}$, whereas the matrix elements of $A_{eff}(x)$ and $\bar{A}_{eff}(x)$ change by factors $\alpha_i^{-1}\alpha_k$ and $\alpha_i^*\alpha_k$, respectively. There exist two ways out of this normalization dilemma. The first one consists in multiplying $\langle\phi_i^L|A_{eff}(x)|\phi_k^R\rangle$ by $\{\langle\phi_i^L|U^{-1}(U^{-1})^\dagger|\phi_i^L\rangle^{-1/2} \langle\phi_k^R|U^+U|\phi_k^R\rangle^{-1/2}\}$, and $\langle\phi_k^R|\bar{A}_{eff}(x)|\phi_k^R\rangle$ by $\{\langle\phi_i^R|U^+U|\phi_i^R\rangle^{-1/2} \langle\phi_k^R|U^+U|\phi_k^R\rangle^{-1/2}\}$. The normalization factors in curly brackets are unity for the correct normalization (10). The resulting ratios are independent of the normalization of $|\phi_i^L\rangle$ and $|\phi_k^R\rangle$ and thus always correct. However, the calculation of the normalization factors involves additional work beyond the calculation of $A_{eff}(x)$ or $\bar{A}_{eff}(x)$. The other way out of this dilemma consists in choosing U unitary, $U^+ = U^{-1}$. Then, $H_{eff}(x)$ is Hermitean, $|\phi_i^L\rangle = |\phi_i^R\rangle = |\phi_i\rangle$, with $\langle\phi_i|\phi_k\rangle = \delta_{ik}$, and all effective operators $A_{eff}(x) = \bar{A}_{eff}(x)$ are Hermitean, too. Normalization problems do not exist.

The first choice was made by Krenziglowa and Kuo [6] who used Brandow's original, non-Hermitean choice for $H_{eff}(x)$, given by [7,8]

$$H_{eff}^B = P_0\, P(x)\, H(x)\, P(x)\, P_0\, \left(P_0 P(x) P_0 \right)^{-1}, \tag{16}$$

so that

$$\mathcal{U}^B = P(x)P_o(P_oP(x)P_o)^{-1} \quad , (\mathcal{U}^B)^{-1} = P_oP(x). \quad (17)$$

[It is obvious that $(U^B)^{-1}U^B = P_o$, and the relation $U^B(U^B)^{-1} = P(x)$ can easily be checked [8] . Moreover, $(U^B)^+ \neq (U^B)^{-1}$] . Brandow's effective interaction has as right-hand eigenvectors simply the projections of ψ_i onto the model space, $|\phi_i^R\rangle = |\phi_i^B\rangle = P_o|\psi_i\rangle$. [H^B_{eff} exists only if the $|\phi_i^R\rangle$ are linearly independent, i.e., if $(P_oP(x)P_o)^{-1}$ exists, which is assumed throughout, cf. the footnote on page 9 . Krenziglowa and Kuo use [6] the transformation U^B and introduce the Hermitean operators $\bar{A}_{eff}(x)$ defined in eq. (15) . They account of the normalization problem by dividing through the second curly bracket introduced above. The normalization integral contains the factor $(U^B)^+U^B$, given by (we put $Q_o = 1 - P_o$)

$$(\mathcal{U}^B)^+\mathcal{U}^B = P_o + (P_oP(x)P_o)^{-1}P_oP(x)Q_oP(x)P_o\,(P_oP(x)P_o)^{-1}$$

$$= P_o + F(I) , \quad (18)$$

where the last equality sign introduces the notation of ref. [6]. Collecting these results, we obtain

$$\langle\psi_i|A|\psi_k\rangle = \langle\phi_i^B|(P_oP(x)P_o)^{-1}P_oP(x)AP(x)P_o(P_oP(x)P_o)^{-1}|\phi_k^B\rangle \cdot$$

$$\cdot \left\{\langle\phi_i^B|P_o + F(I)|\phi_i^B\rangle^{-1/2}\langle\phi_k^B|P_o + F(I)|\phi_k^B\rangle^{-1/2}\right\}. \quad (19)$$

This is the form used in ref. [6] .

The second, Hermitean choice for U was advocated by Brandow [7] in his study of effective operators. His operator U^B has the form[+]

+ Brandow writes this operator in the form $(1+\theta)^{1/2}H^B_{eff}(1+\theta)^{-1/2}$ where H^B_{eff} is given by eq. (16) and remarks that this form is "not evidently Hermitean". The Hermitecity is quite evident from our eq. (21).

$$\tilde{\mathcal{U}}^{B} = P(x)P_o\,(P_oP(x)P_o)^{-1/2} \;,\; (\tilde{\mathcal{U}}^{B})^{-1} = (P_oP(x)P_o)^{-1}P_oP(x) = (\tilde{\mathcal{U}}^{B})^{+}. \quad (20)$$

[Brandow uses the notation $(1+\Theta)^{1/2}$ instead of $(P_oP(x)P_o)^{-1/2}$].

Again, the relations (6) can easily be checked. The new effective Hamiltonian[+]

$$\tilde{H}^{B}_{eff}(x) = (P_o P(x) P_o)^{-1/2} P_o P(x) H(x) P(x) P_o\,(P_o P(x) P_o)^{-1/2} \quad (21)$$

differs from the one in eq. (16). It can be checked that up to and including terms of 3rd order in x, we have, fortunately,

$$\tilde{H}^{B}_{eff}(x) = \frac{1}{2}\left(H^{B}_{eff}(x) + (H^{B}_{eff}(x))^{+}\right). \quad (22)$$

[Eq. (22) is definitely incorrect for terms of 4th order in x!] Since no realistic calculations of H_{eff} have ever been pushed beyond 3rd order in G, we can use eq. (22) to construct $\tilde{H}^{B}_{eff}(x)$ and the associated orthonormal eigenfunctions $|\tilde{\phi}^{B}_{i}(x)\rangle$ from $H^{B}_{eff}(x)$ which has been calculated. With these eigenfunctions, the calculation of matrix elements of effective operators is straightforward, since

$$\langle \Psi_i(x) | A | \Psi_k(x) \rangle$$

$$= \langle \tilde{\phi}^{B}_{i}(x) | (P_oP(x)P_o)^{-1/2} P_oP(x)\,A\,P(x)P_o\,(P_oP(x)P_o)^{-1/2} | \tilde{\phi}^{B}_{k}(x) \rangle. \quad (23)$$

In the perturbation-theoretical evaluation of the effective operators of eq. (23), the expansion of the square-root factors $(P_oP(x)P_o)^{-1/2}$ produces combinatorial factors which are absent, for instance, if the convention (19) is used. Brandow can show that the perturbation expansion of the effective operators (21) consists of linked diagrams only. It is not clear that this is also the case for the expansion of the expression (19).

+ See footnote on page 6.

Considerations quite analogous to the ones presented in this section apply if A_{eff} connects states in two different model spaces which are, for instance, distinguished by spin or parity.

5. Ambiguities of the Hermitean choice.

The Hermitean choice of eqs. (21), (23) for $H_{eff}(x)$ and $A_{eff}(x)$ is not the only one possible, and very many such choices can, in fact, be made. This was apparently first pointed out by Johnson and Baranger [9] . A more restricted and more specific claim, related to our eq. (24), is due to Oberlechner et al. [20]. Let U_1, U_2 be two unitary mappings of S_o onto $S(x)$, and let $H_{1,eff}(x)$ and $H_{2,eff}(x)$ or $A_{1,eff}(x)$ and $A_{2,eff}(x)$ be the associated operators, see eqs. (7) and (13). It is then straightforward to show that

$$H_{2,eff}(x) = M^{+} H_{1,eff}(x) M \quad ; \quad A_{2,eff}(x) = M^{+} A_{1,eff}(x) M, \qquad (24)$$

where $M = U_1 U_2^{+}$ is a unitary mapping of the model space onto itself. In words: All possible Hermitean effective operators can be transformed into each other by a rotation within the model space. Such rotations can be "naturally" defined within the frame of our problem, i.e. without recourse to quantities that would not already appear in the equations given above. For instance, the operator $\exp(iP_o P(x) P_o)$ is unitary on S_o and defines a rotation within the model space, and a multitude of such operators can be constructed. These remarks show that the matrix representation in the model space of $H_{eff}(x)$ and $A_{eff}(x)$ is not unique. This freedom might be useful in attempts to improve the rate of convergence of the perturbation expansion, but this question has not yet been investigated. Conversely, this ambiguity may affect inferences on the existence of three-body forces in $H_{eff}(x)$ which are certainly dependent on the representation chosen.

II. Convergence properties of the perturbation expansions.

6. Branch points. Intruders.

The explicit calculation of the effective interaction, and of the effective operators, cannot be performed with the help of eqs.(8) and (13,15) since these are based on the assumed knowledge of the actually unknown eigenvalues λ_i and eigenfunctions $|\psi_i\rangle$ $i = 1, \ldots, M$. Instead, perturbation expansions for $H_{eff}(x)$ and $A_{eff}(x)$ or $\bar{A}_{eff}(x)$ are used which can formally be characterized as expansions in powers of x. Three questions arise.

(i) Do the series converge ?

(ii) If the series converge, which of the many possible operators $H_{eff}(x)$ and $A_{eff}(x)$ or $\bar{A}_{eff}(x)$ are approximated? More precisely: Which eigenvalues $\lambda_i(x)$ of the full problem are reproduced by $H_{eff}(x)$?

(iii) If the series converge, how many terms are needed to obtain a sufficiently accurate approximation to H_{eff} and to A_{eff} ?

We discuss these questions in turn and devote this and the following section to questions (i) and (ii). To answer question (i), we must find the closest singularities of $H_{eff}(x)$ and $A_{eff}(x)$ in the complex x-plane, since these determine the radius of convergence. Inspection of eq. (8) shows that the x-dependence of $H_{eff}(x)$ has three sources: The linear x-dependence of $H(x)$, the x-dependence of $U(x)$, and the x-dependence of $P(x)$. The first produces no singularities for finite x [in the case of an infinitely dimensional Hilbert space, some care is required, see ref. [8]]. Since $U(x)$ is a one-to-one mapping from S_o to $S(x)$, it can always be chosen [8] to be analytic in the domain of x-values where $S(x)$ is defined and analytic.[+] But $S(x)$ is defined in terms of the projection operator $P(x)$. This shows that the radius r_o of convergence of $H_{eff}(x)$ is determined by the closest singularity of $P(x)$. The same statement applies to $A_{eff}(x)$ and $\bar{A}_{eff}(x)$. Hence, all effective operators acting in the same model space have the same radius of convergence r_o. Moreover, r_o is independent of the (Hermitean or non-Hermitean) representation chosen for the effective operators.

[+] Some choices may introduce additional singularities. This is true, for instance, of the choices (17) or (20) which are singular wherever $(P_oP(x)P_o)^{-1}$ does not exist. However, other choices can easily be found [8] which avoid this problem.

It is determined by the closest singularity of $P(x)$, i.e. by an intrinsic property of the decomposition (1) .

The nature and physical significance of the closest singularity of $P(x)$ can also be exhibited. It follows [8] from analytic perturbation theory, and from a generalization of the Wigner - von Neumann non-crossing rule, that this singularity consists of a pair of complex conjugate branch points, located at x_o and x_o^* . To characterize x_o and x_o^* , let us follow the eigenvalues $\lambda_j(x)$ of $H(x)$ as functions of x . We call the first M of these which for $x = 0$ coincide with the eigenvalues of $P_oH_oP_o$, "internal" eigenvalues, the rest, "external" eigenvalues. The points x_o and x_o^* with $|x_o| = |x_o^*| = r_o$ denote the smallest value of $|x|$ where an internal and an external eigenvalue coincide.

These points of coincidence, x_o and x_o^* , thus mark the places beyond which "intruder" states appear in the spectrum of $H_{eff}(x)$: A state originating, at $x = 0$, from $(1-P_o)H_o(1-P_o)$ moves down into the set of states originating, at $x = 0$, from $P_oH_oP_o$. Such intruder states are characterized by the statement that their wave functions have the main component outside the space S_o . A more detailed and intuitive discussion of such states may be found in ref. [10]. We see that the existence of intruder states, mostly low-lying collective states originating from the next-higher shell, is a safe indication that $r_o < 1$. i.e., that the perturbation series for H_{eff} and A_{eff} diverge at $x = 1$. [Unfortunately, the converse is not true, i.e., the absence of intruder states does not necessarily signal convergence. For instance, intruder states might appear in the spectrum of $H_{eff}(x)$ if the sign of V , i.e. of x , were reversed. The convergence of the perturbation expansion is, on the other hand, determined by <u>all</u> singularities within the unit circle $|x| < 1$.] The existence of intruders is, unhappily, rather typical of many low-lying nuclear spectra. The associated divergence of the series for H_{eff} and A_{eff} have spurred a considerable amount of activity in recent years. This is reviewed in other contributions to this Conference.

We have thus answered question (i) raised at the beginning of this section. Concerning question (ii), the results just summarized show that the perturbation series for $H_{eff}(x)$, if convergent at $x = 1$, must approximate that particular H_{eff} which has as eigenvalues the lowest M eigenvalues of $H(x)$.

It can be shown that the conslusions presented above are not affected by transcribing the perturbation series into a diagram expansion and by the associated cancellation of unlinked diagrams [8].

7. Problems of Double-Counting.

All the conclusions presented in the previous section are based on the decomposition (1) of $H(x)$, and on a perturbation expansion of $H_{eff}(x)$ and $A_{eff}(x)$ in powers of (xV). In actual cases, complications arise because of the introduction of the Bethe-Brueckner G-matrix which by definition sums all two-particle ladders. As a consequence, two-particle ladders of G-matrices are forbidden as they would constitute double-counting of certain diagrams. The expansions for $H_{eff}(x)$ and $A_{eff}(x)$ based on the G-matrix are, therefore, not identical with the expansions obtained by writing $H(x) = H_o + xG$, and expanding the effective operators in powers of (xG). Such an expansion would introduce two-particle ladders of G-matrices. We now ask how our previous conclusions are affected by this double-counting problem.

Perhaps the most satisfactory way to avoid this problem consists in using the double-partitioning technique advocated by Brandow [11] and Brown [12] and used recently by Barrett and collaborators [13] . The total Hilbert space is divided into three orthogonal subspaces, the model space S_o, a space S_1 of low-lying configurations right above S_o , and the remainder S_2 . The G-matrix is determined by summing all two-particle ladders with intermediate states in S_2 . In this way, the short-range or high-momentum components of the nucleon-nucleon interaction are satisfactorily taken into account. On the remaining space, spanned by all vectors in S_o or in S_1 , the Hamiltonian has the form $H(x) = H_o + xG$. No double-counting problems arise in expanding H_{eff} and A_{eff} in powers of (xG), and the conclusions drawn in section 6 remain valid.

A second alternative can best be described by analogy to the average potential U . The latter is often defined diagrammatically [7, 11] as the sum of all on-shell contributions to the mass operator. This prescription leads to the identification of an operator U which is then added to and subtracted from H in such a way that $(T + U) = H_o$ is the new unperturbed Hamiltonian, and $x(V-U)$ the new perturbation. It appears that one might proceed similarly in the case of the double-counting problem, identifying an operator $\mathcal{O}$ which just cancels two-particle ladder contributions of G-matrices. The perturbation would then have the form $x(G + \mathcal{O})$, and our previous conclusions would remain unaltered. This possibility seems not to have been explored, however.

8. How serious are branch-point singularities ?

Intruder states are often collective states, i.e. they consist of a linear combination of many states in the next higher major shell above the model space. Because of cancellation effects, one might, therefore, expect [14] their interaction with the states in the model space to be weak. This means that the perturbation expansion for H_{eff} and A_{eff} , although divergent, will build up the divergence only in high orders of the series, and that the first few terms may still be useful in providing a good approximation to the actual values of H_{eff} and A_{eff}. This can be seen by looking at a recent model calculation by Pittel, Vergados and Vincent [15] . These authors study the 0^+ states in ^{18}O in a space of 171 dimensions and compare "exact" elements of H_{eff} in a 3-dimensional model space with the results of the perturbation series. Because of the presence of an intruder, the latter series diverges. For $n = 1,2,3,4$, the contributions of all terms up to n^{th} order for the $s^2_{1/2}$ diagonal matrix element are -2.171, -2.046, -1.895 and -1,826 MeV, respectively, while the "exact" value is - -1.897 MeV. Clearly, 3rd order perturbation theory is very good. This is also true for the other matrix elements of $H_{eff}(x)$.

Unfortunately, the perturbation series do not provide us with any clue as to where to terminate the expansions, or with any means of estimating the difference between the results obtained in n^{th} order, and the exact numbers. The observations just mentioned are thus not very useful unless they are supplemented by extraneous information concerning, for example, the strength of the coupling between intruder and model space states. Such estimates can be obtained, for instance, from simple collective models of the intruder state [16,17] . Procedures like these leave, however, the realm of perturbation theory discussed in this paper.

9. If convergent, how quickly do the series converge ?

Intruders need not always be present, and the question raised is thus of considerable practical interest. This is true in particular since until the present, practical calculations beyond third order in G seem hopelessly complicated, and since one aims at an accuracy of ~100 keV in the calculation. A general answer to the question seems difficult to obtain, as it depends on the choice of single-particle energies, on the method of calculation (double-partitioning or not), and on properties of the interaction. Moreover, double-counting must be avoided. There exists a number of investigations concerning these points (see refs. [13,18,19,21] and references to earlier work therein.) It is difficult at this time to draw any definite conclusions. Suffice it to say that the tensor force necessitates [18] the inclusion of harmonic oscillator states with $n\hbar\omega$ where $n \approx 10$ or 12, which has never been done yet in any realistic calculations, and that there are indications that fairly high orders in the perturbation series, if convergent, may be needed to obtain an accuracy of better than 100 keV [19] . Recent calculations of the spin-averaged effective interaction further strengthen these observations [21] .

10. Conclusions.

Both the algebraic structure of and the convergence properties of the perturbation expansion for $H_{eff}(x)$ and $A_{eff}(x)$ seem to be reasonably well understood by now. Methods to overcome the singularities and associated divergences have also been developed, as is discussed in other contributions to this Conference. The most sobering aspect appears to be the necessity to include contributions of fairly high order in n , the order of the perturbation series, and/or contributions of considerable size in $\hbar\omega$, the harmonic oscillator excitation energy, for a realistic calculation. The evaluation of such contributions with present techniques seem very hard.

The author is grateful to J. Richert and M. Simbel for helpful comments concerning the preparation of this manuscript.

References.

1 J. Richert, T.H. Schucan, M.H. Simbel, and H.A. Weidenmüller, to be published.

2 I. Talmi, Invited paper given at this Conference, and references therein.

3 J.P. Schiffer, Invited paper given at this Conference.

4 A.D. Jackson, Invited paper given at this Conference.

5 R.L. Becker, Invited paper given at this Conference.

6 E.M. Krenciglowa and T.T.S. Kuo, Nucl.Phys. A240 (1975) 195

7 B.H. Brandow, Rev.Mod.Phys. 39 (1967) 771; Ann. Phys. (N.Y.) 57 (1970) 214.

8 T.H. Schucan and H.A. Weidenmüller, Ann.Phys.(N.Y.) 76(1973)483

9 M.B. Johnson and M. Baranger, Ann.Phys. (N.Y.) 62 (1971) 172

10 H.A. Weidenmüller, International Conf. on Nuclear Structure and Spectroscopy, Amsterdam 1974, H.P. Blok and A.E.L.Dieperink, editors, Scholar's Press, Amsterdam 1974, p.1.

11 B.H. Brandow, Lectures in Theoretical Physics, vol. XIB, ed. K.T. Mahanthappa and W.E. Brittin, Gordon and Breach, New York, 1969, p. 55.

12 G.E. Brown, Rev. Mod. Phys. 43 (1971) 1

13 B.R. Barrett, Nucl.Phys. A221 (1974) 299; B.R. Barrett, W. Weng, E. Osnes and J. Richert, to be published; B.R.Barrett and F.L. Herbert, Contribution to this Conference; F.L.Herbert, Ph.D. Thesis, 1975 (unpublished).

14 M. Vincent and S. Pittel, Phys. Lett. 47B (1973) 327.

15 S. Pittel, J.D. Vergados, and C.M. Vincent, Contribution to this Conference.

16 H.M. Hofmann, S.Y. Lee, J. Richert, H.A. Weidenmüller, and T.H. Schucan, Ann. Phys. (N.Y.) 85 (1974) 410.

17 S. Pittel, preprint.

18 J.P. Vary, P.U. Sauer, and C.W. Wong, Phys.Rev. C7 (1973) 1776

19 Y. Starkand and M. Kirson, Phys. Lett. 55B (1975) 125; H.M. Hofmann, Y. Starkand, and M. Kirson, Contribution to this Conference.

20 G. Oberlechner, F. F. Owono N'Guema and J. Richert, Nuovo Cimento B68 (1970) 23

21 P. Goode and D.S. Koltun, Nucl. Phys. A243 (1975) 44

H.A. WEIDENMUELLER: CONVERGENCE PROPERTIES OF THE PERTURBATION EXPANSION FOR THE EFFECTIVE INTERACTION

Brandow: (Remark) We have just heard that convergence becomes a serious problem when states from (1-P) approach the states in P. I'd like to mention that there is a different but related pathology which can occur when there are many valence particles in a number of different valence shells, if one partitions according to shells instead of according to the many-body unperturbed energies. In this case the lowest unperturbed states in Q_0 may drop below the highest states in P_0, leading to the possibility of vanishing energy denominators. I've suggested a way to deal with this in my Rev. Mod. Phys. article.

Vincent: (Reply to Brandow's remark) I'm not sure that that is a pathology. Convergence may even be better for these cases, if it helps the model and excluded states to avoid crossing each other.

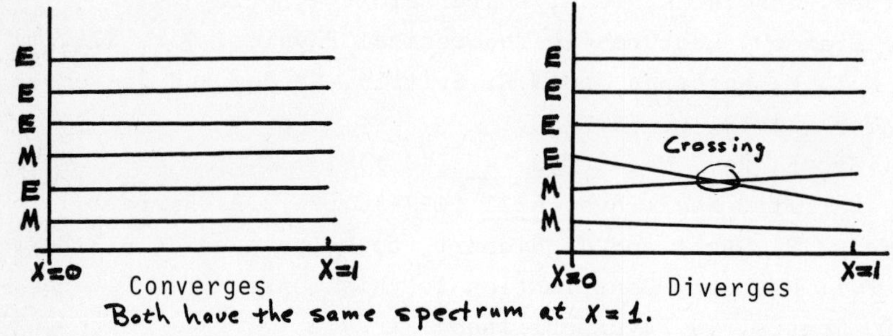

If one is only looking for convergent schemes, what about the Lanczos algorithm? It is guaranteed to converge, and about 17 iterations usually seem to suffice.

Weidenmüller: Concerning the last point, I believe the Lanczos algorithm is useful only if we know the effective interaction.

Harvey: In connection with the problem of intruder states it seems to me that the problem arises not so much from the "intruding" energy but rather that the intruder state couples (no matter how weakly) with the chosen model states. If one could switch off the coupling, the problem of the intruder would disappear. This suggests that one should recognize an approximate quantum number characterizing and distinguishing the intruder states from the normal model states. If this can be done then the original Hamiltonian H can be divided into a part H that preserves the quantum number and a part

H_2 that destroys the quantum number. The Hamiltonian H should then be solved in two parts: first the discovery of the eigensolutions of H_1 (and if perturbation theory is used certainly the intruder state is no longer a problem); and then seeking the eigensolutions of H in terms of those of H_1 by diagonalizing the appropriate energy matrix - i.e. treating the coupling of the intruder to the model space to all orders.

Weidenmüller: I agree, but it is not clear to me how easy it is in any given case to identify the approximate quantum numbers referred to above.

Zamick: Does non-convergence mean that the answer goes to infinity?

Weidenmüller: Yes.

Is There a Universal Relationship Connecting All Two-Body Effective Interactions?**

J. P. Schiffer

Argonne National Laboratory, Argonne, Illinois 60439

and

University of Chicago, Chicago, Illinois 60637

In this talk I would like to summarize the status of a class of information which should have a close connection to an effective interaction of nucleons in nuclei. The data generally come from nuclei in the vicinity of closed shells, in the sense indicated schematically in Fig. 1. Much of what I have to say I covered in a talk four years ago at the Gull Lake Conference on "The Two-Body Force in Nuclei," which is the immediate predecessor of this conference. That paper[1] should be consulted as an introduction to this. The situation since then has not changed qualitatively, though new data have been included here that were obtained since that time.

The data included here are summarized in Table 1, showing that information from ^{6}Li to ^{210}Po are included. Most of the information is derived from transfer reactions, which allows us to test the purity of wave functions, in the sense that one-nucleon transfer reactions measure the one-nucleon components of the wave functions; by requiring that the cross sections satisfy sum rules we test the purity of configurations. To the extent that the two-nucleon configurations are fragmented among the actual states of the nucleus, one sees such fragmentation in the transfer cross sections, and the centroid of fragmented strengths was used in a manner outlined in Fig. 2.

These data are best displayed as "angular distributions," matrix elements plotted as a function of the angle θ_{12}, where

$$\theta_{12} \equiv \arc \cos((J(J+1) - j_1(j_i+1) - j_2(j_2+1))/(2(j_1(j_1+1)j_2(j_2+1))^{\frac{1}{2}}).$$

Such plots are seen in Figs. 3 and 4 for the identical orbit data, with the matrix elements normalized by their (2J + 1)-weighted average,

$\bar{E} \equiv \sum_J [J] E_J / \sum_J [J]$ *, for the identical orbit ($j_1 = j_2$) data. The points for the T = 0 matrix element fall on remarkably smooth curves, indicating an overall pattern as seen in Fig. 3. The points in Fig. 4 show a similarly smooth behavior except for the points corresponding to the 0^+ matrix elements at 180°. This may perhaps be understood by considering the fact that for a short-range interaction the overlap between orbits is complete for coupling to $J^\pi = 0^+$, and these matrix elements are therefore the largest. On either side of 180° the interaction decreases. Now using the 0^+ matrix element of the $(g_{9/2})^2$ multiplet or the $(d_{3/2})^2$ multiplet corresponds to averaging over a larger or smaller angular interval, much as the finite width of an aperture needs to be considered in measuring asymmetries of angular distributions. The $(g_{9/2})^2_0$ matrix element samples the strength of the 180° interaction much more effectively than the $(d_{3/2})^2_0$ one, because it corresponds to a "narrower slit," and the pattern of the 0^+ matrix elements reflects this feature. A more puzzling aspect of Fig. 4 is that all matrix elements with $\theta_{12} < 120^\circ$ are above the axis, in other words, they reflect a repulsive component of the interaction.

The data in which the two orbits are different ($j_1 \neq j_2$) are shown in Fig. 5 for the multiplets where both isospin states are known. One feature we immediately note is that the centroid of the T = 0 matrix elements is always attractive while that for T = 1 matrix elements is repulsive, yet the pattern of downward curvature, a characteristic of both the T = 0 and T = 1 matrix elements, is the signature for a short-range attractive interaction.

In order to allow us to study these "angular distributions" and compare them in more detail we can perform various multipole decompositions. The relevant expressions may be defined as[2]

$$\gamma_{kT} \equiv (-1)^{j_1+j_2} \left[\frac{k}{j_1 j_2}\right]^{\frac{1}{2}} \sum_J (-1)^J [J] W(j_1 j_2 j_1 j_2, Jk) E_{j,T}$$

if one analyzes the two-isospin components separately, or, if one wishes the isospin dependence explicitly, one may define

$$a_{k\tau} \equiv (-1)^{j_1+j_2+1} \left[\frac{k}{j_1 j_2}\right]^{\frac{1}{2}} \sum_{J,T} (-1)^{J+T} W(j_1 j_2 j_1 j_2; Jk)$$
$$\times W(\tfrac{1}{2}\tfrac{1}{2}\tfrac{1}{2}\tfrac{1}{2}, T\tau)[JT] E_{JT}.$$

---

* The [] symbol is used to denote 2J+1; $[J] \equiv (2J+1)$, $[j_1 j_2] \equiv (2j_1+1)(2j_2+1)$.

Note that

$$a_{k0} = \tfrac{1}{2}(\gamma_{k0} + 3\gamma_{k1}) \text{ and } a_{k1} = \tfrac{1}{2}(\gamma_{k1} - \gamma_{k0})$$

while for the mixed isospin (n-p) multiplets, where the filling of shells determines uniquely which orbit may be occupied by a neutron and which by a proton and thus the isospin of the pair is not unique, $E_J^{n-p} = \tfrac{1}{2}(E_J^{T=0} + E_J^{T=1})$ or

$$a_k^{n-p} = \tfrac{1}{2}(\gamma_{k0} + \gamma_{k1}) = \tfrac{1}{2}(a_{k0} - a_{k1}).$$

The multipole decomposition for $j_1 = j_2$ multiplets is not well defined in these terms. The various multipole coefficients may be calculated and allow a quantitative comparison of the data completely analogous to the qualitative comparisons in the "angular distribution plots of Figs. 3—6. Two features may be noted. The odd coefficients reflect the spin dependence of the interaction, and change sign when one of the two orbits is changed from $j = \ell + s$ to $\ell - s$ (or vice versa). The rate of falloff of the even coefficients is a measure of the range of the force, a long-range force would correspond to matrix elements constant with J (or θ), therefore only the monopole coefficients would have finite values. A finte range (compared to the size of the orbits), implies finite values of the even multipole coefficients with k greater than 0.

It is from this last point of view that one should look at Fig. 7, where a_{k1} is plotted; these express the isospin dependence of the two-body interaction as a function of multipolarity. It is qualitatively clear that the isospin dependence has monopole character for the data from lighter nuclei and has higher multipole content (finite range) for the heavier ones. It is thus not unreasonable to think of this effect as an interaction with a specific range that happens to be long compared to the orbits in the lighter nuclei (N = 3 and 4 at A = 34 and N = 5 at A = 90) but is comparable to the orbits in Pb (N = 6 and 7 at A = 208).

These trends in the data lead one quite naturally into attempts to see whether all the data can indeed be fitted by a single interaction. The first attempt at this was reported in Ref. 1 and in a subsequent publication. The improved procedure followed in fitting the data that has been developed in the last two years, with the help of Bill True,[3] and the results form the remainder of this talk.

All the data listed in Table I were included in a least-squares fitting procedure. The observed energy matrix elements were weighted in this fit first of all by the average of a given $(j_1j_2)_T$ multiplet and second by my personal bias regarding the reliability of the data. These weighting factors are given in Table II. Then fits were attempted by using two ranges of the central force, with Yukawa shapes, and one range, fixed at the one-pion exchange value, for the LS and tensor interactions. The central interaction was separated into triplet-singlet, odd-even components so that, in all, the interaction had twelve components. Of these, six determine the T = 0 matrix elements and six the T = 1; if it were not for the mixed-isospin n-p matrix elements the separation would be complete and the fits to the two isospin components could be completely independent.

The first step in this procedure was to determine the extent to which the various components of the force are required to fit the data. The values of χ^2 are shown. For the T = 0 data the triplet-even component of the force fits almost as well as a six-parameter interaction, tensor-even and singlet-odd help only slightly. For T = 1, on the other hand, there is a clear need for both the singlet-even and triplet-odd force and a second range is badly needed, at least for the triplet-odd component, and helps in the singlet-even significantly. The tensor interaction does not help. The LS force is not needed in T = 0 or 1.

Since two ranges of the interaction are needed in the T = 1 interaction, one would like to know what these two ranges are.

The answer to this is in the plot shown in Fig. 8. The value of χ^2 is mapped out as a function of two ranges in the T = 1 interaction and it is clear that as long as two ranges are used, the level of fit is essentially the same. χ^2 increases substantially only on the line representing a single-range interaction. Thus there is no clear preference for any two ranges. The plot of the sum of two Yukawas representing the triplet-odd interaction is shown in Fig. 9. It is apparent that various, drastically different, combinations of ranges give similar total potentials, the specific ranges used for the Yukawa is irrelevant, the general shape is reasonably well defined by the data and the insensitivity to the choice of range seen in Fig. 8 is not, in fact, a gross lack of sensitivity to the shape of the interaction. A similar plot for the singlet-even component is shown in Fig. 10, and for the triplet-even in Fig. 11. Here the need for a second range is less clear from the data.

To quote numbers for the T = 0 interaction presents no great problems. For T = 1 the situation is complicated by the fact that the strengths corresponding to the two ranges are strongly correlated. The matrix elements on the average are small and to reproduce them one needs a cancellation between an attractive interaction and a repulsive one of longer range at nearly the same volume integral. The error matrix for these two components is badly skewed and in order to correctly reflect the uncertainties instead of quoting the two strengths V_1 and V_2 and their errors, it would be more meaningful to quote a combination that is less correlated. Such a combination is

$$V_A \equiv V_1 \text{ and } V_B \equiv V_1 - \alpha V_2$$

where α is chosen such that the average matrix element of V_B is zero. Thus V_B and its error reflect the correction needed from a second term in a given interaction. Table III gives such a set of numbers for V_1 being a Yukawa potential with the one-pion exchange range ($r_1 = 1.415$ fm) and V_2 one with three times that range. It should be stated also that the wave functions used in these calculations were oscillator wave functions with the oscillator constant adjusted such that the rms radius of all the proton orbits in that nucleus be equal to that of the measured charge distribution. The uncertainties reflect a 5% variation in χ^2, they demonstrate the same point as is seen in Table IV that a second range is of some use in the T = 1 case but not in T = 0 and that the tensor force is of some help in T = 0, it is better defined for T = 1 but the value is near zero.

The calculated matrix elements were shown in Figs. 5, 6 and 12 and it is clear that they indeed reproduce the trend in the data reasonably well. No obvious systematic deviations are seen.

To conclude then, we see that it is indeed possible to fit the two-body matrix element data from A = 8 to 208 with a single interaction. The implications of this result, whether it is profound, trivial, or accidental I leave for you to judge.*

*It should be noted that some of these same data have been also fitted by a delta function plus monopole, dipole and quadrupole terms by A. Molinari, M. B. Johnson, H. A. Bethe and W. M. Alberico, Nucl. Phys. A239, 45 (1975). This represents a different parameterization, separating the effects of a short-range interaction from the modifications due to long-range components. All in all, eleven apparently independent parameters seem to be used to fit roughly the same set of data as is treated here, and not including absolute magnitudes. Since the paper only appeared recently and I have not fully understood it, I would not like to comment on it further.

References

** Work performed under the auspices of the U. S. Atomic Energy Commission and the Energy Research and Development Administration.

[1] J. P. Schiffer in The Two-Body Force in Nuclei, ed. by S. M. Austin and G. M. Crawley (Plenum Press, New York 1972) p. 205. Also N. Anantaraman and J. P. Schiffer, Phys. Lett. 37B, 229 (1971).

[2] M. Moinester, J. P. Schiffer and W. P. Alford, Phys. Rev. 179, 984 (1969).

[3] The subroutines for calculating the two-body matrix elements were written by W. W. True. Without his help this work would not have been possible.

Figure Captions

Fig. 1. Schematic definition of the matrix elements extracted from the data.

Fig. 2. Schematic showing the way in which fragmentation of specific states is allowed for. The horizontal displacement represents energy, the vertical lines represent spectroscopic strengths (components of a given configuration).

Fig. 3. $T = 0$ matrix elements for $j_1 = j_2$ orbits plotted against the angle θ_1.

Fig. 4. $T = 1$ matrix elements for $j_1 = j_2$ orbits plotted against the angle θ_1.

Fig. 5. Matrix elements for $j_1 \neq j_2$ with separate isospin components.

Fig. 6. Matrix elements for $j_1 \neq j_2$ with mixed isospin components.

Fig. 7. Even isovector multipole coefficients.

Fig. 8. Map of χ^2 as a function of two ranges in the interaction. The value of 1.415 fm corresponds to one-pion exchange.

Fig. 9. Observed and calculated matrix elements for $j_1 = j_2$ multiplets.

Fig. 10. Plot of the sum of two Yukawa potentials for the triplet-odd interaction. The results from various combinations of ranges are shown together with the Hamada-Johnson interaction. $V r^2$ is plotted to emphasize the contribution to the volume integral.

Fig. 11. Same as Fig. 10 for the singlet-odd interaction.

Fig. 12. Same as Fig. 10 for the triplet-even interaction.

TABLE I. Multiplets Used in Present Work.

	Number of Matrix Elements			Atomic Weight	Reference
	T = 0	T = 1	n-p		
$(1p_{3/2})^2$	2	2		8—16	a
$(1d_{5/2})^2$	3	3		18	b
$(1d_{3/2})^2$	2	2		34	c
$(1d_{3/2}1f_{7/2})$	4	4		34	c,d
$1f_{7/2}2p_{3/2}$			4	50	c
$(1f_{7/2})^2$	4	4		48	c
$(1g_{9/2})^2$	5	5		90	c
$1g_{9/2}2d_{5/2}$	6	6		90	e
$1h_{9/2}1i_{13/2}$	10	10		208	c
$1h_{9/2}2f_{7/2}$	8	8		208	c
$1h_{9/2}2f_{5/2}$			6	208	c
$1h_{9/2}3p_{3/2}$			4	208	c
$1h_{9/2}2g_{9/2}$			10	208	c
Total	44	44	24		

[a] D. Kurath and S. Cohen, Nucl. Phys. 73, 1 (1965).

[b] Deduced from J. L. Wigor, R. Middleton, and P. V. Henka, Phys. Rev. 141, 975 (1966); L. M. Polsky, L. H. Holbrow, and R. Middleton, Phys. Rev. 186, 966 (1969).

[c] See Reference 1.

[d] D. Crozier, Nucl. Phys. A198, 209 (1972).

[e] H. Fann, J. P. Schiffer, and U. Strohbusch, Phys. Lett. 44B, 19 (1973).

TABLE II. **Weights Used in the Least-Squares Fitting.** [a]

Multiplet	Weight (W)			A
	T = 0	T = 1	n-p	
$(1p_{3/2})^2$	0.1	0.1		12
$(1d_{5/2})^2$	0.13	0.22		18
$(1d_{3/2})^2$	0.29	0.35		34
$(1d_{3/2}1f_{7/2})$	0.20	0.64		
$(1f_{7/2})^2$	0.27	0.53		48
$(2p_{3/2}1f_{7/2})$			0.2	50
$(1g_{9/2})^2$	0.90	2.9		90
$(1g_{9/2}2d_{5/2})$	0.57	1.0		90
$(1h_{9/2}1i_{13/2})$, $(1h_{9/2}2f_{7/2})$	1.4	7		208
$(1h_{9/2}2g_{9/2})$			1.4	
$(1h_{9/2}2f_{5/2})$			1.4	208
$(1h_{9/2}3p_{3/2})$			1.4	

[a] $\chi^2 \equiv \sum_i [W_i(E^i_{calc} - E^i_{obs})]^2$

TABLE III. One Interaction Determined from the Fits.[a]

Type of Interaction	V_A(MeV)	V_B(MeV)
T = 0		
Singlet-Odd	140 ± 100	100 ± 300
Triplet-Even	-285 ± 20	60 ± 70
Tensor-Even	-165 ± 70	
T = 1		
Singlet-Even	-85 ± 10	-90 ± 30
Triplet-Odd	30 ± 10	-430 ± 70
Tensor-Odd	10 ± 10	

[a] V_A is the strength of a Yukawa potential of one-pion exchange range required to fit the data, V_B is a difference, as defined in the text, with the second range 3 times the one-pion exchange range. The uncertainties correspond to a 5% change in χ^2.

TABLE IV. Improvements in χ^2 with Various Interactions.[a]

T = 0

Singlet Odd	Triplet Even	Tensor Even	LS Even	χ^2
	√			1.34
	√		√	1.29
√	√			1.27
	√ √			1.27
	√	√		1.26
	√	√	√	1.23
√	√	√		1.19
	√ √	√		1.16
√	√ √	√		1.06
√	√ √	√	√	1.03
√ √	√ √	√	√	1.00

T = 1

Singlet Even	Triplet Odd	Tensor Odd	LS Odd	χ^2
√	√			3.5
√	√	√		3.5
√	√		√	3.4
√ √	√			2.6
√	√ √			1.27
√	√ √	√		1.26
√ √	√ √			1.03
√ √	√ √	√	√	1.00

[a] The values of χ^2 are normalized to 1 for the 12 parameter fit. The checkmarks show when a given interaction was used, two checkmarks denote two separate ranges.

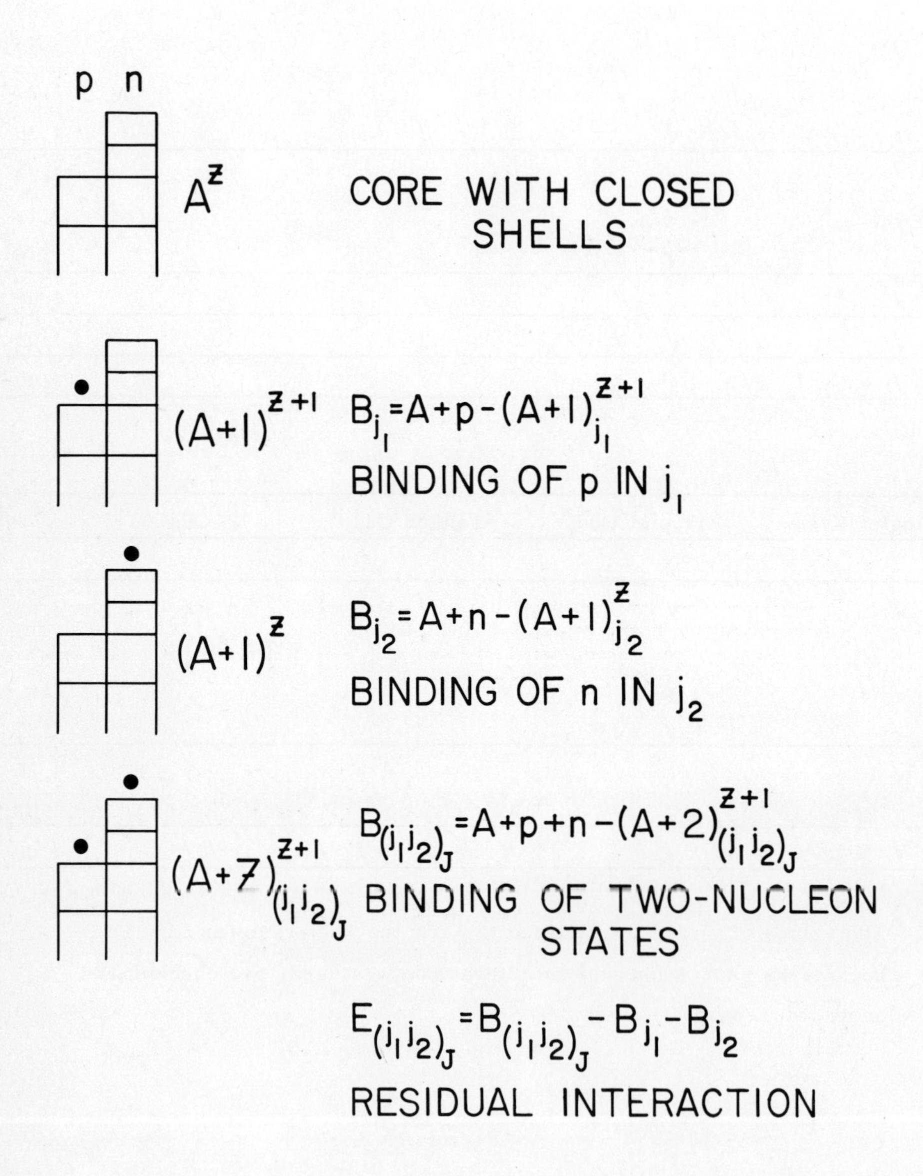
p n
A^{Z}
CORE WITH CLOSED SHELLS
$(A+1)^{Z+1}$
$B_{j_1} = A + p - (A+1)^{Z+1}_{j_1}$
BINDING OF p IN j_1
$(A+1)^{Z}$
$B_{j_2} = A + n - (A+1)^{Z}_{j_2}$
BINDING OF n IN j_2
$(A+Z)^{Z+1}_{(j_1 j_2)_J}$
$B_{(j_1 j_2)_J} = A + p + n - (A+2)^{Z+1}_{(j_1 j_2)_J}$
BINDING OF TWO-NUCLEON STATES
$E_{(j_1 j_2)_J} = B_{(j_1 j_2)_J} - B_{j_1} - B_{j_2}$
RESIDUAL INTERACTION

Figure 1

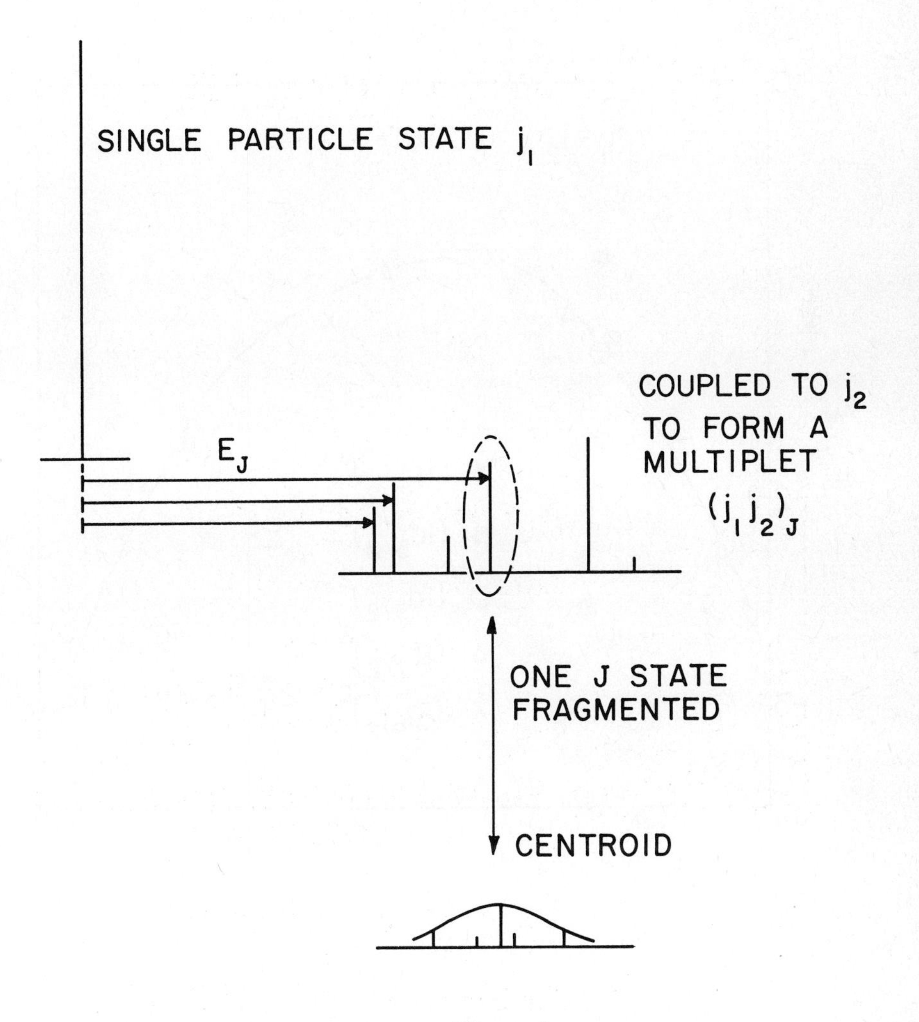
SINGLE PARTICLE STATE j_1
E_J
COUPLED TO j_2
TO FORM A
MULTIPLET
$(j_1 j_2)_J$
ONE J STATE
FRAGMENTED
CENTROID

Figure 2

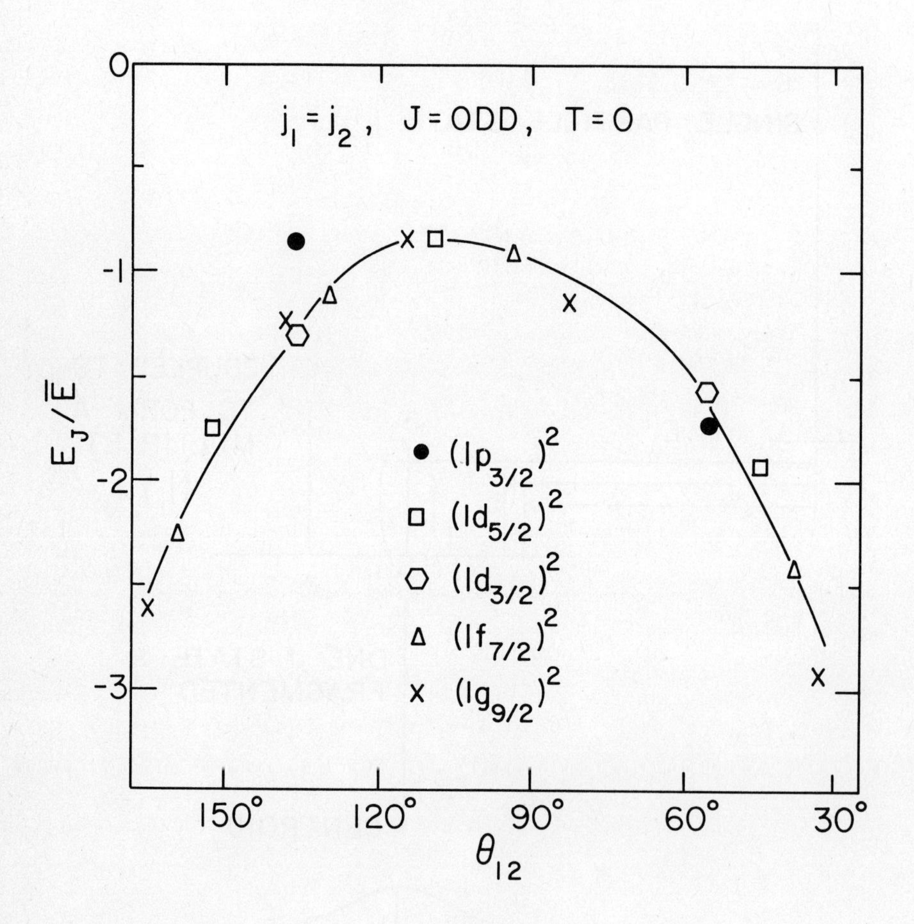

$j_1 = j_2$, J = ODD, T = 0
$E_J/\bar{E}$
θ_{12}
0
-1
-2
-3
150°
120°
90°
60°
30°
$(1p_{3/2})^2$
$(1d_{5/2})^2$
$(1d_{3/2})^2$
$(1f_{7/2})^2$
$(1g_{9/2})^2$

Figure 3

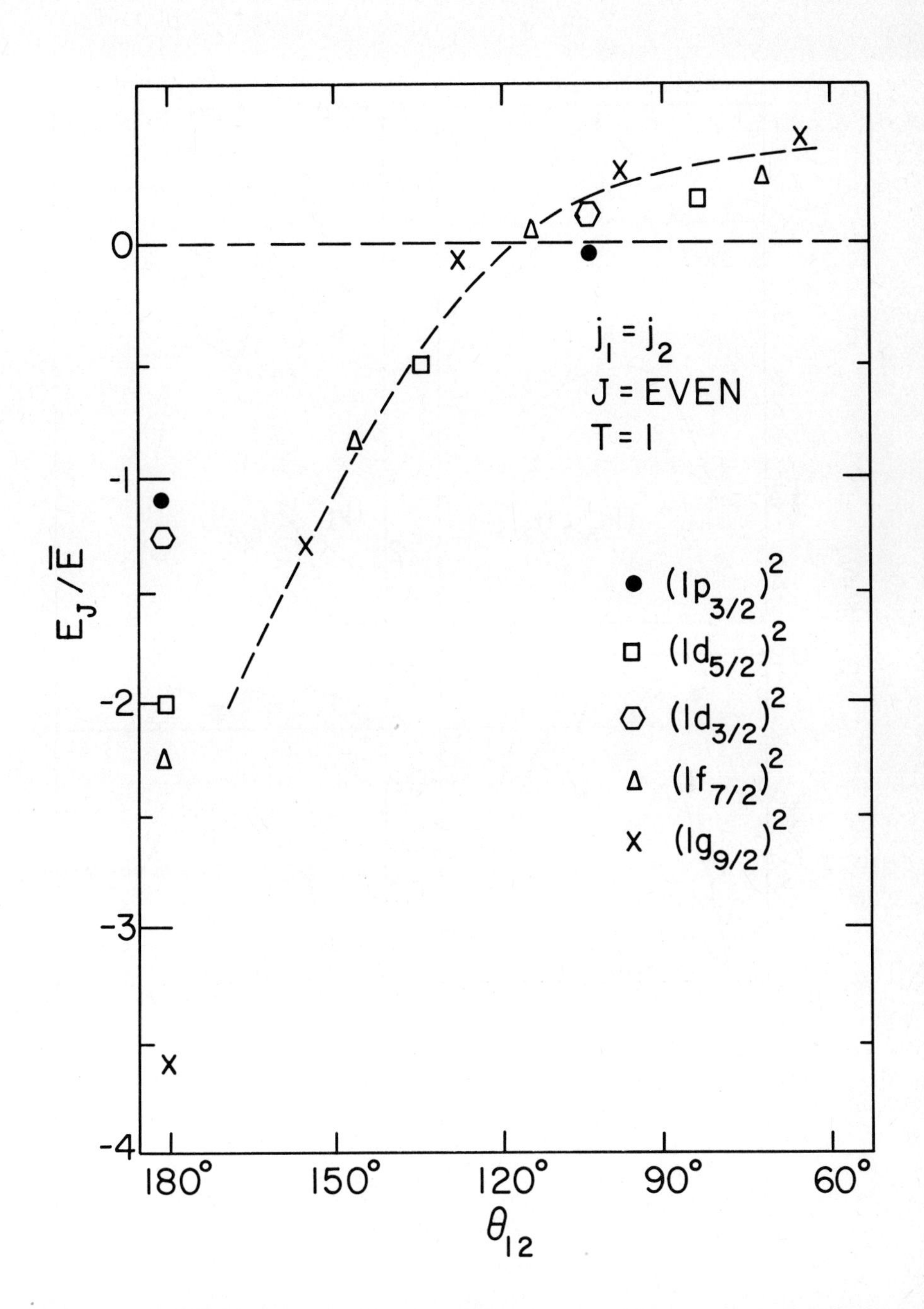

0
-1
-2
-3
-4
180°
150°
120°
90°
60°
$E_J/\bar{E}$
θ_{12}
$j_1 = j_2$
J = EVEN
T = 1
$(1p_{3/2})^2$
$(1d_{5/2})^2$
$(1d_{3/2})^2$
$(1f_{7/2})^2$
$(1g_{9/2})^2$

Figure 4

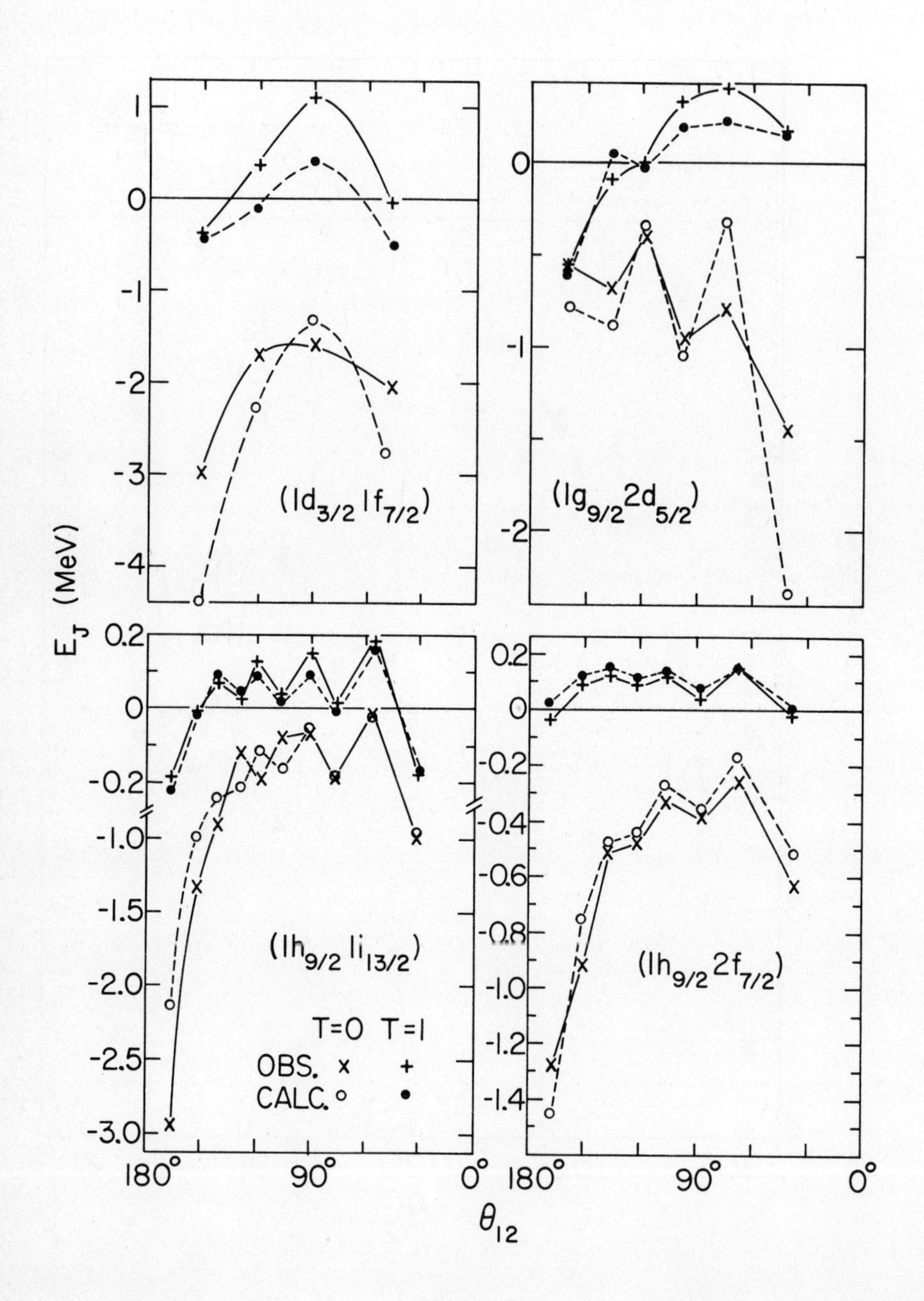

1
0
-1
-2
-3
-4
$(1d_{3/2}1f_{7/2})$
0
-1
-2
$(1g_{9/2}2d_{5/2})$
E_J (MeV)
0.2
0
-0.2
-1.0
-1.5
-2.0
-2.5
-3.0
$(1h_{9/2}1i_{13/2})$
T=0 T=1
OBS. x +
CALC. o •
0.2
0
-0.2
-0.4
-0.6
-0.8
-1.0
-1.2
-1.4
$(1h_{9/2}2f_{7/2})$
180°
90°
0°
180°
90°
0°
θ_{12}

Figure 5

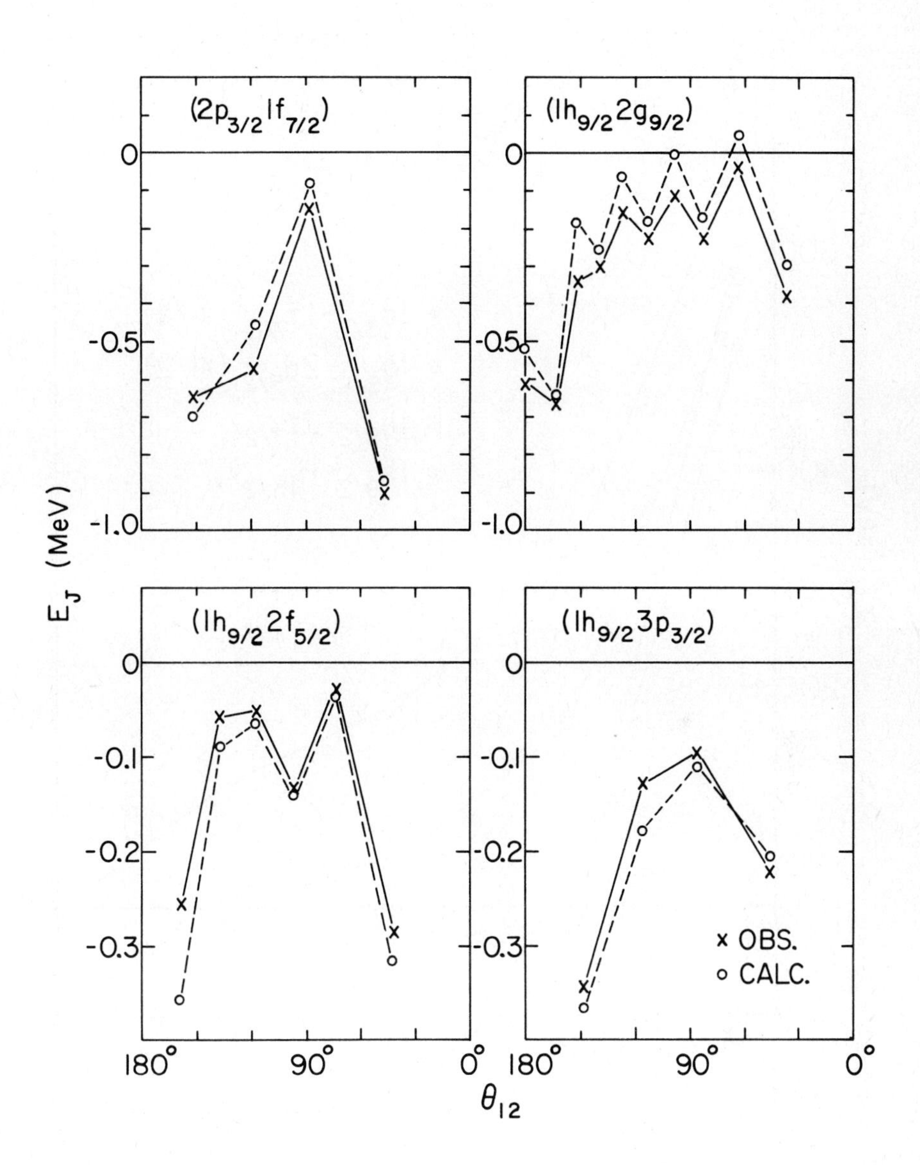

$(2p_{3/2}1f_{7/2})$
$(1h_{9/2}2g_{9/2})$
$(1h_{9/2}2f_{5/2})$
$(1h_{9/2}3p_{3/2})$
0
-0.5
-1.0
-0.1
-0.2
-0.3
E_J (MeV)
180°
90°
0°
θ_{12}
× OBS.
○ CALC.

Figure 6

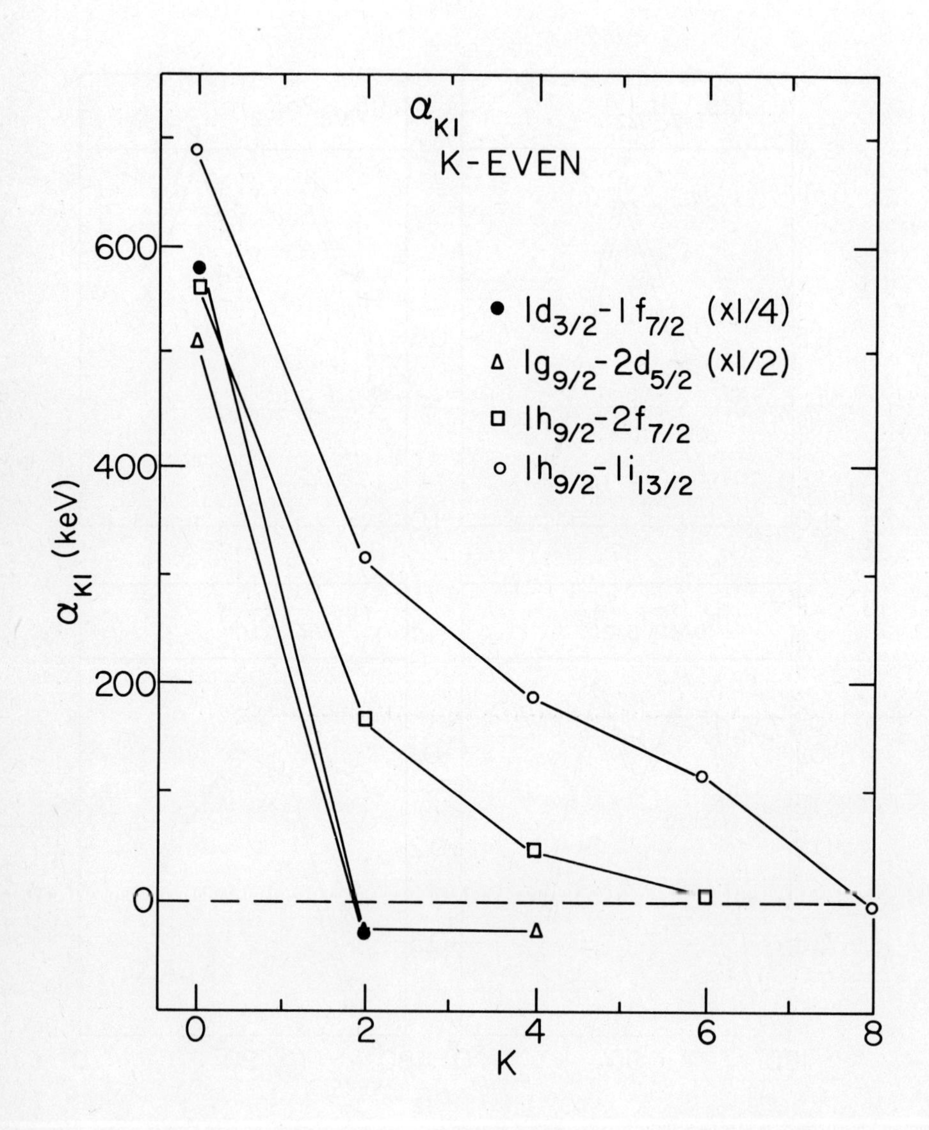

α_{K1}
K-EVEN
$1d_{3/2}-1f_{7/2}$ (x1/4)
$1g_{9/2}-2d_{5/2}$ (x1/2)
$1h_{9/2}-2f_{7/2}$
$1h_{9/2}-1i_{13/2}$
α_{K1} (keV)
600
400
200
0
0
2
4
6
8
K

Figure 7

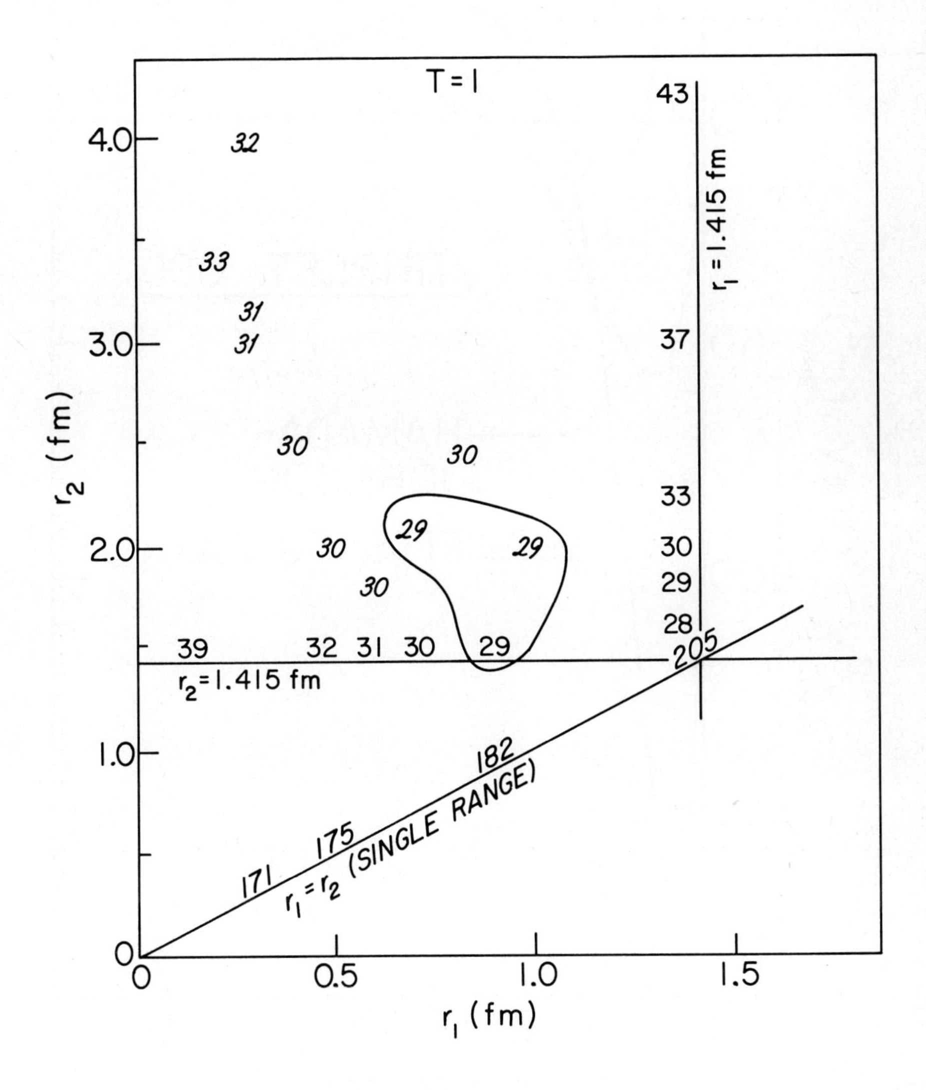
T = 1
43
32
r_1 = 1.415 fm
33
31
31
37
30
30
33
29
29
30
30
30
29
28
39
32
31
30
29
205
r_2=1.415 fm
182
$r_1 = r_2$ (SINGLE RANGE)
175
171
4.0
3.0
2.0
1.0
0
r_2 (fm)
0
0.5
1.0
1.5
r_1 (fm)

Figure 8

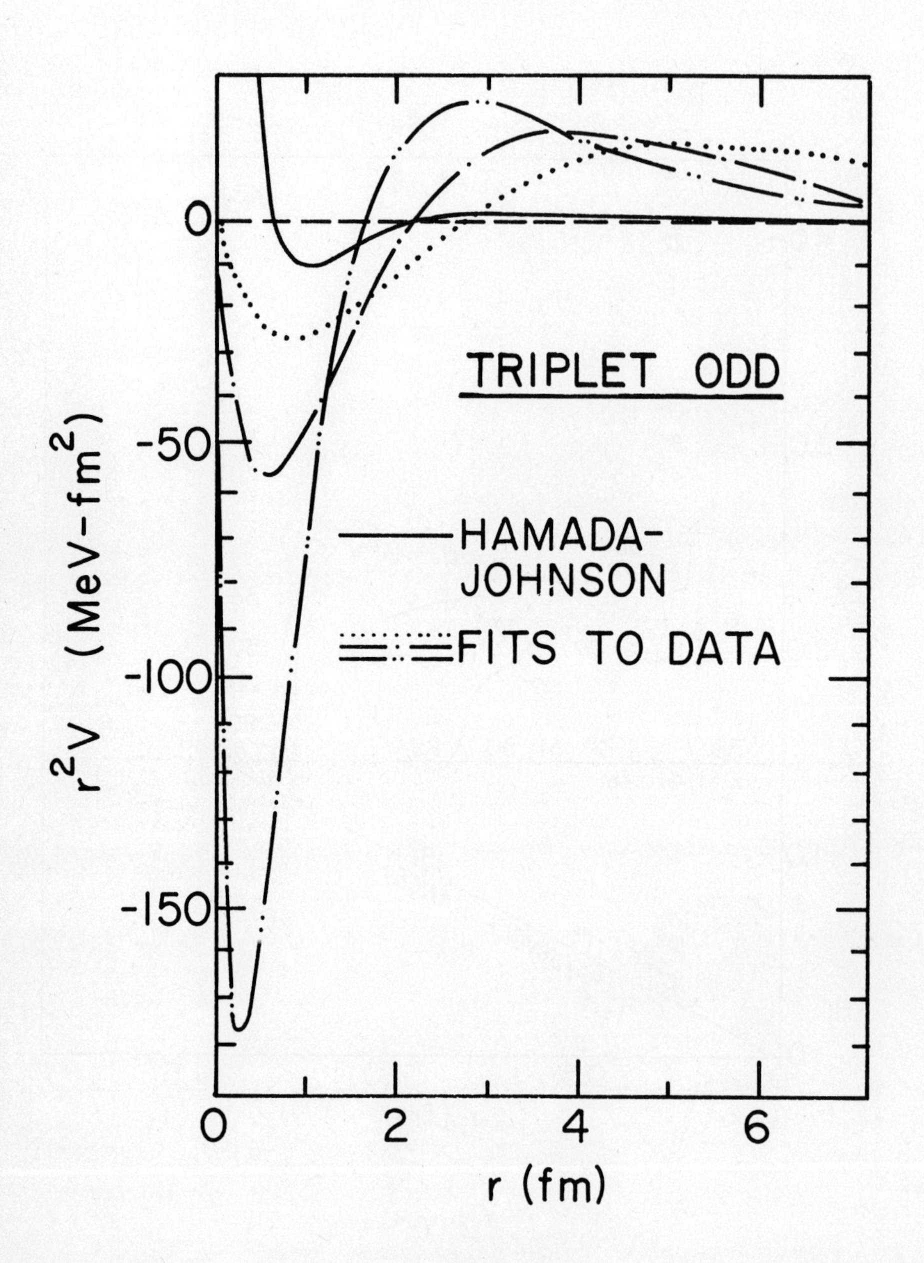

TRIPLET ODD
HAMADA-JOHNSON
FITS TO DATA
r^2V (MeV-fm^2)
0
-50
-100
-150
0
2
4
6
r (fm)

Figure 9

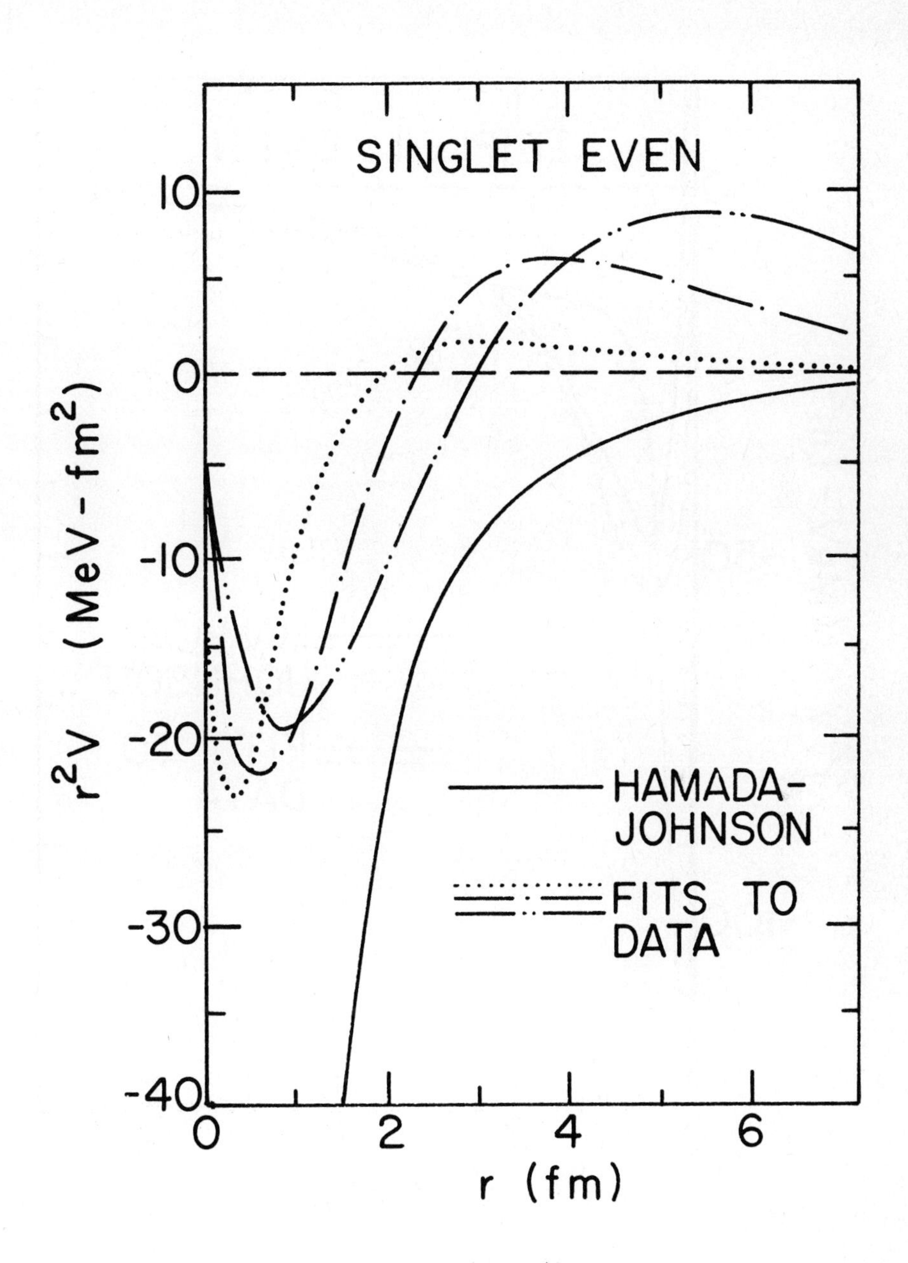
SINGLET EVEN
10
0
-10
-20
-30
-40
r^2V (MeV - fm^2)
0
2
4
6
r (fm)
HAMADA-
JOHNSON
FITS TO
DATA

Figure 10

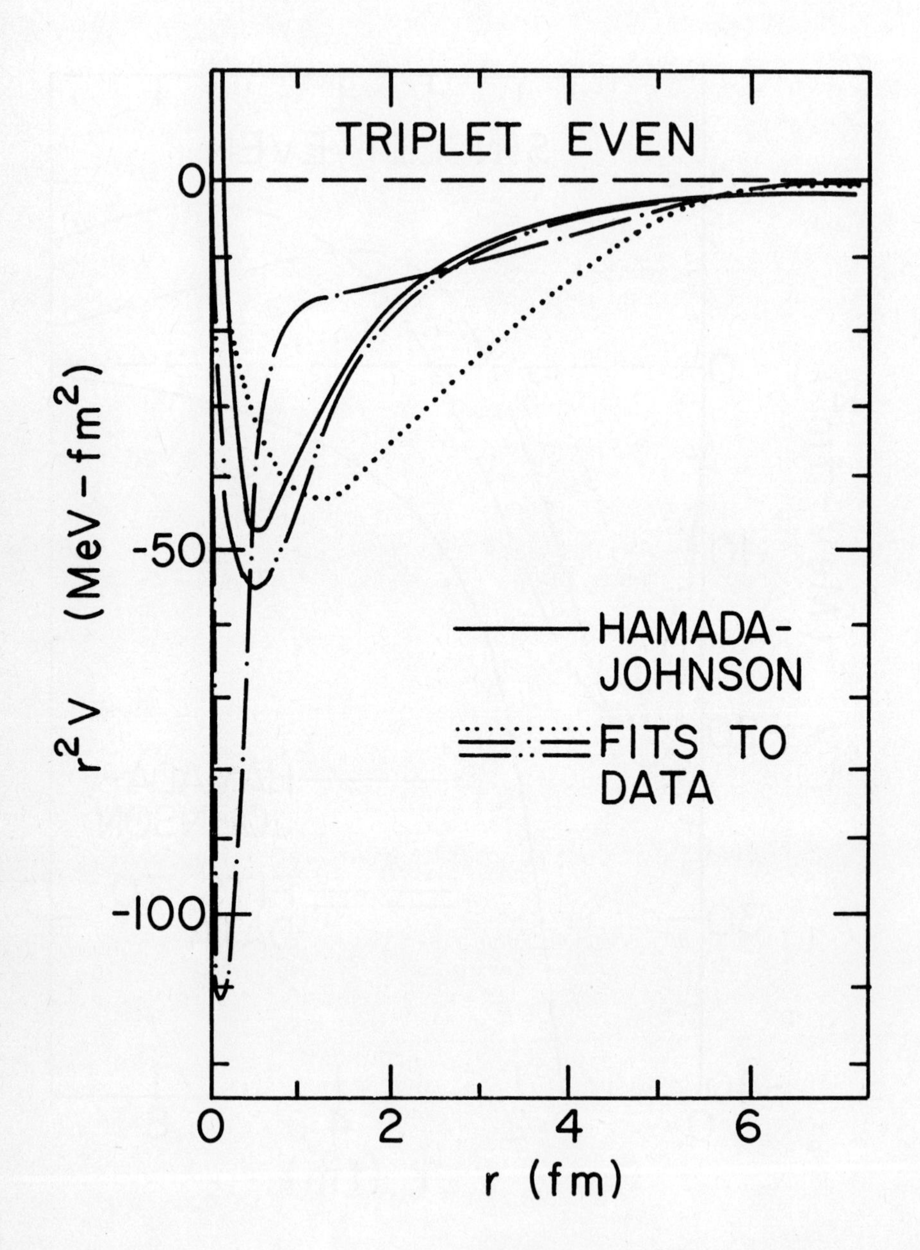
TRIPLET EVEN
0
-50
-100
r^2V (MeV - fm^2)
HAMADA-JOHNSON
FITS TO DATA
0
2
4
6
r (fm)

Figure 11

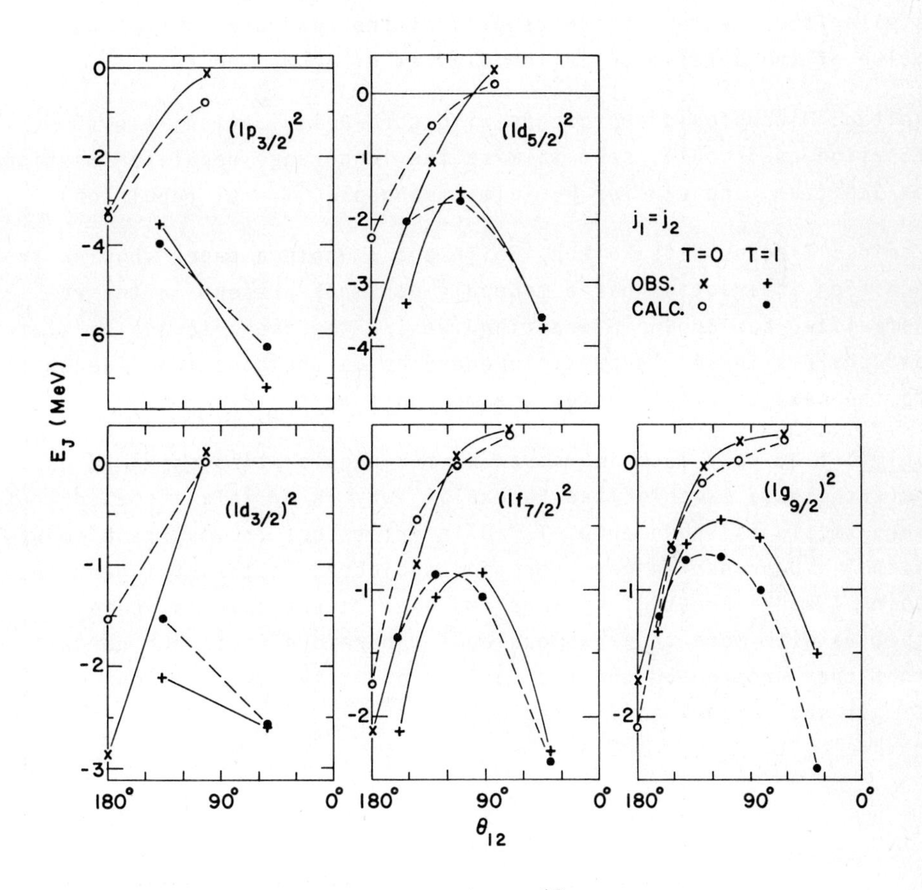

$(1p_{3/2})^2$
$(1d_{5/2})^2$
$(1d_{3/2})^2$
$(1f_{7/2})^2$
$(1g_{9/2})^2$
$j_1 = j_2$
T=0 T=1
OBS.
CALC.
E_J (MeV)
θ_{12}
180°
90°
0°

Figure 12

J.P. SCHIFFER: IS THERE AN UNIVERSAL RELATIONSHIP CONNECTING ALL TWO-BODY EFFECTIVE INTERACTIONS?

Johnson: I would like to mention some results of a calculation by Molinari, Bethe, Alberico and myself. We attempted to analyze your data in a semiclassical scheme which is somewhat more closely related to the theoretical framework about which we have heard so much during the last few days, than is your scheme. We represented the short-range nucleon interaction by a delta function and the long-range core-polarization by a quadrupole term. Our fits to the data reproduce the trends in most cases well. We found some close similarities between the strengths of the quadrupole term and the value of the core-polarization diagram of Brown and Kuo.

Koltun: Microscopic approaches with core-polarization, e.q. G function quadrupole, seem to miss the long-range repulsion mentioned by Schiffer, and earlier by Talmi (monopole, T = 0, repulsion).

Zamick: I agree with Koltun. Golin and I wrote a paper showing the Schiffer interaction has a monopole part not present in the re-normalized Kuo-Brown interaction. We use the isobaric analog states and charges in single-particle energies as one goes from one shell to the next as evidence for the monopole part.

Talmi: I am very much impressed by the regularities shown to us. Nevertheless, I would like to explain why in my talk I stated that very little is known about T = 0 interactions. The matrix elements with T = 0 presented here were not subjected to the stringent tests which I spoke about. They were not used in the calculation of spectra with more than two nucleons. Therefore it is not certain that they are indeed the two-nucleon effective interactions of simple shell-model configurations.

RELEVANT ASPECTS OF STATISTICAL SPECTROSCOPY*

J. B. French
Department of Physics and Astronomy
University of Rochester
Rochester, N.Y. 14627

I shall discuss three aspects of statistical spectroscopy which are relevant to the purposes of this conference. These are:

1. Something about the information content of complex spectra.
2. Procedures for spectroscopy in huge model spaces, which should be useful in effective-interaction theory.
3. Practical ways of identifying and calculating measurable parameters of the effective Hamiltonian and other operators, and of comparing different effective Hamiltonians.

1. Information Content of Complex Spectra

This matter has been of major interest in the slow-neutron domain, particularly for the long sequences of observed $1/2^+$ levels. Apart from an absolute partial level density, suggestions have been made about two kinds of physical information which might be extracted. The first is about hidden symmetries which would have a major effect on the energy-level fluctuations (deviations from uniformity); the second is about the k-body nature of the interaction, which would be successful if there were different fluctuation predictions for the GOE and EGOE cases discussed ahead.

It might reasonably be asked what this has to do with the effective interaction which is more concerned with low-energy observations. However it has gradually become clear that the high-energy fluctuation pattern really extends over the entire spectrum, right into the ground-state domain, so that the subject is really of some relevance to this conference. I mention only one case of low-energy fluctuations, a remarkable demonstration[1] that, if all the nuclei are considered to form an ensemble, then a Poisson behavior is found for the lowest-order spacings irrespective of the exact quantum numbers, and a Wigner "repulsion" for the spacings between pairs of the same (J,π).

---

*Supported in part by the U.S. ERDA.

Let us now consider, to begin with, the case of m non-interacting particles distributed over N single-particle states, the spectrum of which follows by elementary arithmetic. The smoothed spectrum will turn out in most cases to be close to Gaussian and for a very simple reason. If we can ignore the Pauli blocking effects (i.e., if $m \ll N$) the m-particle eigenvalue density or spectrum, $\rho_m(x)$, is simply the convolution of ρ_{m-1} and ρ_1; thus

$$\rho_m(x) = \int \rho_{m-1}(y)\rho_1(x-y)dy \equiv \rho_{m-1} \otimes \rho_1 = \rho_1 \otimes \rho_1 \otimes \ldots \otimes \rho_1 \tag{1}$$

and is therefore the (m-1)-fold convolution of the 1-particle density with itself. But the most elementary central limit theorem (CLT) tells us that practically any function becomes Gaussian under repeated convolution, and so the result follows. We can exclude the exceptional cases where the single-particle spectrum is so exotic that an m-value $\gtrsim N$ would be required in order to generate Gaussian.

Obviously the statistically smoothed density carries very little information, so that, as we have proceeded from the 1-particle to the many-particle model space, we have either lost a lot of information or have "forced" it into the microscopic energy-level fluctuations. At first sight we might argue that the CLT should eventually smooth away the energy-level fluctuations as well, but this is not necessarily so. For, as we increase particle number in order to take greater advantage of the CLT, the level spacing (the natural unit for measuring the fluctuations) decreases rapidly. In fact even for the simple case of non-interacting particles it is not obvious at sight whether, with increasing particle number, the fluctuations will die down, will diverge in amplitude, or will be stable.

For interacting particles the fluctuations generated by a given Hamiltonian are, *a forteriori*, also difficult to deal with. This is analogous to the difficulty, or impossibility, in classical statistical mechanics of following the representative point in phase space as a complicated system develops in time. But now, unlike the situation with non-interacting particles, things are no longer trivial for the density, which is not now simply determined by a CLT. The essential requirements for the applicability of the CLT is that we should be dealing with the combination of *statistically-independent additive* variables; the Pauli effects disturb the statistical independence but this is often a minor effect, easily corrected for; more important is the non-additivity of the energy when we add a particle to a system. As a practical solution to the density problem for a given H we can calculate the low-order moments (up to fourth-order say) and make a reasonable inference about the density from them. But that leaves unsolved the general density problem and is of course no help with the fluctuations.

One gets around these difficulties, as in statistical mechanics, by introducing an ensemble of Hamiltonians however instead of states. The procedure then is schematically:

$$\begin{array}{lcl} \text{Ensemble Average} & \Longrightarrow & \text{Characteristic Form} \\ \text{Asymptotic Variance} & \Longrightarrow & \text{Ergodicity} \\ \text{Covariance} & \Longrightarrow & \text{2-Point Fluctuations} \end{array} \tag{2}$$

The point of the first two lines is that an ensemble-averaged quantity may be taken as relevant to a single characteristic H as long as its variance over the ensemble is small (or, better, $\to 0$ in some natural limit). The third line corresponds to the fact that the two-point measures of fluctuations are by definition those which follow from the covariance of the ensemble density, $\overline{\rho(x)\rho(x')}-\overline{\rho(x)}\,\overline{\rho(x')}$ (in which the bar denotes ensemble averaging). We shall see that there is a beautiful procedure for passing from the covariance to an elegant picture of the fluctuating density.

A standard ensemble is the Gaussian Orthogonal Ensemble (GOE) of real symmetric matrices of large or asymptotic dimensionality, in which the matrix elements are statistically independent and distributed about zero with essentially identical Gaussian distributions. The invariance of this ensemble under orthogonal transformations is of major importance in the "classical" way (Wigner, Mehta, Dyson and many others[2]) of dealing with it, but in fact, for the measurable quantities, neither orthogonality nor the Gaussian form is at all necessary (symmetry about zero is of consequence, however). There are two disconcerting features of this ensemble. The first is that, when used in a many-particle model space, it describes only simultaneous interactions among *all* the particles; the second is that, except for time-reversal invariance, all its properties follow from the *measure* implied by its construction, so that physical notions play only a minimal role.

If, on the other hand, we permit a GOE defined in the k-particle space to operate in the m-particle space we have, especially when $m \gg k$, a system in which the particle structure does play a role, not all the properties follow from the measure, and the interactions are of definite k-body nature[3]. We may specify that the H's preserve J (and T where relevant), the new ensemble being then describable, for $k=2$, as TBRE (two-body random ensemble); or we may ignore the (J,T) restriction and speak of the Embedded Gaussian Orthogonal Ensemble (EGOE) which is analytically more tractable.

We shall consider results for GOE ($m=k$) and for EGOE with $m \gg k$, agreeing always however that $m \ll N$ (dilute systems!). Let us normalize the matrix-elements of

$$H = \sum W_{\alpha\beta}\psi_\alpha(k)\psi_\beta^{+}(k) \tag{3}$$

(where $\psi_\alpha(k)$ creates k particles in state α) so that, for $\alpha \neq \beta$, $\overline{W^2_{\alpha\beta}} = \binom{N}{k}^{-1}$

(and $\overline{W^2_{\alpha\alpha}} = 2\binom{N}{k}^{-1}$). Then describing the densities via moments

$$M_p = \binom{N}{m}^{-1} \mathrm{Tr}(H^p) \equiv \langle H^p \rangle^m \tag{4}$$

whose averages vanish unless p=even, one finds easily that

$$\overline{M}_{2\nu}(m=k) = (\nu+1)^{-1}\binom{2\nu}{\nu} \Longrightarrow \overline{\rho}(x) = (2\pi)^{-1}(4-x^2)^{\frac{1}{2}} \tag{5}$$

$$\overline{M_{2\nu}}(m>>k) = (2\nu-1)!!\binom{m}{k}^{\nu} \Longrightarrow \overline{\rho}(x) \equiv G(0,\sigma^2=\binom{m}{k}) \tag{6}$$

the first of which is Wigner's semicircle[2] while the second is Gaussian with centroid and variance as indicated. It will be important to understand the genesis of the Gaussian result[4]. Let us agree that m>>pk; in order that moments should not vanish via $\overline{W}_{\alpha\beta} = 0$ we need pairwise associations of H's, the transformation effected by one H being undone by its partner (quaternary and higher-order associations are of vanishing weight in the large-N limit). Then for two H's standing together

$$H^2 \to \underbrace{H\,H} = \sum W^2_{\alpha\beta}\psi_\alpha\psi^+_\beta\psi_\beta\psi^+_\alpha \Longrightarrow \binom{n}{k} \tag{7}$$

where the last step follows on ensemble averaging, using the results that $\sum \psi_\alpha\psi^+_\alpha = \binom{n}{k}$, $\sum \psi^+_\beta\psi_\beta = \binom{N-n}{k} \to \binom{N}{k}$, where n is the number operator. However, since there are many particles, we can take for granted that H operators in different correlated pairs act on disjoint sets of particles so that they effectively "commute", each complete pairwise association gives $\binom{m}{k}^{\nu}$ and the number of associations is simply the number of binary associations of 2ν objects, *viz* $(2\nu-1)!!$. The result is that we do find Gaussian for the ensemble-averaged density, even without the operation of an elementary CLT. One can easily study the rate at which the semicircular → Gaussian transition proceeds as we increase particle number; in usual situations 6-8 particles should be enough for an excellent Gaussian density.

To deal with ergodicity and with the fluctuations let us consider the ensemble average of the product of two moments, $\overline{M_pM_q}$. Once again we need consider only binary correlations but now, as we shall see, there is a fundamental classification according to the number of *cross-linked* operator pairs (in which one member comes from M_p and one from M_q). We have, suppressing the m-dependence, that

$$\overline{M_pM_q} = \sum_{\zeta=0}^{(p,q)_<} g_p^{(\zeta)}g_q^{(\zeta)}I^{(\zeta)} \tag{8}$$

where the sum is parity restricted, the $\zeta=0$ term gives simply $\overline{M}_p\overline{M}_q$ (so that the moment covariance is given by $\sum$), the $g^{(\zeta)}$ quantities are independent of N in the large-N limit, and the $I^{(\zeta)}, \zeta \geq 1$ which will play the role of spectral intensities, are given by fully-cross-linked averages,

$$I^{(\zeta)} = \underbrace{\langle H^{\zeta}\rangle^{m} \langle H^{\zeta}\rangle^{m}}_{(\zeta)} \Longrightarrow 2\zeta \binom{m}{k}^{2} \binom{N}{k}^{-2} \tag{9}$$

whose evaluation, as given, is valid for large N and follows by an elementary argument.

We see immediately from (9) that we do have an ergodic behavior for the moments, whose variance goes to zero for large N, and hence for the density. Thus as $N \to \infty$ the chance goes to zero that a chosen Hamiltonian will give a density which deviates from the ensemble average.

For the fluctuations the significant step is to turn things around and to regard the $g_p^{(\zeta)}$, *for fixed* ζ *and varying* p, as the moments of a function $\bar{\rho}(x)G_{\zeta}(x)$ which, since it has non-vanishing moments only of orders $\geq \zeta$, must oscillate more and more rapidly the higher the cross-linkage order. But now we have immediately a representation of the ensemble density in terms of a sequence of statistically uncorrelated elementary excitations characterized by ζ (which is essentially the inverse wave length).

$$\begin{aligned} \rho(x) &= \bar{\rho}(x) \{1 + \sum_{\zeta > 1} R_{\zeta} G_{\zeta}(x)\} \\ \int \bar{\rho}(x)\, G_{\zeta}(x)\, x^{p} dx &= g_{p}^{(\zeta)} \\ \overline{R_{\zeta}} = 0 \; ; \quad \overline{R_{\zeta} R_{\zeta'}} &= I^{(\zeta)} \end{aligned} \tag{10}$$

from which we can calculate easily all those fluctuation measures which depend only on the two-point correlation function.

For the GOE we have $g_p^{(\zeta)} = \binom{p}{\frac{p-\zeta}{2}}$ and then

$$\begin{aligned} \pi\bar{\rho}(x) G_{2\nu}(x) &= (-1)^{\nu} (4-x^{2})^{-\frac{1}{2}} \operatorname{Cos} 2\nu\phi \\ \pi\bar{\rho}(x) G_{2\nu+1}(x) &= (-1)^{\nu} (4-x^{2})^{-\frac{1}{2}} \operatorname{Sin} (2\nu+1)\phi \end{aligned} \tag{11}$$

where $\phi \equiv \phi(x) = \operatorname{Sin}^{-1}(x/2)$ and is identifiable on the semicircle as the angle between the radial coordinate and the normal through the center. We then for example find immediately that the mean squared deviation of the number of levels in an interval is given by a cosine integral,

$$\Sigma^2(n) = \frac{4}{\pi^2}\int_0^{d/2} \frac{d\nu}{\nu}\sin^2\nu(\phi_2-\phi_1) \xrightarrow{\overline{n}\to\infty} \frac{2}{\pi^2}\ln\overline{n} \qquad (12)$$

where $d = \binom{N}{k}$, and from this follows the remarkable long- and short-range spectrum rigidity, first discovered for the central region of the GOE by Dyson and Mehta[5]. We find in fact that the central-region results apply over the entire spectrum and, as is clear from the derivation, do not call for orthogonality of the ensemble, nor for Gaussian matrix-element distributions.

While the classical GOE method seems entirely intractable for EGOE the covariance method used above extends easily. For $\zeta << m/k$ the same argument (following (7) above), which led there to a Gaussian density, now gives Hermite-polynomial excitations, defined by $g_p^{(\zeta)}(m) = (p-\zeta-1)!!\binom{p}{\zeta}\binom{m}{k}^{(p-\zeta)/2}$, whose intensity however falls off so rapidly with increasing ζ that they generate no short-range fluctuations at all. However for fixed particle number the free association of self-linked operator pairs which generates the Gaussian moment must fail when $\zeta \sim m/k$ (for then there are not enough particles for free commutations). For excitations of higher order, and hence for the short-range fluctuations (which have $\zeta_{max} = \binom{N}{m}$), a transition occurs to a system of inhibited binary associations in which very many H's act on the same particles. The GOE, in which *all* H's act on the same particles is the limiting case of such a system. The details of this transition have not yet been worked out, but, in the light of Monte-Carlo experience, it seems likely that the EGOE fluctuations will be very close to GOE. This would give an explanation for many shell-model and Monte-Carlo results in which no differences have been detected between GOE and EGOE fluctuations, and would rule out the study of energy-level fluctuations as a means of learning about the k-body nature of the interaction. We could say then that the CLT is ineffective in modifying the short-range fluctuations and that the GOE fluctuations, though derived with an unrealistic ensemble, would represent the fluctuations to be expected in all complex spectra and at all excitations, being then of a very general nature but carrying little or no information.

We have taken for granted that our spectra are "pure", i.e. do not involve interwoven sets of states with different exact symmetries; such sets when taken separately would display the standard fluctuations but, since there is no "repulsion" between them, this is not true when all are taken together. This has a calculable effect on the fluctuations at high excitation energy, but, more important for our purpose, a different and larger one in the ground-state domain. In the case of angular momentum, for example, a J-dependence of the centroids and widths may produce a large variation in the fixed-J ground-state energies (giving an extended yrast spectrum which *does* carry information, though in limited amount).

It is in terms of such "end effects" that we may think of the standard shell-model analyses of low-lying energies; in practice this seems particularly successful for odd-even nuclei. On the other hand there are also CLT's on groups which might permit a more sophisticated analysis of the mixed spectra.

2. Spectroscopy in Huge Spaces

Let us now leave ensembles and turn to spectroscopy as defined by a single Hamiltonian. The true level density is, of course, not Gaussian but increases exponentially with the energy. Moreover only a small part of the low-lying tail region of the Gaussian can reasonably correspond to Hamiltonian eigenstates, say the first 10 MeV or so in a $(ds)^{12}$ spectrum (which would span about 100 MeV), since at higher energies we would encounter excitation from lower orbits and into higher ones. If our interest is with higher excitation energies we must enlarge the model space by taking more single-particle states into account (allowing, to begin with, excitations from 1p and into 2p-1f in the example). But then the spectrum span grows and the centroid rises very rapidly. We can appreciate now the importance of the spectrum rigidity, for in a sense we are tying everything to a centroid which is far above the domain adequately treated in the model space.

But there is a limit to how far we can go with this, for we must recognize that the CLT procedure is a strong-interaction one (in which the matrix elements enter exponentially). It requires for its validity a kind of *chaos* which cannot properly represent the weak interaction of very distant states. We can see this easily by considering the simple example of m particles in two orbits with single-particle splitting Δ; the particles are interacting but, to begin with, there is no interaction between orbits. We now turn on a pairing interaction $G+G^+$ (where G promotes two particles from orbit 1 to orbit 2) and ask for the ground-state shift.

In perturbation theory (as long as $2\Delta >> \sigma \equiv \sigma(m,0)$, the width of the lowest configuration) we have

$$\Delta E_g^{(pert.)} \simeq - \frac{1}{2\Delta} \langle g|G^+G|g\rangle \qquad (13)$$

and in Gaussian statistical theory, where we take account of the pairing contribution to σ^2, the very different result

$$\Delta E_g^{(stat.)} \simeq - \frac{(E-E_g)}{2\sigma^2} \langle g|G^+G|g\rangle \qquad (14)$$

where E is the (m,0) centroid. When the interaction is weak (large Δ) the distribution of the ground-state configuration is really *bimodal*, a small part of the intensity occuring at $\sim 2\Delta$ above the ground state, and then it is improper to apply the statistical method which takes for granted that the second moment defines a realistic "width" for the density of the ground-state configuration. When the interaction is strong the perturbation result is in obvious error and the statistical treatment appropriate.

The statistical and perturbation methods are complementary and join together to yield an effective procedure for a large model space (of dimensionality >5000 say). We partition the single-particle space into "orbits" and the many-particle space into *interacting* subspaces according to the configurations defined thereby. We would naturally choose orbits whose centroids are reasonably separated so that the many-particle-configuration centroids themselves span a fair fraction of the many-particle spectrum. The interaction between a pair of subspaces we treat statistically, in terms of partial widths (and higher-moment contributions in some cases) if the subspace centroids are close and the interaction therefore strong, or perturbatively for distant subspaces. The procedure is shown schematically in Fig. 1 where, as indicated, a rough criterion for the nature of the interaction is defined by the ratio of the partial width connecting two subspaces

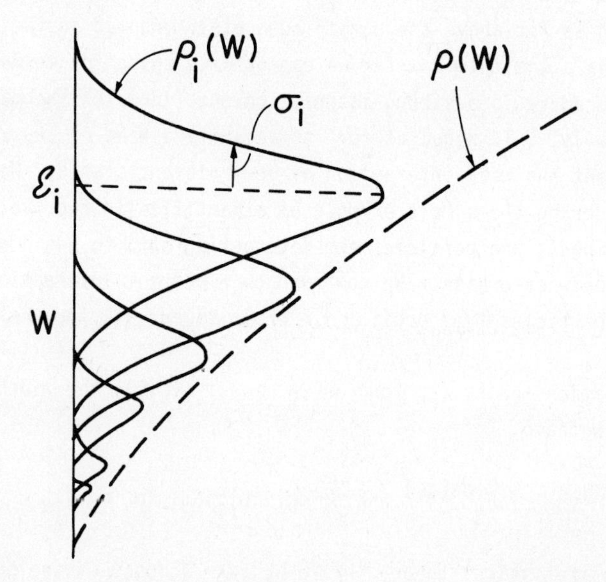

Fig. 1. The decomposition of the density $\rho(W)$ into partial densities $\rho_i(W)$ corresponding to a set of interacting subspaces. We have $\sigma_i^2 = \sum_j \sigma_{ij}^2 = \sum_j \langle HP_jH\rangle^i$, where P_j is a projection operator for subspace j, and the sum is over the subspaces which strongly interact with i. As a very rough criterion, j interacts weakly with i if $\sigma_{ij} \ll |E_i - E_j|$.

and the centroid difference between them. A better procedure in fact is to calculate the (partial) third moments for each subspace, and hence the "skewness" which is very sensitive to small contributions at large distances (corresponding to an H^3 sequence in which the first H excites a particle or pair to a distant orbit, the second H contributes a (large) single-particle-energy factor and the third de-excites). The distant subspaces which make large contributions to the skewness should then be ignored or their effects treated via perturbation theory.

With this procedure, which has been used a great deal (mostly ignoring the pertubative effects) it is usually satisfactory to treat the subspace distributions as Gaussian, so that, for each subspace, we calculate a centroid and width. Even if there are thousands of subspaces the computing operations involved are trivial so that this procedure is entirely practical. We are now of course making use of more information than in the simple Gaussian case, though by no means as much as we might think. For, on the one hand, the centroids are fixed by a low-order polynomial in the configuration partition numbers, while, on the other, it is an empirical fact, which derives from a "uniformity" of the standard effective interactions, that the widths of subspaces which differ only by configuration are closely equal (±5-10%).

Although there is no formal proof of it, it is our belief that the techniques described here are applicable in indefinitely large model spaces. In order that they should constitute a general method of spectroscopy in huge or unbounded model spaces two extensions are required. The first is to subspaces with specified exact symmetries (and indeed often for broken symmetries as well). Much of this has been done but there has been a difficulty in specifying the angular momentum; it may be that this problem is now solved. The other extension would treat the lowest-lying subspaces by matrix methods so that we would have a ternary rather than binary subspace classification; but nothing effective has been done about this. We comment ahead on the relevance of all this to effective-interaction theory.

3. Expectation Values, Sum Rules and Strengths

All of the results for the eigenvalue density emerge in terms of explicit functions of the Hamiltonian matrix elements so that we can explore easily what feature of H is responsible for a certain behavior of the density. But going well beyond that, and in analogy with partition-function techniques of statistical mechanics, we can test the response of the system under infinitesimal variations of H and in that way can calculate expectation values of operators (including sum-rule quantities) as well as strength distributions of various kinds.

We see from Fig. 2 that in the CLT limit, in which both H and $(H + \alpha K)$ have Gaussian spectra (the latter to first order in α), the expectation value $K(W) \equiv \langle W|K|W\rangle$, in a Hamiltonian eigenstate, of an operator K, varies linearly with the energy:

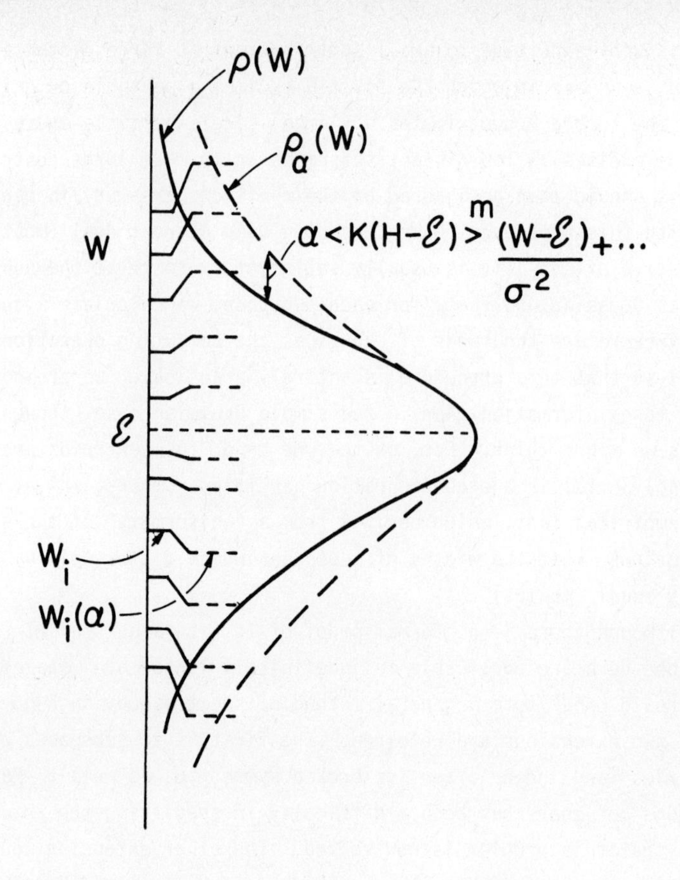

Fig. 2. The expectation values $\langle W|K|W\rangle \equiv K(W) = \left(\frac{\partial W(\alpha)}{\partial\alpha}\right)_{\alpha=0}$, as determined by the response of the system under $H \to H_\alpha = H + \alpha K$ for infinitesimal α. The figure is for the CLT limit, assuming that $\langle K\rangle^m = 0$, so that the only response (to within fluctuations) is a linear scale variation, $\sigma^2 = \langle (H-E)^2\rangle^m \to \langle (H-E+\alpha K)^2\rangle^m$, and then the level shifts are linear in the energy as shown. Allowing for a non-zero value of $\langle K\rangle^m$ gives a (constant) centroid shift as well.

$$K(W) = \langle K\rangle^m + \langle K(H-E)\rangle^m \frac{(W-E)}{\sigma^2} \tag{15}$$

where E and σ are the density centroid and width. This result can of course be correct only to within fluctuations which, as we now know, are not eliminated by the CLT. Implicit in (15) is a (unitary) geometry in which operator magnitudes are expressed in terms of the standard unitary norm as

$$\{\| K \|\}^2 = d^{-1}\, Tr^{(m)}(K^+K) \equiv \langle K^+K\rangle^m \tag{16}$$

where d is the model-space dimensionality. This, coupled with the CLT result that the centroid and width dominate the behavior of operators, tells us that we have a geometry, whose effectiveness is guaranteed by the CLT, and which gives precise definitions of orthogonality, projection and so forth. We see immediately that (15) simply transcribes into the energy "space" a model-space scalar product, or correlation coefficient, for the two "centered" operators, $K - \langle K \rangle$, $H - E$.

It is easy to write the exact result, of which (15) gives the first two terms. Introducing the orthonormal polynomials defined by the density as weight function, we find

$$K(W) = \sum_{\mu} \langle K\, P_{\mu}(H) \rangle^{m}\, P_{\mu}(W) \tag{17}$$

The terms of order $\nu \geq 2$, which derive from the shape variation of the density as we test its response, are of course strongly inhibited by the CLT, the higher the order the stronger the inhibition, so that the series is very rapidly convergent.

An interesting application, and one which promises to inform us about one particular aspect of effective interactions, is to ground-state occupancies, which are measurable by single-nucleon transfer. We find that $\langle n_s (H-E) \rangle^m$ is essentially the induced single-particle energy, for state #s, i.e., the modification of the (traceless) s.p.e. which is due to the interaction of one particle in state #s with all the others. Explicitly we have, in terms of the parameters of H,

$$n_s(W) = \frac{m}{N} \{1 + \frac{N-m}{N-1}\, \xi_s(m)\, \frac{W-E}{\sigma^2}\} \tag{18}$$

$$\begin{aligned} \xi_s(m) &= (e_s - \frac{1}{N} \sum e_i) + (m-1)(N-2)^{-1} \{ \sum_j W_{sjsj} - N^{-1} \sum_{i,j} W_{ijij} \} \\ &= (e_s - \frac{1}{N} \sum e_i) + (m-1)(N-2)^{-1} \lambda_s \end{aligned} \tag{19}$$

where then λ_s, which measures the s.p.e. shift, is the induced s.p.e. in the one-hole system.

Different interactions in common use, in the (ds) shell for example, give considerably different induced s.p.e., different slopes therefore for the occupancy curves, and different ground-state occupancies. We would expect a correlation with the ground-state occupancies and this is well demonstrated in Fig. 3. Obviously we have here, even in quite complicated systems, isolated a parameter of the effective interaction which is susceptible both to measurement and to calculation.

As another example[7] we show the NEW sum-rule quantity as calculated via (15), and various extensions of it, for E2 transitions in the (rather unlikely) case

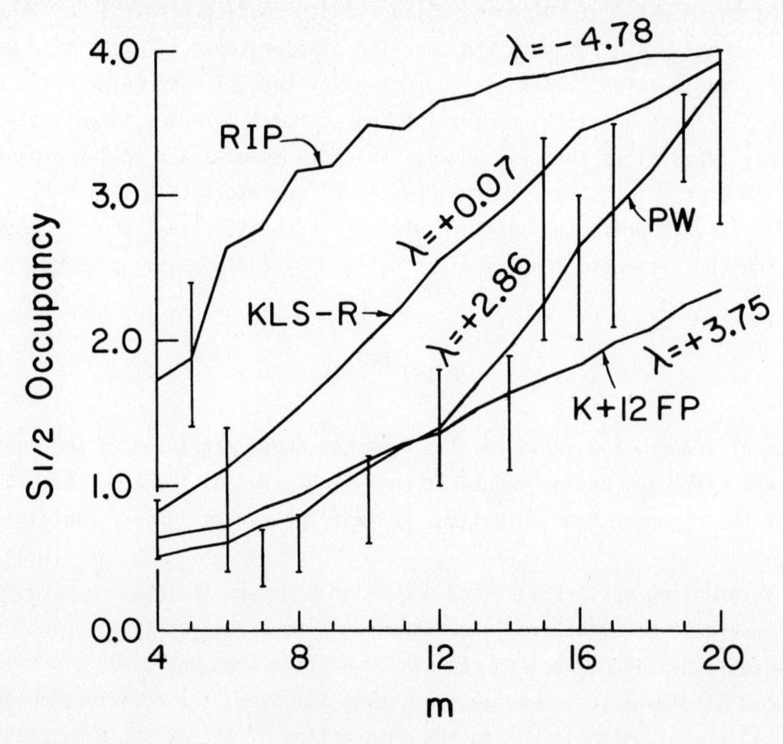

Fig. 3. The $s_{1/2}$ ground-state occupancies for $(ds)^m$ calculated throughout the (ds)shell (A=20-36) for four interactions and compared with experiment (vertical bars). Given in MeV for each interaction is the induced s.p.e. parameter λ (19). Results are from Potbhare and Pandya[6].

$\{(ds)^6: J=2, T=0 \to J=0, T=0\}$, for which shell-model results are also given. Agreements are excellent reminding us that total strength measurements would determine for us a Q·Q *vs.* H correlation coefficient, or, if you wish, tell us how much Q·Q is "contained in" H. The general solution for that of course is

$$\hat{H} = \beta \hat{Q}\cdot\hat{Q} \; ; \qquad \beta = \frac{\langle \hat{Q}\cdot\hat{Q} \; \hat{H} \rangle^m}{\langle (\hat{Q}\cdot\hat{Q})^2 \rangle^m} \tag{20}$$

Where we have written $\hat{O} = O - \langle O \rangle$. Similarly we can make further decompositions of H and can study for example how large are the contributions which the various "parts" of H make to the energy of the system in various regions of the spectrum.

We may be interested in the differences between two effective interactions, for example between a "bare" interaction and a renormalized version of it. Eq.(15) and its extensions, when applied to (H-γH') and to $(H-\gamma H')^2$ can yield an excellent comparison between them and tell us, *inter alia*, at what energies the differences show up; all that we need is a table of traces of low-order powers and products

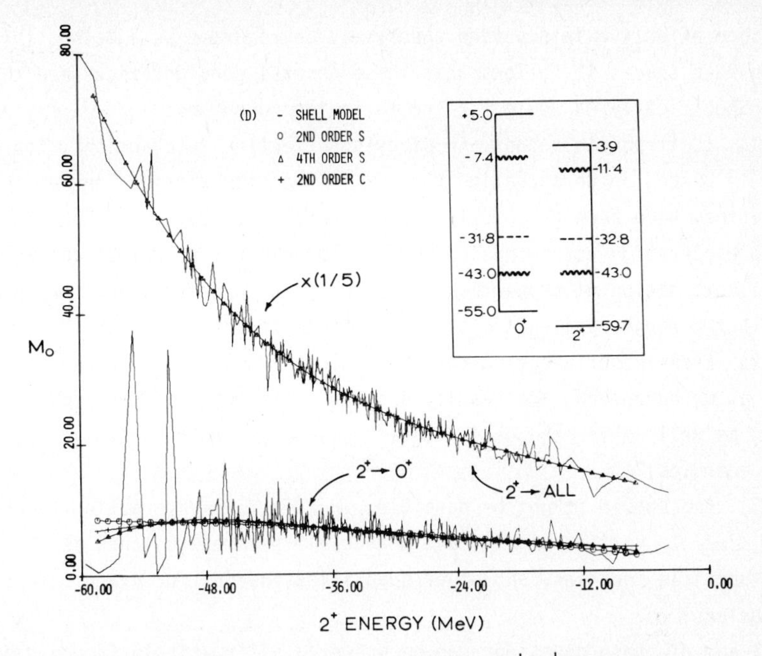

Fig. 4. The NEW sum-rule quantity for the $(2^+\to 0^+)$ and the $(2^+\to$ all J) E2 transitions in the T=0 states of $(ds)^6$. The shell-model energies for the highest and lowest states, the scalar centroids (---) and the highest and lowest configuration centroids are given in the inset. The strength is normalized to $Q\cdot Q=4G-3L^2$, where G, the SU(3) Casimir operator has eigenvalues $(\lambda+\mu+3)(\lambda+\mu)-\lambda\mu$. The fluctuating lines give the detailed shell-model results and the continuous lines those derived from various extensions of (15).

of H and H', methods and programs for the evaluation of which are available.

Corresponding to (15,17) there are CLT and exact expansions for strength distributions but these I shall not discuss here, except to remark that the CLT strengths (squares of matrix elements of excitation operators) are linear in both the initial and final energies. For expectation-value and strength fluctuations there are also two-point-correlation measures, which have however not yet been applied in any serious way.

4. Conclusions

Much of complex spectroscopy is dominated by four characteristic forms, Gaussian densities (or natural superpositions of them), expectation values and strengths which are respectively linear and bilinear in the energy, and fluctuations which are close to those determined by the GOE. There is a long-range rigidity in the distributions, and the fluctuations are small. Relatively little information is to be extracted so that the structures encountered are much simpler than would seem at first sight.

Since effective-interaction theory may be regarded as a method for spectroscopy in huge spaces it follows that these general considerations are relevant here also. Specifically we would ask the following questions:

(1) Is it possible that, in studying effective interactions, too much reliance is placed on perturbation theory? PT has its glories, and at this conference they have been gracefully exhibited for us. But should one ignore the CLT whose operation determines so much of the general structure and which has an unusual combination of properties, being elegant, powerful, and forgiving (in the sense that a rough application of it is liable to seize upon the truth)?

(2) Even in perturbation theory, could one not make good use of the results which, as we have seen, are readily available for the configuration centroids and widths, as well as for the basis-state occupancies (which are given at all excitation energies)?

(3) And should it not be possible, in effective-interaction theory, to calculate *directly* various significant measurable parameters, such as the induced single-particle energies, and other quantities, useful for example in comparing Hamiltonians?

If any of these questions can be answered affirmatively it could be that statistical spectroscopy will indeed make a contribution to the subject of this conference.

Acknowledgements

Much of the work of §1 has been done with K. K. Mon and is as yet unpublished. For the rest I have drawn freely on the work of friends and colleagues in many laboratories.

References

1. E. Cota, J.Flores, P.A.Mello and E.Yépez, Phys. Lett. 53B (1974) 32.
2. E.P.Wigner, SIAM Review 9 (1967) 1; *Statistical Theories of Spectra: Fluctuations*, ed. C.E.Porter (Academic Press, N.Y.1965); M.L.Mehta, in *Statistical Properties of Nuclei*, ed. J.B.Garg (Plenum Press, N.Y.1972); M. L.Mehta, *Random Matrices* (Academic Press, N.Y. 1967).
3. J.B.French and S.S.M.Wong, Phys. Lett. 33B (1970) 449; 35B (1971) 5; O.Bohigas and J.Flores, Phys. Lett. 34B (1971) 261; 35B (1971) 383.
4. K.K.Mon and J.B.French, University of Rochester Report COO-2171-46 (to be published).
5. F.J.Dyson and M.L.Mehta, J.Math. Phys. 4 (1963) 701, reprinted in Porter (2).
6. V.Potbhare and S.P.Pandya, to be published.
7. J.P.Draayer,J.B.French,V.Potbhare and S.S.M.Wong, Phys. Lett. 55B (1975)349.

J.B. FRENCH: WHAT FEATURES OF NUCLEAR SPECTRA ARE DIFFERENT FROM WHAT WE EXPECT BASED ON STATISTICS?

Zamick: You have an association of two spectra. How do you know that the wave functions of two levels that you associate are nearly the same?

French: You are discussing the relationship between a detailed spectrum and one which is calculated by making use of the first few moments only (via a Gram-Charlier expansion). There is no freedom at all in the association; the smoothed density determines the distribution function and hence the corresponding spectrum (by a method first used by Ratcliff). One does not know about the wave-functions because the methods don't deal with them but only with trace-like information (which could however be calculated from wave functions if they were available).

Zamick: Do you think your techniques will be able to distinguish between intermediate structure and Ericson Fluctuations?

French: I regard this kind of thing as by far the most interesting aspect of fluctuations in nuclear spectroscopy but we have only quite recently learned how to deal with fluctuations and nothing has yet been done along this line. I cannot predict what would emerge.

Vary: You have told us there is a great reduction in the actual information content of complex spectra. You did not go so far as to tell us how many parameters of information are left. For example, the width and location of the distributions, and the information contained in higher moments must be obtained somehow. How many of these are meaningfully determined by the data and how do we go about determining them theoretically?

French: I have attempted to show that the information content is very much smaller than one might expect; for example, in the centroids and widths, but, as I stated, each of these quantities obeys a few-parameter equation. It is important also not to take for granted that higher moments do yield new information, with a Gaussian (or any other predetermined two parameter) distribution, for example, that is not true. Your question is quite important but I know no general method for counting the number of pieces of information.

Weidenmüller: On the one hand, there are these very elegant and powerful statistical techniques that you described, on the other, we discuss in this conference methods of calculating nuclear spectra with perturbation and diagonalization techniques. This leads to the important question, related to Vary's remark: is there a statistically meaningful way of extracting from a nuclear spectrum those pieces of information which show deviations from statistics and must therefore be accounted for in terms of dynamical models?

French: It is hard but there are clearly some situations where one can gain real information. As far as spectra are concerned you must remember that I have taken systems with no exact-symmetry admixtures; for example, I have not dealt with states of different angular momentum. It is clear here that variations with J in the centroids and widths reflect themselves particularly in the yrast spectrum, so that from low-lying spectra we can of course learn a fair amount. I should have added that it would be good to extend the system of weakly and strongly interacting subspaces in order to treat the lowest ones by matrix methods. There would be some redundancy here but that would cause no harm. Finally one can learn new things, as I indicated in discussing occupancies, from sum-rule quantities, and in fact from the other aspects of strength distributions.

Harvey: Implicit in this approach is the assumption of a single-particle potential. Is there not some physical information in this single-particle spectrum? Can one interpret the content of your talk as an analysis of what complex spectrum can tell us over and above the physics already contained in the single-particle model?

French: For one thing it should be understood that by different methods, I am really doing standard shell-model problems and I make the same assumptions as other people who do these things. But you may be suggesting that the difficulty in extracting new information from complicated spectra is related to the fact that the single-particle spectrum really determines much that can be determined about the Hamiltonian. That seems to me to be a very interesting proposition.

INFINITE PARTIAL SUMMATIONS

DONALD W.L. SPRUNG
Physics Department, McMaster University, Hamilton, Ontario
Canada

This talk is intended to be a brief review of those aspects of the effective interaction problem that can be grouped under the heading of infinite partial summations of the perturbation series. To begin with, I will attempt to put this subject into the perspective of the conference, without (I hope) trespassing too far onto the territory occupied by previous speakers. After a brief mention of the classic examples of infinite summations, I turn to the effective interaction problem for two extra core particles. Their direct interaction is summed to produce the G matrix, while their indirect interaction through the core is summed in a variety of ways under the heading of core polarization. Then I mention the work of Weidenmüller and coworkers, specifically the use of Padé approximants as a method of summing the series. Finally I discuss the shell model diagonalization method of Rowe et al. Only a brief outline is given of the analogous work on the effective charge problem.

Background The working hypothesis of nuclear physics is that only two body forces are important. In the nucleus however one is free to introduce an external one body potential U in the shell model sense. Then

$$\begin{aligned} H = T + v &= (T+U) + (v-U) \\ &= H_o + V \end{aligned} \tag{1.1}$$

is divided into an unperturbed Hamiltonian H_o and a perturbation V. In principle, U is arbitrary, but one hopes by choosing it cleverly to minimize the effects of V. If the two body force were weak, it would be reasonable to choose U in the Hartree-Fock sense; with matrix elements

$$\langle p|U|q\rangle = \sum_m \langle pm|v|qm-mq\rangle n_m \tag{1.2}$$

where n_m is the occupation of the orbital $|m\rangle$. One would choose these orbitals to diagonalize H_o.

Hartree-Fock theory is in fact the most familiar example of an infinite partial summation, as pointed out by Thouless (Tho 61a). Consider an arbitrary Goldstone diagram containing a particle (or hole) line labelled by $|p\rangle$. In this line one may make any number of 'hole-bubble' insertions as illustrated in (1.3). Each such insertion multiplies the contribution of the graph by a factor

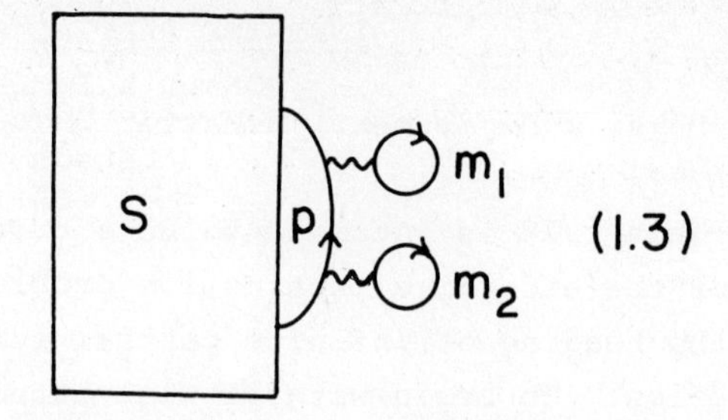

$$\sum_{m_i} \mp \frac{\langle pm_i|v|pm_i - m_i p\rangle}{E_s \pm T_p} = \pm \frac{\langle p|U|p\rangle}{E_s \pm T_p} \qquad (1.4)$$

where E_s is the 'spectator' energy from the rest of the diagram and T_p is the unperturbed energy of the orbital p. The sum of all such repeated insertions is a geometric series whose sum is

$$\frac{1}{E_s \pm (T_p + \langle p|U|p\rangle)} = \frac{1}{E_s \pm T_p} \mp \frac{1}{E_s \pm T_p} \langle pUp\rangle \frac{1}{E_s \pm T_p} + \ldots \qquad (1.5)$$

That is, the sum of all hole-bubble insertions converts the energy of orbital $|p\rangle$ into $\varepsilon_p = T_p + \langle p|U|p\rangle$. Since this is true for any particle (or hole) line, ε_p will appear everywhere.

Of course, the geometric series converges only if $\left|\frac{\langle p|U|p\rangle}{E_s \pm T_p}\right| < 1$. In the sense of perturbation theory, where the perturbation $\lambda V(x)$ is turned on, $(\lambda=0 \to \lambda=1)$ we can take the sum as an analytic continuation of the series beyond its radius of convergence.

There are two other classic examples of infinite partial summations. The Brueckner theory introduces a G-matrix

$$G(\omega_o) = v - v \frac{q}{H_o{}' - \omega_o} G(\omega_o) \qquad (1.6)$$

(1.7)

which is the sum of ladder diagrams, in which a given pair of particles interacts repeatedly via the two body potential. $G(\omega_o)$ is a function of the 'starting energy' ω_o, representing the energy available to the pair in the medium. The Pauli operator q forces the intermediate states to lie outside the Fermi sea. H_o' defines the energy of the two body intermediate states. G represents the self consistent pair interaction to all orders, in the presence of other particles, whose influence is expressed in q and $e = H_o' - \omega_o$.

The third classic example of an infinite sum of diagrams is offered by the TDA or RPA series. These were introduced in nuclear physics by Brown and Bolsterli (BB 59) and Thouless (Tho 61b) to explain the position of the giant dipole resonance. The absorption of a gamma ray is a one body operator, so must proceed by creating a particle-hole (p-h) excitation of the nucleus; an energy of order $\hbar\omega$. The resonance occurs at a higher energy. The explanation of Brown and Bolsterli was that this p-h excitation will interact via the nucleon-nucleon force, creating other p-h pairs of the same energy. This gives a secular problem to be diagonalized, yielding eigenstates which are coherent linear combinations of the degenerate p-h excitations. One of these states was shown, in a schematic model, to capture all the dipole absorption strength, and at the same time this 'dipole state' was greatly shifted in energy, by the trace of the particle-hole interaction matrix. In terms of diagrams, the TDA or RPA theories correspond to the sum of repeated particle-hole diagrams.

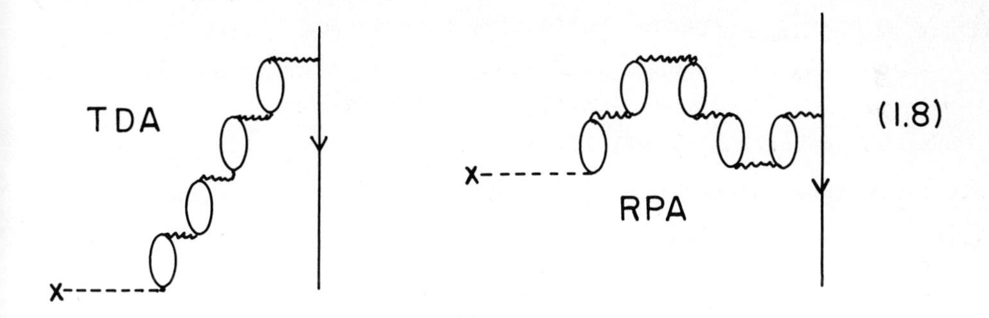

General Theory In most practical calculations, the one body potential U of (1.1) is chosen as an harmonic oscillator potential, so the unperturbed eigenstates of H_o are many-body oscillator wave functions:

$$H_o|\Phi_i\rangle = E_i^o|\Phi_i\rangle \qquad (2.1)$$

In the effective interaction problem we are concerned with a nucleus consisting of a closed core plus a small number n (usually 2) of active nucleons in valence orbits. Perturbation techniques such as those of Brandow (Bra 67) relate the unperturbed states $|\Phi_i\rangle$ to the exact solutions $H|\Psi\rangle = E|\Psi\rangle$ (2.2)

In this approach one fixes attention on a certain subset 'd' of the unperturbed states $|\Phi_i\rangle$, with projector

$$P = \sum_{i\varepsilon d} |\Phi_i\rangle\langle\Phi_i| \qquad (2.3)$$

The end result of the perturbation theory is to produce a secular equation

$$P[\Sigma\, \varepsilon_k - E_o^V - \Delta E^V + \mathcal{V}]|\Psi_D\rangle = 0 \qquad (2.4)$$

for the projection $|\Psi_D\rangle = P|\Psi\rangle$ (2.5)

of the true eigenstate $|\Psi\rangle$ onto the model space 'd'. In this secular equation the true single particle energies ε_k of the particle or hole orbitals enter. E_o^V is the unperturbed energy of the particles in valence orbitals and ΔE^V is the perturbation. The energy independent effective interaction $\mathcal{V}$:

$$\mathcal{V} = V - [V \frac{Q}{H_o^V - E_o^V} \mathcal{V}]_{\text{linked folded}} \qquad (2.6)$$

is obtained as the sum of all linked folded diagrams with n particle lines (the valence nucleons) entering at the bottom and leaving at the top of the diagram. H_o^V is the unperturbed Hamiltonian after the unperturbed core energy is subtracted out. The expansion is of Rayleigh - Schroedinger type, with only unperturbed energies occurring in the denominator. Since the states $|\Psi_D\rangle$ are projections onto a subspace, they are not orthonormal, and this is reflected in having a non-hermitian effective interaction $\mathcal{V}$.

Because the N-N force is a balance between strongly attractive and repulsive elements, a perturbation theory in V is certain to diverge. In the simpler problem of nuclear matter, the remedy devised to overcome this difficulty is to reorder the series into a series in the number of interacting particles, a linked cluster expansion. The two particle cluster brings in the Brueckner G-matrix, (1.6). This can also be done for the expansion (2.6). The result is to interpret all perturbation theory diagrams as containing G-interactions rather than v-interaction vertices, and omit all diagrams containing two or more successive two body interactions between the same pair (not delineated by an interaction elsewhere in the diagram) because these have already been included in G. Schematically we can say

$$\mathcal{V} = G - \left[G \frac{Q'}{H_o^V - E_o^V} \mathcal{V}\right]_{\text{linked folded.}} \qquad (2.7)$$

The only complication in this procedure is to see that when G is used at a particular place in a particular diagram, the energy denominator used in calculating G is the same as the one that occurs in the ladder sum which G is replacing. This means we must, in (2.7) and (1.6) identify

$$H_o^V - E_o^V = H_o' - \omega_o \qquad (2.8)$$

which defines the proper starting energy to use. However if H_o^V is an oscillator Hamiltonian while H_o' uses plane wave energies for the intermediate states, some compromise must be made. This is a source of uncertainty in the results of actual calculations. There is also the question of making the Pauli operators q and Q consistent. The more recent methods of calculating G-matrix elements for finite nuclei [BHM 71] are better in this respect, in making clear just what propagator is being used.

In nuclear matter, the next step is to sum all three body cluster diagrams, using Bethe-Faddeev-Day techniques [Bet 65, Day 66]. This is necessary because the perturbation series in powers of G is still divergent. The third order diagrams alone are quite misleading, compared to the complete three body cluster. It is believed that the convergence parameter of the cluster expansion is the 'wound integral'

κ, which is roughly equal to 1/6 for nuclear matter at normal density. Thus the energy per particle due to n-body clusters should be of order κ^2 times the two body potential energy of order 40 MeV. For the three particle case, special considerations reduce this to only about 1 MeV per particle (Dah 69).

Specific Cases Most, if not all, of the effective interaction calculations have been carried out for the case of two valence particles beyond a closed shell, mainly for the case of ^{18}O, ^{18}F and ^{42}Ca, ^{42}Sc. I suppose for this reason little has been said about evaluating the three body cluster contribution. It is in the very nature of the procedure that for three valence particles, the effective force $\mathcal{V}$ will come out as a three body interaction which is not simply the sum of three pair-wise interactions. A careful evaluation of three body cluster diagrams will be required.

Beyond the introduction of the G-matrix, calculation of the effective interaction has been guided by the classical model of the valence particles interacting indirectly by exciting vibrations of the core. In a microscopic model, the simplest such process is the single particle-hole bubble graph considered by G. Bertsch (Ber 65).

This process was added to the G-matrix by Kuo and Brown (KB66) and was found to be an important part of the force, giving it some of the pairing and quadrupole character of the phenomenological P and Q force. By analogy with the giant dipole resonance which is also explained as an excitation of particle-hole pairs, one would expect that repeated particle hole bubbles should be important. These are summed up by the TDA or RPA series, as done by Osnes and Warke (OW 69). In this work the RPA equations were solved for the phonon wave functions

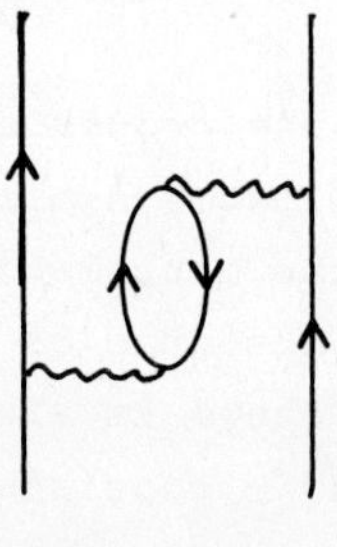

(3.1)

$$|nJT\rangle = C^{+}(nJT)|\tilde{0}\rangle = \sum_{ph} [X_{ph}(nJT)A^{+}(phJT) - (-)^{p-h} Y_{ph}(nJT)A^{+}(hpJT)]|\tilde{0}\rangle \tag{3.2}$$

and then these collective states were used to evaluate the core polarization diagram.

In this procedure the intermediate state energy denominators are of the form $[\varepsilon_c - \varepsilon_a - \omega(nJ''T'')]$ with $\omega(nJ''T'')$ being the phonon eigenvalue. The corresponding formulae for TDA are obtained by omitting the Y amplitudes from the calculation. Bertsch's calculation corresponds to identifying each particle-hole state as being one of the modes (n). Since the experimental energies ε_c, ε_a were used both here and in solving the RPA equations, this calculation is not quite equivalent to others using oscillator energies. Aside from this difference, the method would be equivalent to Kirson's work to be described below, as shown by Ellis and Siegel [ES 70].

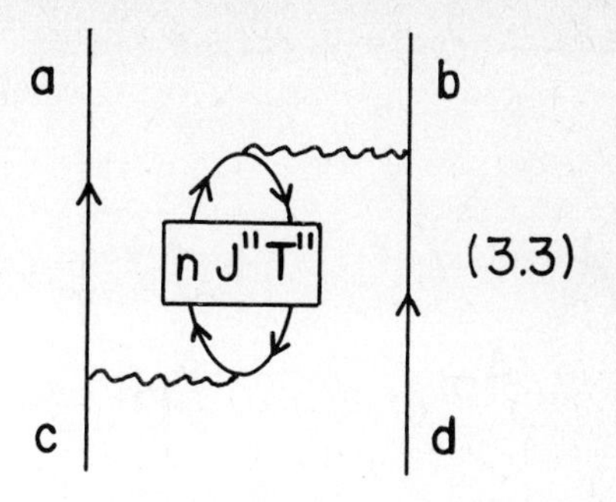

When the resulting effective interaction is diagonalized to give the excited states of ^{18}O, ^{18}F, it is found that, in going from ph to TDA to RPA, the ground state is strongly depressed; an example is shown in (3.4). The same result

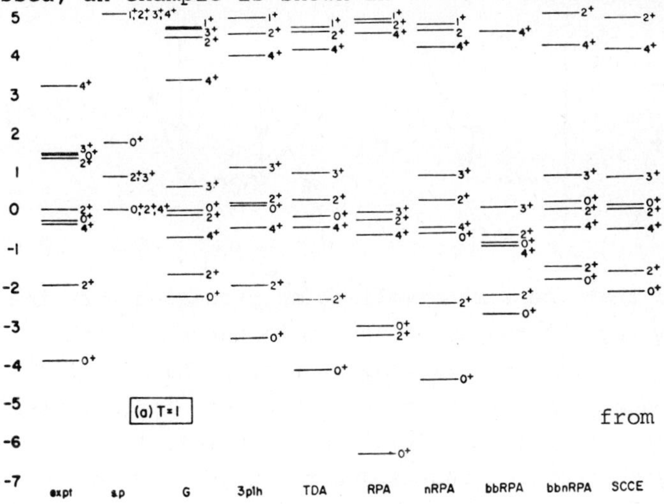

(a) T=1

expt sp G 3p1h TDA RPA nRPA bbRPA bbnRPA SCCE

(3.4)

from Kir 71

seems to follow for other forces (Zam 69, ES 70, KZ 70]. Presumably the lesson to be drawn from this circumstance is that the series of graphs which has been summed does not include all the important physical processes. Physical intuition was called upon again for guidance as to further series to be summed. Blomquist and Kuo (BK 69) included corrections to the bare particle-hole vertices, which occur in the TDA or RPA series.

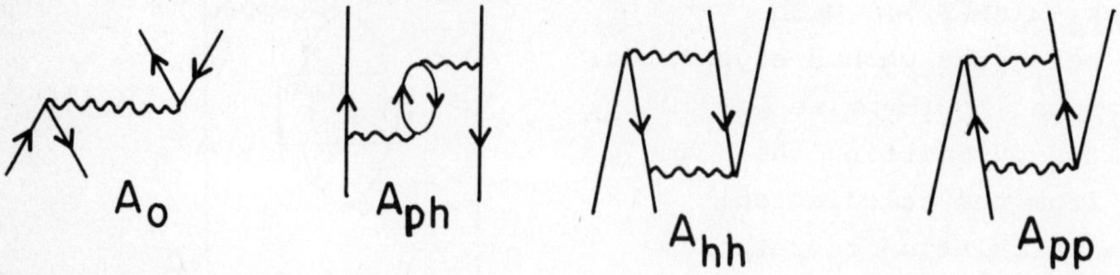

A_o A_{ph} A_{hh} A_{pp} (3.5)

Kuo and Osnes (KO 74) showed that these 'self screening' corrections damp the collectivity of the core vibrations.

It seems generally agreed (Kir 70, BK 75) that the pp ladder designated as A_{pp} represents an overcounting, because the G matrix interaction should already include this graph.

Kirson and Zamick (KZ 70) classed the possible additional graphs into two groups. Propagator renormalization includes processes which renormalize the propagation of a particle hole pair, such as the processes shown above. Vertex renormalization consists of all processes which connect a valence particle (or hole) to a particle-hole pair such as

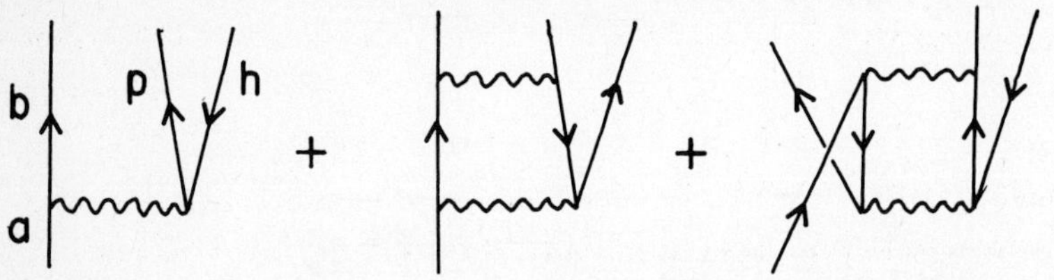

(3.6)

They noted that these renormalization processes were individually large, but the two types tended to cancel, so they proposed to calculate them to all orders. Kirson (Kir 71, Kir 74) undertook such a calculation for ^{18}O. The sum of all vertex corrections is taken to define a "Black Box" vertex. The sum of all propagator corrections is described as "nested" propagators. It is easiest to explain Kirson's work if we follow him in defining five convenient vertices [he includes certain phase and renormalization factors so that in graphs containing sequences of interactions, most phase factors are absorbed into the definitions].

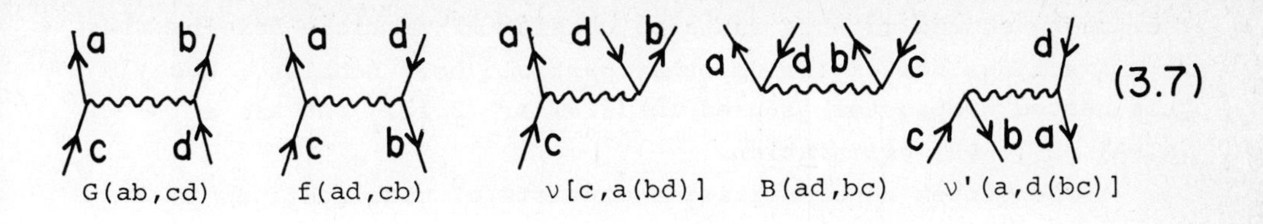

Since these are defined as antisymmetrised matrix elements, I should really use the Hugenholtz notation and pinch the "interaction line" into a dot, but I draw them in Goldstone form out of habit.

The simplest core polarization process (3.1) gives an effective interaction

$$\mathcal{V}_{ph} = G + \nu \frac{1}{e} \nu \tag{3.8}$$

The TDA sum is represented by

$$\begin{aligned}\mathcal{V}_{TDA} &= G + \nu \frac{1}{e} \nu + \nu \frac{1}{e} f \frac{1}{e} \nu + \nu \frac{1}{e} f \frac{1}{e} f \frac{1}{e} \nu + \ldots \\ &= G + \nu \frac{1}{e+f} \nu\end{aligned} \tag{3.9}$$

Kirson uses a particular version of RPA in which the secular equation takes the form

$$[\varepsilon+f - B \frac{1}{\varepsilon+f} B]|\chi\rangle = E|\chi\rangle \tag{3.10}$$

and the effective interaction is

$$\mathcal{V}_{RPA} = G + \nu[A - BA^{-1}B]^{-1}[\nu - BA^{-1}\tilde{\nu}] \tag{3.11}$$

with

$$A = \varepsilon+f.$$

Screening or "nesting" corrections are modifications to the particle-hole interaction, described by

$$f_\infty = f + \nu \frac{1}{\varepsilon+f_\infty} \nu' \tag{3.12}$$

Graphically, this gives

(3.13)

The graphs summed in this instance consist of particle-hole bubbles in TDA strings nested inside other particle hole bubbles. When this nested propagator is used in (3.9) or (3.11), one has a [nTDA] or [nRPA] calculation.

The vertex renormalization consists of modifications to the p-2p1h vertex ν or its time reversal mate ν'. These are called the 'black box' vertex and may be computed as follows:

$$\nu_{bb} = \nu + \nu \frac{1}{e+f} f + \tilde{\nu} \frac{1}{e+f} B \qquad (3.14)$$

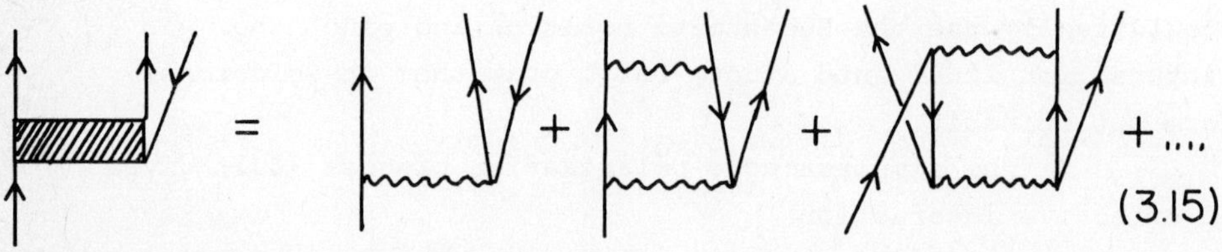

$$= \quad + \quad + \quad + \ldots \qquad (3.15)$$

and similarly for ν'_{bb}. Calculations including this effect are called BBTDA, BBRPA. In a subsequent paper Kirson (Kir 74) considered renormalizations to the B vertex of the same type:

$$B_{bb} = B + B \frac{1}{\varepsilon+f} f$$

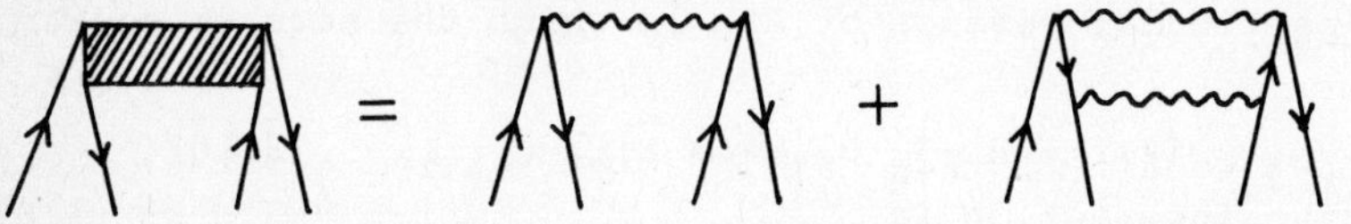

$$= \quad + \quad + \ldots\ldots \qquad (3.16)$$

Whether this should be counted as part of the propagator renormalization is a moot point. In the most complete calculation, which Kirson called SCCE (self-consistent coupled equations), all of these processes are taken together:

$$\left.\begin{aligned}
f_\infty &= f + \nu_\infty \frac{1}{\varepsilon+f_\infty} \nu'_\infty \\
\nu_\infty &= \nu + \nu_\infty \frac{1}{\varepsilon+f_\infty} f_\infty + \tilde{\nu}_\infty \frac{1}{\varepsilon+f_\infty} B_\infty \\
\nu'_\infty &= \nu' + \nu'_\infty \frac{1}{\varepsilon+f_\infty} f_\infty + \tilde{\nu}'_\infty \frac{1}{\varepsilon+f_\infty} B_\infty \\
B_\infty &= B + B_\infty \frac{1}{\varepsilon+f_\infty} f_\infty
\end{aligned}\right\} \qquad (3.17)$$

This means that the renormalization corrections to each vertex are calculated using the most completely renormalized vertices and propagators. These equations are solved by an interative procedure. Unfortunately, the net result of including all the corrections summed by Kirson, was that the resulting effective interaction was almost unchanged from the bare G matrix. There was almost complete cancellation of all the correction terms.

Some insight into this result is afforded by the schematic model of Brown and Bolsterli (BB 59), which assumes a separable interaction

$$G = f = \lambda|D\rangle\langle D| \quad , \quad B = \mu|D\rangle\langle D| \tag{3.18}$$

where $|D\rangle$ is a ket in particle hole space. Assuming the unperturbed energies to be degenerate simplifies the results further. In any case, a measure of the force strength is

$$\lambda\langle D|\frac{1}{\varepsilon}|D\rangle \equiv \frac{\lambda}{c} \tag{3.19}$$

When the force is strong relative to the p-h energy ε, c will be small, and $\frac{\lambda}{c}$ will be larger. In this model

$$\begin{aligned} V_{ph} &= |D\rangle\lambda\langle D| + |D\rangle\lambda\langle D|\frac{1}{\varepsilon}|D\rangle\lambda\langle D| \\ &= |D\rangle\lambda\langle D|\ (1 + \frac{\lambda}{c}) \end{aligned} \tag{3.20}$$

which shows an amplification factor $S_{ph} = 1 + \frac{\lambda}{c}$ relative to the unperturbed force. One similarly finds (SJ 72)

$$S_{TDA} = \frac{1+2\frac{\lambda}{c}}{1+\frac{\lambda}{c}} \quad \text{and} \quad S_{RPA} = \frac{1+3\frac{\lambda}{c} + 2(\lambda^2-\mu^2)/c^2}{1+2\frac{\lambda}{c} + (\lambda^2-\mu^2)/c^2} \tag{3.21}$$

For very weak forces ($\frac{\lambda}{c} \to 0$) these agree but for strong forces they behave quite differently. For attractive forces ($\frac{\lambda}{c} < 0$) there is a reduction in coupling strength, and in the TDA or RPA cases, a singularity at the critical force strength (where the perturbed energy E would go to zero and become imaginary).

In the case of screening or nesting, (3.12) we can solve by making the ansatz $f_\infty = |D\rangle\kappa\langle D|$. Then

$$\kappa = \lambda + \lambda^2\langle D|\ \frac{1}{\varepsilon+f_\infty}\ |D\rangle \tag{3.22}$$

or

$$\frac{\kappa}{c} = \frac{\lambda}{c} + (\frac{\lambda}{c})^2\ \frac{1}{1+\frac{\kappa}{c}} \tag{3.23}$$

$$\approx \frac{\lambda}{c}\ (1 + \frac{\lambda}{c}\ \ldots) \text{ for weak forces.}$$

The effect of screening, for attractive forces, is to <u>reduce</u> the effective coupling constant from $\frac{\lambda}{c}$ to $\frac{\kappa}{c}$. Even if the force strength $\frac{\lambda}{c}$ were close to the critical value, (where $S_{RPA} = \infty$). The reduced strength $\frac{\kappa}{c}$ will give a finite renormalization.

The vertex corrections can be handled similarly. The relevant ansatz is that $\lambda \to \lambda t, \mu \to \mu t$ with the same proportionality constant t. We found (SJ 72) that

$$t = (1 + \frac{\kappa}{c})/(1 - \frac{\mu^2}{\lambda c}) , \qquad (3.24)$$

which is <1 for attractive forces, and an amplification factor

$$S = \frac{1 + \frac{2\kappa}{c} + \frac{\lambda t^2}{c} + [(\frac{\kappa}{c})^2 + \frac{\lambda \kappa t^2}{c^2} - (\frac{\mu}{c})^2 (1+t^2)]}{1 + \frac{2\kappa}{c} + [(\frac{\kappa}{c})^2 - (\frac{\mu}{c})^2]} . \qquad (3.25)$$

For realistic forces, $\mu \approx \frac{1}{2} \lambda$.

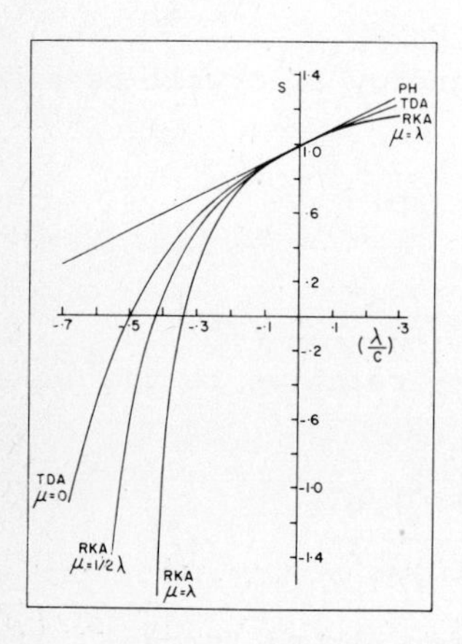

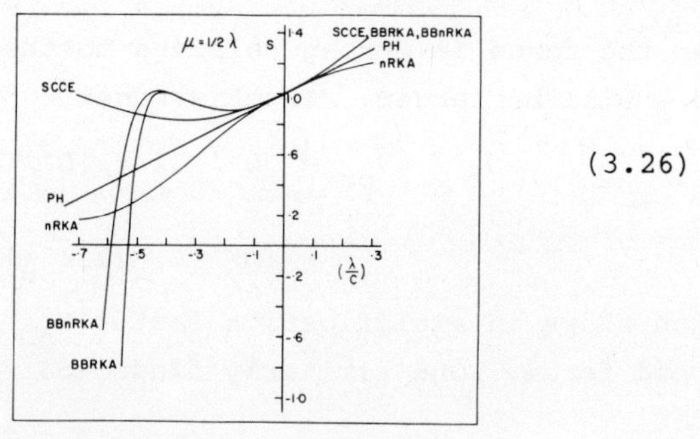

Amplification factor S of the p–h interaction calculated in various ways for nRPA, bbRPA, bbnRPA, and SCCE under the assumption $\mu = (1/2)\lambda$.

(3.26)

In 3.26 this amplification factor is plotted vs. $\frac{\lambda}{c}$ for several variants of Kirson's partial summations. It is seen that for attractive forces there is a broad region of attractive force strengths $(-.5 < \frac{\lambda}{c} < 0)$ over which the amplification factor S remains close to 1, meaning that the renormalized force is nearly equal to the bare force. This applies to the BBRPA, BBnRPA and SCCE theories.

Kirson's calculations for ^{18}O were repeated by Jopko and were carried out also for ^{42}Ca (JS 73, Jop 74). The main conclusions are completely parallel in the two cases. This is disappointing because the bare G-matrix gives unsatisfactory spectra for ^{18}O, ^{42}Ca and something like the simple p-h core polarization correction is required to correct this. Features of the calculation which may be responsible for this are the limitation of all these calculations to only $2\hbar\omega$ or $1\hbar\omega$ excitations. It was shown by Vary, Sauer and Wong (VSW 73] that to properly take account of the tensor force which is strong either in the bare NN force or in the G matrix, one has to include up to 12 $\hbar\omega$ excitations. Another feature is the use of G matrix elements, generally those of Kuo and Brown, calculated by approximate techniques. Those of Barrett,Hewitt and McCarthy (BHM 71) are much more reliable and allow one to take account of various treatments of the low-lying two-body intermediate state spectrum in calculating G. (This is one way of approximately including three body cluster effects in the G-matrix - see for example Grangé (Gra 75)). Variations here are known to have a big effect on the spectra. There is the certainty that many more diagrams remain to be summed. Based on a calculation by Goode (Goo 71), Barrett and Kirson have conjectured (BK 73) that additional vertex corrections may be useful. Finally there is a question whether the use of an harmonic oscillator H_o is sufficiently realistic. Use of Hartree Fock wave functions (EO 72, Row 73) and of Woods Saxon wave functions (KLS 69) have been shown to affect particle-hole calculations. From another point of view, Ellis and Mavromatis (EM 71) have shown that when oscillator wave functions are used, there are sizeable contributions from (v-U) insertions, which would be absent in a HF basis.

Convergence In a series of papers, Weidenmüller and collaborators have studied the analytic properties of the perturbation series (SW 72, SW 73). They conclude that, in all practical cases, the perturbation series must diverge, because the set of lowest lying perturbed states will not correspond to the the lowest states of the unperturbed problem. For example, in ^{18}O there can be deformed 4p-2h states which lie very low in energy, among the 2p shell model states. These papers have been very important, both in clarifying the problem and in stimulating work to overcome the difficulty. Hofmann et al. (HLRW 73, HLRW 74) considered a number of methods of reordering the series, and concluded that Padé approximants were the most promising. Krenciglowa et al. (Kre+ 73, KK 74) have also used Padé techniques in this connection.

The divergence of the perturbation series has been discussed by Ellis and Osnes (EO 73), Vincent and Pittel (VP 74) and P. Schaefer (Sch 74), for a simple two channel model space Hamiltonian. The general conclusion of all these papers that by summing the perturbation series one can determine not the lowest lying perturbed states but rather those perturbed states which have the greatest overlap with the model space. The problem considered by the above authors is

$$H = H_o + xV = \begin{pmatrix} e+u_o x & \gamma u_o x \\ \gamma u_o x & 0 \end{pmatrix} \quad (4.1)$$

The exact eigenvalues are

$$\lambda_\pm(x) = \left(\frac{e+u_o x}{2}\right)\left[1 \pm \left\{1 + \frac{4\gamma^2 u_o{}^2 x^2}{(e+u_o x)^2}\right\}^{1/2}\right] \quad (4.2)$$

Because of the square root, there are two branch points x_B and x_B'. As the coupling between channels is turned on ($\gamma = 0 \to \infty$) these travel around a circle B in the complex x-plane, starting at $x_p = -\frac{e}{u_o}$ and finishing at the origin as illustrated: (I assume attractive forces so $u_o<0$).

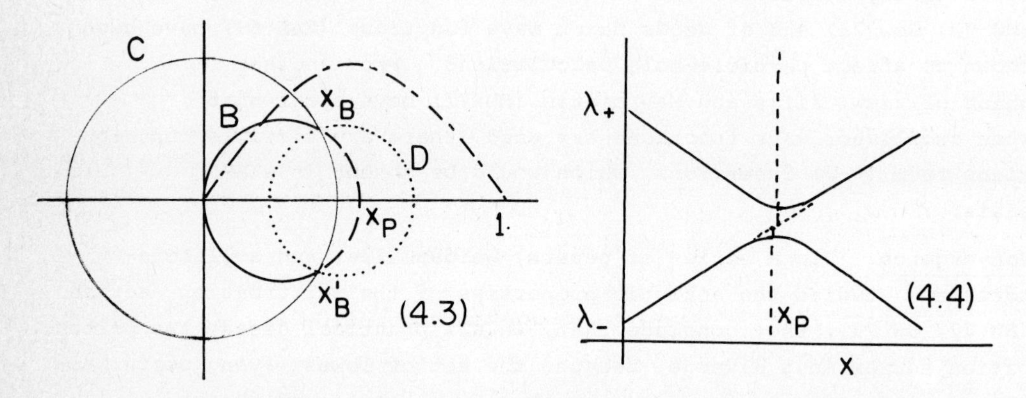

$\lambda_\pm$ are two branches of an analytic function. It is convenient to draw the Riemann cut along an arc of the circle B, joining the two branch points (heavy dashed line). Since there are no other singularities, Taylor's theorem tells us that the power series for $\lambda_\pm$ will converge inside the circle C, and will clearly diverge at x=1.

Suppose we selected $\lambda_-(x)$ as our initial state, and let x(real) increase from 0 to 1. The divergence of the power series is seen (4.4) to be due to the sharp curvature of $\lambda_-(x)$ in the vicinity of the branch cut. At x_p, in 4.3, we drop down onto the lower Riemann sheet and continue to follow the branch $\lambda_-(x)$.

It was noted by Ellis and Osnes that the folded diagram

perturbation series is equivalent to a continued fraction expansion of the eigenvalues. If this is truncated at any finite order, one has a Padé approximant. These converge on the upper sheet of the Riemann surface to an analytic continuation of the function whose power series converges inside C. (However, the convergence is poor inside the "orthogonal circle" D.) So, if we begin with $\lambda_-(x)$ near x=0, and adopt the Padé method, we will arrive at x=1 with a perfectly well defined value. To remain on the top sheet, the path followed must be one (dotted line 4.3) which does not cross the cut. But at x=1, because we are on the top sheet, we must have the other value, namely $\lambda_+(x)$. The dotted line in 4.4 is intended to give an impression how one changes from λ_- to λ_+ near x_p.

The paper of Ellis and Osnes discusses in more detail how the continued fraction expansion converges. That of Vincent and Pittel adopts the viewpoint that if one stays far enough away from the cut, its influence on the computed values will be bounded. For this reason, a Padé approximant based on low orders of the perturbation series may be a useful approximation to the exact analytic continuation.

Shell Model Approach Lo Iudice, Rowe and Wong (LRW 71, LRW 74) attempted to cut the Gordian knot by a fresh approach. They chose to do an exact diagonalization of the G-matrix in the space of two particle, and of 3p-1h ($2\hbar\omega$) excitations; using shell model techniques. In principle the method sums a very large class of diagrams, which are constructed using the vertices G, f, and ν of (3.7). However it does not include the RPA type strings, and various other graphs included by Kirson. They found much more satisfactory results, with the ^{18}O levels only slightly shifted from the positions found in a lowest order (G+ph) core polarization calculation. Their eigenvectors had only small 3p-1h components, and showed little collectivity. These differences from the perturbative calculations were ascribed in part to the use of experimental single particle energies in the diagonalization, rather than simple $2\hbar\omega$ energies.

This point has been investigated by Starkand and Kirson (SK 75). They repeated the LRW calculation using $\hbar\omega$ energies, and found only small changes. However, they then carried out the calculation in a slightly different manner, using experimental energies for the 2p states (P space) and $\hbar\omega$ energies for the 3p-1h states (Q space). These calculations are for the $J^{\pi}T = 0^{+}1$ states of ^{18}O. For diagonal matrix elements they now found sizeable differences, of order 35% when $\hbar\omega$ = 14 MeV, For off diagonal matrix elements the differences are much smaller. For $\hbar\omega$ = 17.5 MeV the shift is smaller. This shows that the single particle energies entering into intermediate states can affect the results significantly.

The diagrams summed by LRW are classified by Starkand and Kirson as follows:

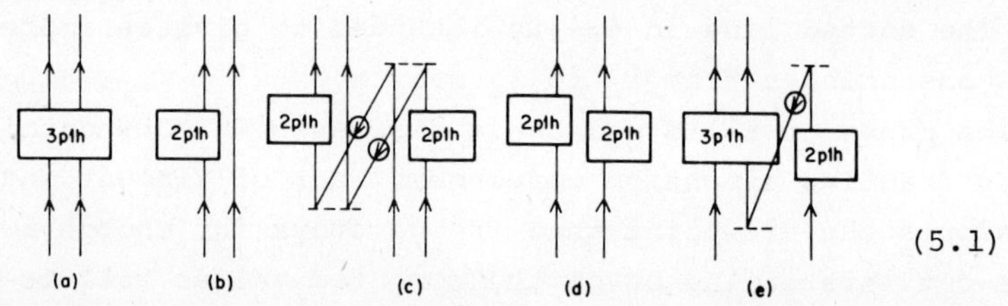

(5.1)

Schematic structure of perturbation theory diagrams contained implicitly in the extended shell-model calculation: (a) general two-body diagrams; (b) one-body contributions, with one spectator nucleon; (c) folded disconnected diagrams; (d) non folded disconnected diagrams (not included in the 2p + 3p1h diagonalization); (e) folded diagrams modifying incoming single-particle energies in energy denominators.

They noted that the folded disconnected diagrams (c) occur in the LRW approach but not in the perturbation theory, where they would be cancelled by the non-folded disconnected diagrams (d). When the contribution of these diagrams is removed, the SK results are much closer to the LRW results, and also close to the third order perturbation theory. This does not necessarily imply that perturbation theory really would converge, only that the higher order terms ultimately cancel each other out. The techniques of (SK 75) will allow a more detailed comparison between the LRW method and the usual perturbation technique. Similar ideas have been advanced by Ellis (Ell 75).

The general conclusion to be drawn from this brief survey, is that we do not yet have a satisfactory understanding of the effective interaction, based on infinite partial summations of the perturbation series. Diagrams beyond those already considered may well be important. The Shell Model method, supplemented by the insights of Ellis, Starkand and Kirson, is probably the most promising avenue for further work.

Effective Charge The problem of deducing the shell model effective interaction can be regarded as a special case of the effective operator problem for a general operator t. For the effective interaction, t is the Hamiltonian H and is a two body operator. The effective interaction $\mathcal{V}$ then contains zero-body, one-body, two-body, three-body, etc. terms. The one body terms are the hole-bubble insertions. Also of great interest is the case when we take t to be an electromagnetic transition operator. Most attention has been devoted to E2 transitions near ^{16}O and ^{40}Ca. The effective operator $\tilde{t}$ is defined on the subspace 'd' of (2.3). It is to have the property that

$$<\Psi_\alpha|t|\Psi_\beta> = <\Psi_\alpha|P\tilde{t}P|\Psi_\beta> = <\Psi_{D\alpha}|\tilde{t}|\Psi_{D\beta}> \qquad (6.1)$$

i.e. the correct matrix elements of t will be obtained if the effective operator $\tilde{t}$ is used with the projected states $|\Psi_D>$. It has been emphasized by Harvey and Khanna (HK 70, HK70b, KLH 71) that the renormalization of the effective operator must be consistent with that of the effective interaction.

In the first of the papers cited, a general formalism for $\tilde{t}$ is developed, and diagrams are introduced for the evaluation of the one body and two body parts of $\tilde{t}$, when t is a one body operator and the interaction QVP connecting the model space to the excluded space is taken to be a two body operator at worst. The one-body part of $\tilde{t}$ can be looked at in nuclei with one active particle such as ^{17}O and ^{41}Ca. For nuclei with two active nucleons, the two-body part of $\tilde{t}$ can come into play. Evidence for the existence of a two-body operator was studied in HK 70b and KHSJ 72, in reference to data on A=18 and A=42 nuclei. Estimates of the size of the two-body terms were made, based on a quadrupole force (Q.Q) for the coupling potential QVP, and also using more realistic forces. This is completely analogous to the existence of an effective three body force in the effective interaction $\mathcal{V}$, which we mentioned above (3.1).

Transition rates are expressed by a number, the B(E2) value in units e^2fm^4. When the experimental value for a transition in ^{17}O or ^{17}F does not agree with the single-particle model, the discrepancy can be expressed by ascribing an effective charge different from e to the odd nucleon. For a neutron, the effective charge e_n must be due entirely to core polarization processes. For a proton, the charge can be written as $1+e_p$, so e_p

also refers to the polarization charge. One finds that generally speaking e_p, e_n are both about 0.5. First order calculations of the effective charge, using the diagrams

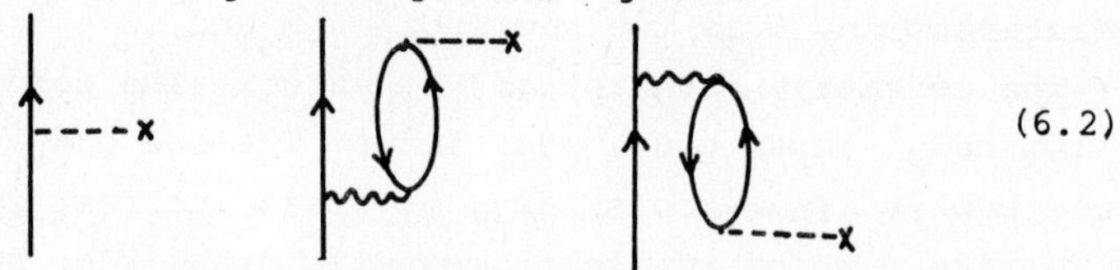

(6.2)

were carried out by Dieperink and Brussard (DB 69). Siegel and Zamick (SZ 70) carried out the first-order, second-order, the TDA and RPA calculations, using the Kallio-Koltveit and Kuo-Brown forces. They found a big increase in the effective charges over the perturbation theory results, with experimental values lying somewhere between the TDA and RPA results. Khanna et al (KLH 71) carried out similar calculations using the Kahana, Lee and Scott separable potential (KLS 69). Kuo and Osnes (KO 73) have studied the effect of vertex corrections of the type eq. (3.5) on the effective charge. As in the effective interaction problem, they found that the vertex corrections severely reduce the collective enhancements of the TDA and RPA theories, and these two calculations give similar results. As a bonus, it was found that the A=17 and A=41 nuclei now behaved quite similarly, and the results were less sensitive to the assumed single particle energies. The problem is that with the bare p-h interaction, the A=41 calculations were close to the instability (zero eigenvalue for RPA calculations), so really large enhancements could be obtained. The vertex corrections give an effectively weaker p-h force, as discussed earlier. In all of these papers it is emphasized that the experimental effective charge is also model dependent, being sensitive to the use of harmonic oscillator as opposed to Woods-Saxon wave functions. Thus, comparison to experiment cannot be carried out in detail.

Jopko and Khanna (JK 73) have calculated the one-body and two-body effective charge for the A=18 and A=42 systems using the various approximations introduced by Kirson (Kir 71) for the effective interaction (see fig. 3.4 for a list). Unfortunately this work has not been published.

In conclusion, I would like to thank the National Research Council of Canada for continued research support under operating grant A-3198.

References

BB 59 G.E. Brown and M.Bolsterli Phys. Rev. Lett. 3 (1959) 72.

Ber 65 G. Bertsch, Nucl. Phys. 74 (1965) 234.

Bet 65 H.A. Bethe Phys. Rev. 138 (1965) B804.

BHM 71 Barrett, Hewitt and McCarthy Phys. Rev. C3 (1971) 1137.

BK 69 J. Blomquist and T.T.S. Kuo Phys. Lett. 29B (1969) 544-547.

BK 73 B.R. Barrett and M.W. Kirson Adv. in Nucl. Phys. 6 (1973) 219.

BK 75 B.R. Barrett and M.W. Kirson Phys. Lett. 55B (1975) 129-133.

Bra 67 B.H. Brandow Rev. Mod. Phys. 39 (1967) 771-828.

Dah 69 T.K. Dahlblom Acta Akad. Aboensis ser.B, 29 #6 (1969).

Day 66 B.D. Day Phys. Rev. 151 (1966) 826.

Ell 75 P.J. Ellis Phys. Lett. 56B (1975) 232-236.

EM 71 P.J. Ellis and H.A. Mavromatis Nucl. Phys. A175 (1971) 309.

EO 72 P.J. Ellis and E. Osnes Phys. Lett. 41B (1972) 97-101.

EO 73 P.J. Ellis and E. Osnes Phys. Lett. 45B (1973) 425-428.

ES 70 P.J. Ellis and S.Siegel Nucl. Phys. A152 (1970) 547-560.

Goo 71 P. Goode Nucl. Phys. A172 (1971) 66.

Goo 75 P. Goode Nucl. Phys. A241 (1975) 311-317.

Gra 75 P. Grangé, preprint, Strasbourg.

HLRW 73 H.M. Hoffman,S.Y.Lee, J. Richert and H.A. Weidenmüller Phys. Lett. 45B (1973) 421.

HLRW 74 ———————————— Ann. Phys. 85 (1974) 410-437.

Jop 74 A. M. Jopko, Ph.D. Thesis, McMaster University (1974).

JS 73 A.M.Jopko and D.W.L.Sprung Can. J. Phys. 51 (1973) 2275-2282.

KB 66 T.T.S. Kuo and G.E.Brown Nucl. Phys. 85 (1966) 40-86.

Kir 70 M.W.Kirson Phys. Lett. 32B (1970) 33-36.

Kir 71 M.W. Kirson Ann. Phys. 66 (1971) 624-650; erratum 68,556.

Kir 74 M.W.Kirson Ann. Phys. 82 (1974) 345-368

KLS 69 S. Kahana, H.C. Lee and C.K. Scott, Phys. Rev. 185 (1969) 1378.

KK 74 E.M.Krenciglowa and T.T.S.Kuo Nucl. Phys. A235 (1974) 171-189.

KO 74 T.T.S. Kuo and E.Osnes Nucl. Phys. A226 (1974) 204-218.

Kre+ 73 Krenciglowa, Kuo, Osnes and Giraud, Phys. Lett. 47B (1973) 322.

KZ 70 M.W.Kirson and L. Zamick Ann. Phys. (N.Y.60 (1970) 188.

LRW 71 N.LoIudice, D.J.Rowe and S.S.M. Wong Nucl. Phys. A219 (1974) 171-189.

LRW 74 ———————————— Nucl. Phys. A219 (1974) 171-189.

OW 69 E. Osnes and C.S. Warke Phys. Lett. B30 (1969) 306-310.

Row 73 D.J. Rowe Phys. Lett. 44B (1973) 155.

Sch 74 Paul A. Schaefer Ann. Phys. 87 (1974) 375-416.

SJ 72 D.W.L. Sprung and A.M. Jopko Can. J. Phys. 50 (1972) 2768

SK 75 Y. Starkand and M.W. Kirson Phys. Lett. 55B (1975) 125-128.

SW 72 T.H. Shucan and H.A. Weidenmüller Ann.Phys. 73 (1972) 108.

SW 73 ——————————————— Ann. Phys. 76 (1973) 483.

Tho 61a D.J.Thouless, "Quantum Mechanics of Many Body Systems", Academic Press, London, p. 48.

Tho 61b D.J. Thouless Nucl. Phys. 22 (1961) 78.

VP 73 C.M. Vincent and S. Pittel Phys. Lett. 47B (1973) 327.

VSW 73 J.P. Vary, P.V. Sauer and C.W. Wong Phys. Rev. C7 (1973) 1776.

Zam 69 L. Zamick Phys. Rev. Lett. 23 (1969) 1406-9.

Additional References (Effective Charges)

DB 69 A.E.L. Dieperink and P.J. Brussard Nucl.Phys. A129 (1969) 33-44.

HK 70a M. Harvey and F.C. Khanna Nucl. Phys. A152 (1970) 588-608.

HK 70b M. Harvey and F.C. Khanna Nucl. Phys. A155 (1970) 337-361.

KHSJ 72 F.C. Khanna, M. Harvey, D.W.L. Sprung and A.M. Jopko, in The Two Body Force in Nuclei", Plenum Press (1972) p. 229

KJ 73 F. C. Khanna and A.M. Jopko unpublished.

KLH 71 F. C. Khanna, H. C. Lee and M. Harvey. Nucl. Phys. A164 (1971) 612-630.

KLS 69 S. Kahana, H. C. Lee and C. K. Scott, Phys. Rev. 180 (1969) 956.

KO 73 T.T.S. Kuo and E. Osnes Nucl. Phys. A205(1973) 1-19.

SZ 70 S. Siegel and L. Zamick Nucl. Phys. A145 (1970) 89-128.

D.W.L. SPRUNG: INFINITE PARTIAL SUMMATIONS

Negele: Are there any counting or Pauli problems in the self-consistent coupled equations?

Sprung: Kirson assures me that there are no overcounting problems in the effective interaction calculation. However, there is the usual neglect of the Pauli principle in solving the RPA equations.

Zamick: I am surprised that you get large screening corrections with the schematic force λDD. This is like a long range Copenhagen force. They use it and show that there are strong RPA correlations. But from time to time they calculate screening corrections and show that they are small - of order $\frac{1}{A}$, and thus justify ignoring them.

Sprung: From (3.36) of the text, you can see that screened RPA (nRPA) is very different from RPA, but in conjunction with the BB vertex, it has only a small additional effect, so nBBRPA = BBRPA. I think this parallel's Kirson's calculation pretty well.

Kümmel: You correctly remarked that three-body effects are important. This completely agrees with the results of our theory. However, there is another point which to me seems to be relevant. The calculations you have described here roughly correspond to solving the Bethe-Faddeev equation with only a few terms of the inhomogenity taken into account and most of them left out. I believe that the other terms (besides RPA with screening, etc.) together are equally important since as often in such a situation cancellations occur. Without them there is no hope for reliable results.

Sprung: The comments you refer to do not appear in the prepared text. I mentioned that Grangé has repeated Dahlblom's three-body calculation in nuclear matter, and in solving for the G-matrix, put in a potential energy for particle states. This is attractive for low lying states and the magnitude is in reasonable accord with Prof. Kümmel's finding that a constant -8 MeV will do. This makes the two-body G-matrix more attractive, as is well known, since the aim is to make G reproduce both the two and three-body binding energy in nuclear matter. This has not been done for finite nuclei, and anyway the additional attraction gained in this way may not be enough to help. Certainly it is better to solve the three-body problem if one can.

PADE APPROXIMANTS AND THE CALCULATION OF EFFECTIVE INTERACTIONS

Thomas H. Schucan +
Swiss Institute for Nuclear Research
CH-5234 Villigen, Switzerland

1. INTRODUCTION

The analytic properties of the effective interaction in nuclei have become increasingly well understood in the last few years. It has been found[1,2] that the corresponding series expansion diverges in most practi-ctical applications due to the occurence of low lying collective states. It is the purpose of this paper to review and discuss an approximation scheme that has been used to rearrange this series with the aim to overcome the difficulties connected with its divergence.

The main features of the divergence problem can be described in a very simple schematic model consisting of two states. In the next section this model is reviewed and it is shown that a continued fraction expansion can be used to calculate that eigenstate that has the larger overlap with the model space. An extension of this method is obtained by the use of Padé approximants (P.A.). These are defined in sect.3 and applied to the effective interaction, to related matrices and to matrix elements in sects. 4-6. The remaining sections are devoted to some mathematical properties of the P.A. that are important in view of these applications.

2. SCHEMATIC MODEL

The simplest model[2,3] which exhibits the main effects of an intruder state on the convergence of a series expansion consists of two states. One of these represents the model-space states, the other one the collective (intruder) states. The Hamiltonian is given by

$$H(x) = H_0 + xV = \begin{pmatrix} 3x & \gamma x \\ \gamma x & 4-3x \end{pmatrix} . \tag{1}$$

The eigenvalues of $H(x)$ are

$$\lambda_{\substack{2\\1}}(x) = 2 \pm \sqrt{(2-3x)^2 + \gamma^2 x^2} , \tag{2a}$$

while the series expansion in x of the model space state is given by

$$\lambda_p(x) = 3x - .25\gamma^2 x^2 - .375\gamma^2 x^3 \pm \ldots \quad (2b)$$

Since we consider a one-dimensional model space, this expression equals the effective interaction $W(x)$. The radius of convergence of this series is determined by the location of the two branch points $\bar{x}_o = 2/(3 \pm i\gamma)$. These eigenvalues are displayed in Fig.1 for $\gamma = .2$.

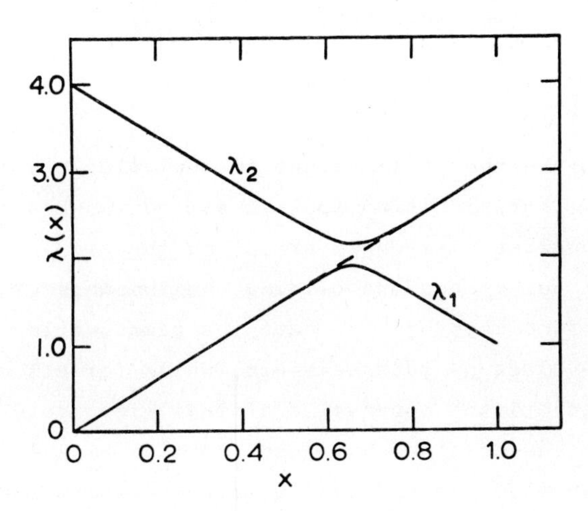

Fig.1 : The eigenvalues of the schematic model. The exact eigenvalues are given by the solid lines while the dashed line represents the perturbation expansion of $\lambda_p(x)$ through third order in x.

We see that $\lambda_p \approx \lambda_1$, for small x, while $\lambda_p \approx \lambda_2$ near x = 1 and that the switch occurs at a crossing point which lies near x = 2/3. On the other hand $W(x)$ is by its definition, always equal to $\lambda_1(x)$[2]. Apparently we need a prescription for reordering the series $W(x)$ in such a way that it approaches λ_2 rather than λ_1, to the right of the crossing point.

This prescription is achieved by expressing $\lambda_p(x)$ in the form of a continued fraction expansion

$$\lambda_p(x) - 3x = \cfrac{\gamma^2 x^2}{6x - 4 + \cfrac{\gamma^2 x^2}{6x - 4 + \cfrac{\gamma^2 x^2}{6x - 4 + \ldots}}} \quad . \quad (3)$$

This is a periodic continued fraction generated by the linear fractional transformation

$$s(w) = \frac{\gamma^2 x^2}{6x - 4 + w}, \tag{4}$$

the fixed points s_1 and s_2 of which are given by the two roots of the quadratic equation $s(w) = w$. Such a continued fraction expansion converges[4], for $\gamma x \neq 0$, if and only if s_1 and s_2 are finite numbers satisfying one of the two conditions

$$\text{I)} \quad s_1 = s_2$$
$$\text{II)} \quad |s_2| > |s_1| \,, \quad s_2 \neq 0 \,. \tag{5}$$

If the expansion converges it converges to the value s_1. In the complex x-plane, the first condition applies at the two branch points while the second applies everywhere except on the arc S of a circle through the branch points and the origin. Furthermore it can be shown that the two eigenfunctions of $H(x)$ have the same overlap with the model space for x-values on this same arc, while for all other values of x, s_1 corresponds to the eigenvalue of that eigenstate which has the larger overlap. Hence, the continued fraction expansion (3) converges to the eigenvalue with maximal overlap (M.O.) whenever the overlaps are different. If the same function $\lambda_p(x)$ is considered in the $1/x$ -plane, the arc S is transformed into a straight line connecting the two branch points.

3. DEFINITION OF PADE APPROXIMANTS

Let $F(z)$ be an operator-valued function given by its Taylor series[5,6]

$$F(z) = \sum_{n=0}^{\infty} A_n z^n . \tag{6}$$

The $[K/L]$ Padé approximant to this function is defined by

$$[K/L]\,F(z) = R_K(z)\, S_L^{-1}(z), \tag{7a}$$

where the operators R_K and S_L are polynomials in z of degree K and L, respectively. These operators are uniquely defined by

$$F(z)\,S_L(z) - R_K(z) = O(z^{K+L+1}), \tag{7b}$$

$$S_L(0) = 1 . \tag{7c}$$

They can be evaluated from the first K+L terms of the power series (6). As an example we give

$$[2/1]\,F(z) = A_0 + z A_1 + z^2 A_2 \left(1 - z A_2^{-1} A_3\right)^{-1} . \tag{8}$$

Although the P.A. defined in this way are obviously designed to reproduce meromorphic functions, they have been used in many fields of physics and applied mathematics to approximate functions with branch points[7]. It might be worthwhile mentioning at this point that the original approximation scheme proposed by Padé[8] was considerably more general and could possibly be adapted to functions with branch points in a more efficient way (I thank D.D.Warner for pointing this out to me).

The result of the Lth step in the continued fraction expansion (3) is equal to $[L+1/L]\,\lambda_p(x)$. The discussion in the previous section thus shows that the series of $[L+1/L]$ approximants to a scalar function with two branch points converges with increasing L for all points x not lying on the arc S. It was noted by Baker[5] that the zeros and poles of diagonal P.A. also accumulate on this arc. Generalizing this to functions with more branch points Baker conjectured that the series of diagonal P.A. converges everywhere outside of a polygon formed by such arcs through each pair of branch points. In the next three sections we find empirically that the domain of convergence seems to be much larger than the one conjectured by Baker. The mathematical convergence properties are further discussed in sect. 7.

4. PADE APPROXIMANTS TO THE EFFECTIVE INTERACTION

Let us now consider the effective interaction W(x) in a model space with dimension M and with an associated projection operator P. The continued fraction expansion in the schematic model can be generalized by forming $[L+1/L]$ Padé approximants to the operator W(x) in this case[9,10]. In analogy to the schematic model we term the results of such an approach successful if the eigenvalues of $PH_oP+W(x)$ agree sufficiently well with those M eigenvalues of the full Hamiltonian $H(x)$, the wave functions of which have the largest overlap (M.O.) with the model space.

For a degenerate model space ($PH_oP = \varepsilon_o$) the first few terms of the power series expansion of $W(x)$ are given by[2,10]

$$W(x) = \sum_{n=1}^{\infty} W_n x^n = x\, PVP + x^2\, PVQDQVP + \\ + x^3 (PVQDQVQDQVP - PVQD^2QVP\,PVP) + \ldots \tag{9}$$

with

$$D = (\varepsilon_o Q - QH_oQ)^{-1}, \tag{9a}$$

$$Q = 1 - P \,. \tag{9b}$$

In applications to realistic cases the first three coefficients of this series are available at the most. From these [2/1] $W(x)$ can be constructed according to eq.(8). The success of such an approximation can be judged by a comparison with the experimental level energies or with exact eigenvalues in solvable models with small dimensions. Since our criterion of success is connected with the model space overlap of the eigenfunctions, we choose the latter method.

It is easy to see that W_2 is singular for all models with a single state in the Q-space. The simplest non-trivial examples are thus given by models with two model-space states and two Q-states. In Fig. 2 the result of such a calculation[9] is shown for the Hamiltonian

$$H(x) = \begin{pmatrix} 1 & 5x & 0 & 5x \\ 5x & 1+25x & 5x & 0 \\ 0 & 5x & 3-5x & x \\ 5x & 0 & x & 3.5-5x \end{pmatrix}. \tag{10}$$

In this figure the exact eigenvalues of $H(x)$ (solidlines) are compared with the results obtained from a perturbation expansion of $W(x)$, truncated after the third order (dotted lines), and with [2/1] $W(x)$ (dashed lines). For $0 \lesssim x \lesssim 0.05$, the eigenvalues obtained with P.A. coincide with the lowest eigenvalues of $H(x)$, while for $0.08 \lesssim x \lesssim 0.3$ they reproduce the lowest and the highest eigenvalue. In each of these intervals the P.A. results approximate the two eigenstates which have the largest overlap with the model space. For $0.3 \lesssim x$, one of the P.A. eigenvalues approximates the largest eigenvalue of $H(x)$ while the other one has a singularity. It is thus shown that the application of a low order P.A. is successful in this example whenever the eigenvalues do not display a strong x-dependence.

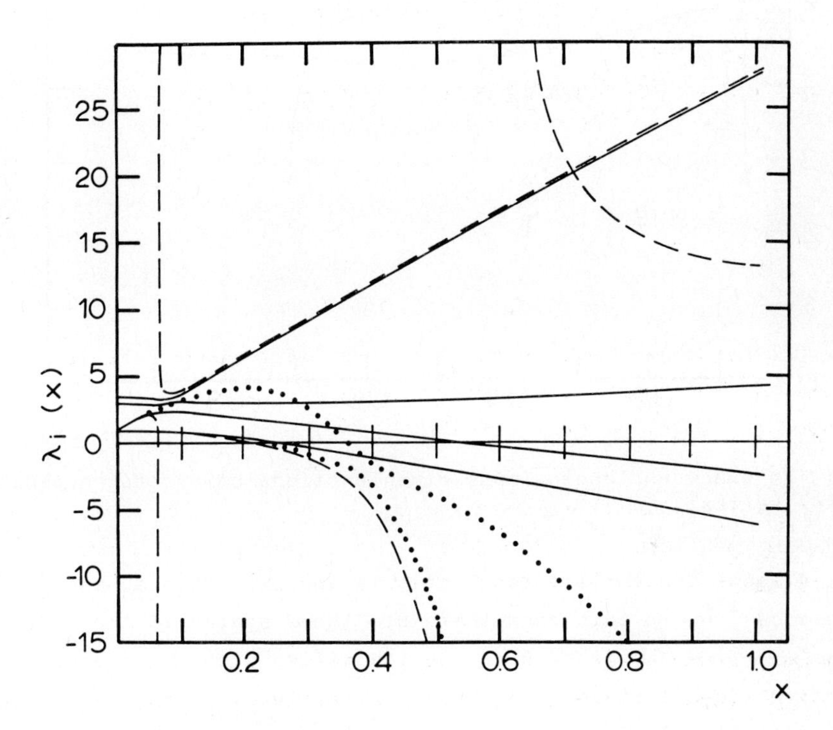

Fig.2 : The exact and approximate eigenvalues as functions of x. See explanations in the text.

However, we note that even near the singularities of the P.A., one of the two eigenvalues provides a good approximation. The same properties of $[2/1]$ $W(x)$ are reflected in the wave functions shown in Fig.3. This figure shows the ratio of the expansion coefficients in the model space. Of the two eigenfunctions of $H(x)$ which have the largest overlap with the model space, the one belonging to the smaller eigenvalue is shown. It is seen that the P.A. result is reasonably successful while the result obtained with the perturbation expansion breaks down sooner than for the corresponding eigenvalue. The detailed discussion of this and similar examples[9] shows that the singularities of the P.A. indicate where some of the branch points of $W(x)$ are and, more important, which levels in the model space are strongly coupled to intruder states.

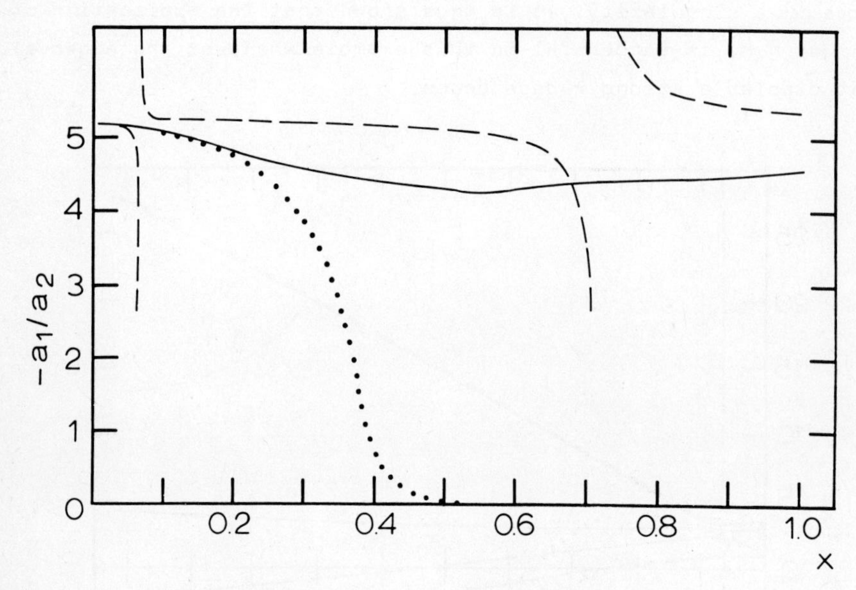

Fig.3 : The exact and approximate eigenfunctions of the eigenvalue defined in the text.

Fig. 4 shows the application[11] of this approximation scheme to the 0^+ levels of ^{18}O. In this case there are three states in the model space. The singularity at $x \approx 0.6$ can be identified with the crossing of a collective (4p-2h) state with the uppermost level in the model space. The identification of the singularity at $x \approx 0.15$ with a level crossing would imply that the collective intruder state lies below the two-particle states at x=1. Moreover, the spectrum of PH_oP is well separated from QH_oQ and it is thus very unlikely that a crossing takes place near the origin. It seems that this pole is spurious in the sense that it is not related to a crossing of levels. The resulting spectrum at x=1 is in good agreement with two of the three lowest experimental eigenvalues (also indicated in the figure), while the collective state is clearly not reproduced.

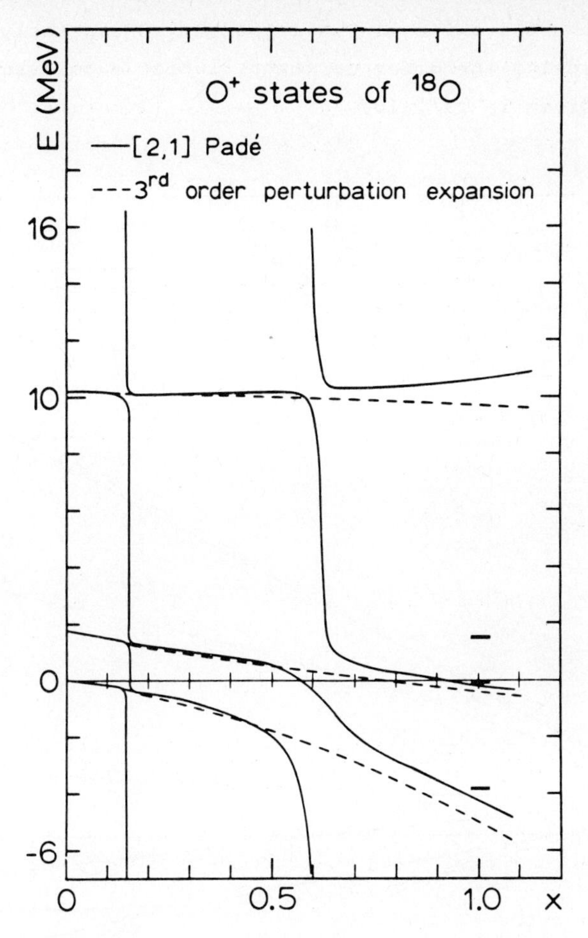

Fig.4 : Comparison of P.A. results with third order perturbation expansion for the 0^+ levels of ^{18}O.

The conditions for the spuriosity of a P.A. pole near the origin have been further investigated in ref.10. It is shown there that a spurious pole of $[2/1]$ W(x) near the origin occurs whenever the coupling of one model-space state to the Q-space is almost linearly dependent on the coupling of the other states in the model space to the Q-space. An example is shown in Fig. 5 for the model Hamiltonian indicated in the figure. It is seen that the pole of $[2/1]$ W(x) near the origin is not associated with a crossing of levels, but it merely reflects that W_2 is almost singular. In the same figure the results obtained with higher order P.A. are also shown. It is seen that their poles lie near crossing points and that the spurious pole of $[2/1]$ W(x) is eliminated in the higher-order approximants. It is also seen that the success of the approximation is increased considerably in this example by increasing the order of the P.A. More examples of higher order P.A. are given in ref. 10.

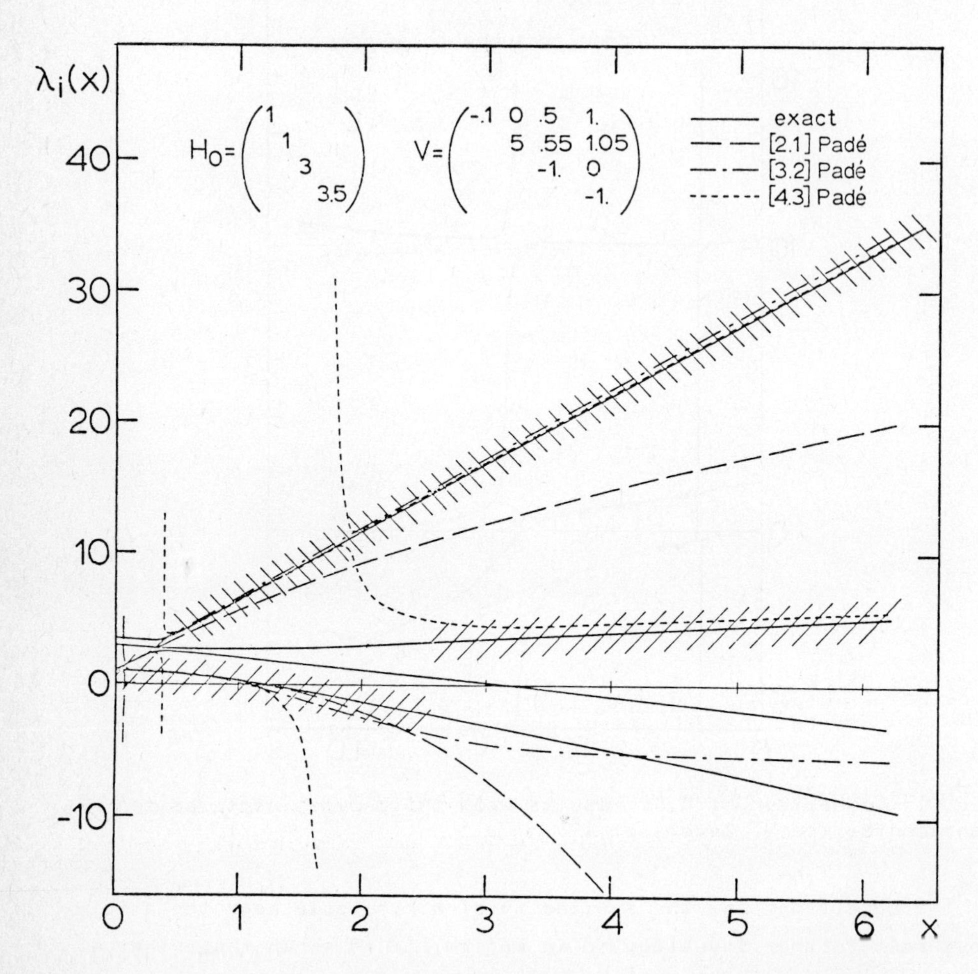

Fig.5 : Exact and approximate eigenvalues of $H(x)$ as functions of x. Note the spurious pole of $[2/1]$ $W(x)$ near the origin.

5. PADE APPROXIMANTS TO RELATED MATRICES

According to the definition (9) the effective interaction $W(x)$ is not Hermitean. While this corresponds to the choice which is commonly used, it has many undesirable features. The first is that the approximated eigenvalues of $PH_0P+W(x)$ can be complex. An example of this is shown in Fig. 6, where the eigenvalues obtained from [2/1] $W(x)$ are complex for $0.70 \leq x \leq 0.77$.

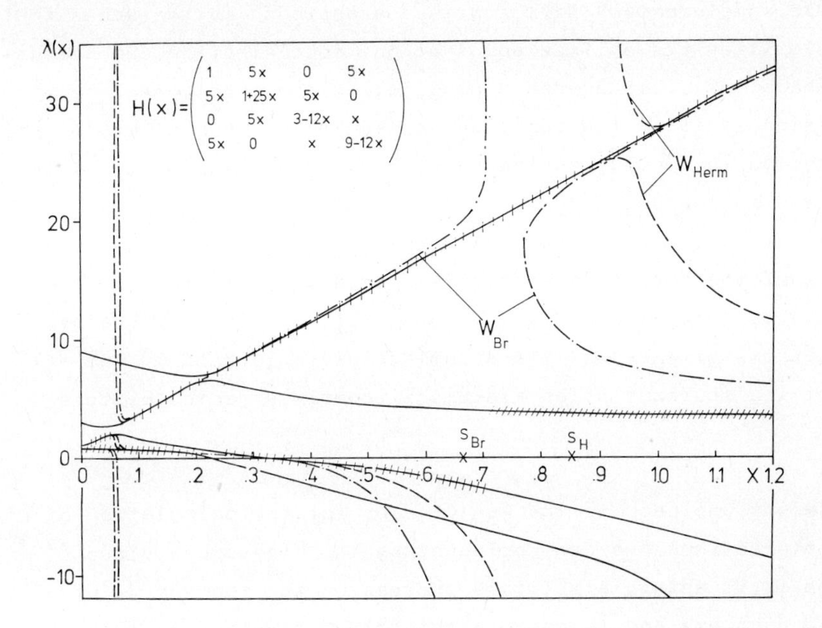

Fig.6 : The exact and approximate eigenvalues as functions of x. The exact eigenvalues are given by the solid lines, while the results obtained with [2/1] $W(x)$ and [2/1] $W_H(x)$ are shown by a dot-dashed and dashed lines, respectively. The two exact M.O. eigenvalues and two of the singularities are also indicated in the figure.

Secondly, the eigenfunctions of $PH_0P+W(x)$ are not orthonormal and can thus not be used to calculate matrix elements of effective operators between different states[12]. Thirdly, P.A. to Hermitean operators are also Hermitean and this facilitates the discussion of their convergence properties considerably.

A prescription for the construction of Hermitean effective interactions can be given[2] in terms of a general operator T and is discussed elsewhere in this conference[13]. The simplest possible choice is T = 0. This choice leads to a Hermitean effective interaction which is, through third order in x, connected with $W(x)$ by

$$W_H(x) = \frac{1}{2} \left[W(x) + W(x)^{\dagger} \right] + O(x^4) . \tag{11}$$

The eigenvalues of $PH_oP+ [2/1] W_H(x)$ are also shown in Fig. 6. It is seen that there are no complex eigenvalues in this case and that the results are complementary to the previous approach in the sense that the singularities are shifted and that one approximation can be successful where the other one has a singularity and vice versa. Apart from the singularities the two methods seem to be equivalent. In particular the approximation to the lower M.O. eigenvalue at $x = 1$ is not improved by the use of W_H.

The use of the Hermitean effective interaction provides a further possibility to check whether a pole of $[2/1] W(x)$ is spurious or not. A reevaluation for $W_H(x)$ of the arguments given in ref. 10 for $W(x)$ shows that the spurious poles are significantly affected by this change.

A different approach to the use of P.A. for the calculation of effective interactions has been proposed by the Stonybrook group[14,15]. Their idea is to obtain a reliable expression for the sum Q_E of all non-folded diagrams and to compute the folded diagrams in terms of this function and its derivatives with respect to the energy. By comparing $[2/1] Q_E(x)$ with the exact values of $Q_E(x)$ in model examples, it is found that the P.A. converge (in a numerical sense) very rapidly. An improved algorythm for the derivation of the folded diagrams from Q_E has been tested[15] in a 14-dimensional model with a 4-dimensional model space. This algorythm used in conjunction with the exact Q_E was found to converge numerically and to reproduce all eigenvalues of the full Hamiltonian, the selection depending on the starting energy. If Q_E is replaced by $[2/1] Q_E(x)$ in the same example, the algorythm is found to be more rapidly convergent. The dependence on the starting energy is significantly changed and not all the eigenvalues are reproduced anymore.

6. PADE APPROXIMANTS TO MATRIX ELEMENTS

The method of P.A. which has been applied to $W(x)$ in sect.4, could also be applied to each matrix element separately. However, the set of $[K/L] W_{ik}(x)$ $(i,k=1,\dots,M)$ has altogether $L\cdot M^2$ poles while $[K/L] W(x)$ has only $L\cdot M$ poles. Hence, the domain of applicability of the former approximation is, in general, more restricted due to the occurrence of a larger number of poles. Moreover, the invariance of $W(x)$ under orthogonal transformations within the model space is respected by P.A. to the matrix $W(x)$ while this is not the case for the matrix formed by the approximants $[K/L]\ W_{ik}$. These assertions are supported by numerical evidence which shows that the matrix approximants give much better results if PVP has large off-diagonal matrix elements or if the coupling to the Q-space is strong[10].

A more elaborate scheme involving P.A. to matrix elements has been suggested by Lee and Pittel[16] to account for a single intruder state strongly coupled to a two-dimensional model space. In this case the two pairs of complex conjugate branch points are decoupled in the effective interaction if the latter is considered in that transformed basis, in which PVP is diagonal. If the orthgonal transformation O is defined by

$$(O\,PVP\,O^{+})_{ik} = \tilde{V}_i\,\delta_{ik}, \tag{12}$$

then each of the two columns of the transformed effective interaction

$$\tilde{W}(x) = O\,W(x)\,O^{T} \tag{13}$$

is dominated by one pair of branch points only. From this fact Lee and Pittel conclude that matrix element P.A. should be calculated to $W(x)$. This conclusion is corroborated by numerical evidence which shows that this method leads to better results than the matrix element P.A. to $W(x)$. In particular the exact effective interaction is better reproduced and the poles of various low order P.A. to the matrix elements of a given column of $W(x)$ seem to lie near the Baker arc through the corresponding two branch points indeed.

Unfortunately, this approximation can not be compared to the results obtained by a matrix P.A. in these cases, since W_2 is singular for any model with a single Q-state. In order to obtain a comparison we have applied the same method to the model shown in Fig. 6. The eigenvalues of $(PH_oP)_{ik} + [2/1]\ W_{ik}(x)$ are displayed and compared with the exact eigenvalues in Fig. 7. The four singularities correspond to the poles of the $[2/1]$ approximants to $W_{21}, W_{11}, W_{12}, W_{22}$, respectively, in ascending order. It is seen that, apart from the singularities, the resulting eigenvalues agree considerably better with the exact results than the ones obtained in Fig. 6 with matrix P.A.

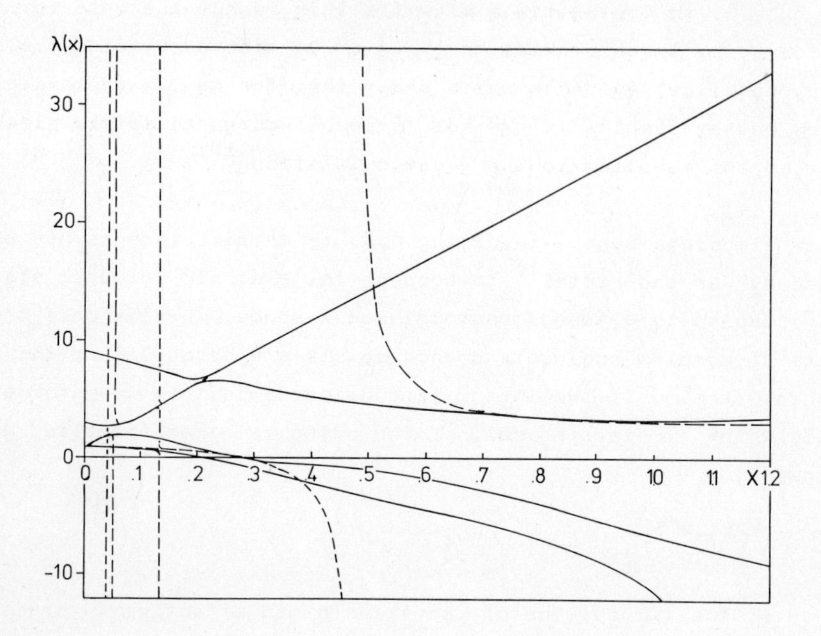

Fig.7 : The exact and approximate eigenvalues of the model defined in Fig.6. The dashed lines represent the results obtained by using the [2/1] P.A. in the transformed basis. Except near the singularities these lines also represent the results obtained with [2/1] $W_H(x)$ and with [2/1] $\lambda_i(x)$ (see text).

of the same order. The repelling of eigenvalues near the singularities is too narrow to be shown in model cases in the figure. The same calculation can also be performed with $W_H(x)$ instead of $W(x)$. In this case the two poles due to the off-diagonal matrix elements are replaced by a double pole at $x = 0.06$. Otherwise, the results are not affected by this substitution. Encouraging as this result looks, we have to keep in mind that the justification for this method has been given for models with one intruder state only. Hence, the present result might be due to a numerical coincidence. This method has thus to be further investigated both in other examples as well as with respect to its mathematical justification.

7. CONVERGENCE OF PADE APPROXIMANTS

It would exceed the scope of this paper to assemble the present knowledge about convergence properties of sequences of Padé approximants. Many theorems and conjectures can be found in the literature[5-7]. Instead I intend to discuss two recent theorems of convergence which seems to me to be sufficiently relevant for our applications as well as representative for the present status of the art. The central notion in both theorems is the transfinite diameter or the capacity of a set of points[17].

Given a bounded closed set E in the complex plane. The transfinite diameter $d_o(E)$ of this set is a measure of how far apart the points of E could get on the average. The maximum of the (geometric) average distance of n points in E is given by

$$d_n(E) = \mathrm{Max}\left\{ \prod_{1\leq j<k\leq n} |z_j - z_k| \right\}^{\frac{2}{n(n-1)}} \qquad (14)$$

It can be shown that the sequence $\{d_n(E)\}$ is decreasing, so that

$$d_o(E) = \lim_{n\to\infty} d_n(E)$$

exists. It is obvious that the transfinite diameter is at most equal to the topological diameter of E and usually considerably less. Thus for a circular disk it equals the radius, for a line segment it is a quarter of the length.

A related notion can be derived in terms of electrostatics. Let the (discrete or continuous) set E of points in the (two-dimensional) plane be furnished with a unit charge. Denoting the distribution of this charge on E by $\rho(x,y)$, we obtain the potential energy

$$I(\rho, E) = \int_E\int_E \log \frac{1}{|\vec{r}-\vec{r}\,'|} \rho(\vec{r})\rho(\vec{r}\,') d^2\vec{r}\, d^2\vec{r}\,' . \qquad (15)$$

Depending on the geometric configuration of E, this energy is either infinite for all possible distributions $\rho(\vec{r})$, or it has a finite minimum

$$V(E) = \inf \quad I(\rho, E) . \tag{16}$$

If the boundary of E is sufficiently smooth, there is a charge distribution $\rho_0(r)$ with

$$I(\rho_0, E) = V(E) . \tag{17}$$

Clearly $\rho_0(\vec{r})$ equals the equilibrium distribution of a unit charge on E, if E is an isolated conductor. Accordingly the quantity

$$\text{cap } E = \exp\left[-V(E)\right] \tag{18}$$

is defined as the (logarithmic) capacity of the set E. It is determined by the geometric configuration of E. For a countable set E the capacity vanishes. If the two variables x,y are represented by a complex variable z, it has been shown[17] that cap E is equal to the transfinite diameter $d_0(E)$ defined above.

Recently Pommerenke[18] and Nuttall[19] have shown that the convergence of sequences of Padé approximants can be described best in terms of convergence in capacity. Both authors discuss the convergence criteria in the (1/z)-plane. Accordingly, they define the [K/L] P.A. to a function f(z), analytic near the point at infinity, by

$$[K/L] f(z) = V_K(z^{-1}) / W_L(z^{-1}), \tag{19}$$

where the polynomials V_K and W_L are defined in analogy to eq. (7). A simplified version of Pommerenkes theorem states that the sequence of diagonal P.A. to a meromorphic function f (z) "converges in capacity" to f(z). This means that, for $\varepsilon > 0, \eta > 0, r > 0$. there exists a number L_0 such that

$$| [L/L] f(z) - f(z) | < \varepsilon^L \qquad (L > L_0) \tag{20}$$

for $|z| \leq r$, $z \in E_L$, where cap $E_L < \eta$. The theorem thus states that the function can be approximated to any desired accuracy for all values of z lying outside of an exceptional set, the capacity of which can be made as small as desired. In its general version the theorem is stated for non-diagonal P.A. and the function f(z) can have a set of essential singularities of zero capacity.

Nuttall[19] has investigated the convergence of diagonal P.A. for a special class of functions with 2ℓ branch points. He considered functions of the type

$$G(z) = \prod_{i=1}^{\ell-1} (z-c_i) \left[\prod_{j=1}^{2\ell} (z-a_j) \right]^{-1/2} , \tag{21}$$

where a_j and c_i are complex parameters. If these parameters satisfy two additional conditions, then an exceptional set S is defined as the union of ℓ finite Jordan arcs, whose endpoints are chosen from the branch points a_j in such a way that each point a_j is the end of one and only one arc in S. The definition of S is made unique by requiring that its capacity is minimal. It is thus uniquely defined by the location of the branch points. The complement S' of S is simply connected. Nuttall's convergence theorem states that the sequence of $[L/L]$ Padé approximants to $G(z)$ converges in capacity as $L \to \infty$ in any closed, bounded region of S'. Nuttall also shows that the two additional conditions on the parameters are rather weak. He speculates that the result might be applicable to a larger class of functions with branch points.

In the case of a function with two branch points, it is easy to show that the set S defined above is equal to the straight line joining the two branch points. We have thus the same convergence domain that we have established before for $[L+1/L]$ $f(z)$ for functions with two branch points (see sect.2). It can also be shown[19] that the set S is contained within any convex polygon that includes all the branch points. Nuttall's result is thus consistent with Baker's conjecture stated in sect.3, but it is much more restrictive.

8. CORRECTION TERMS

In practical applications the knowledge of the perturbation series for the effective interaction is usually limited to three terms. This implies that it is probably much more important to investigate the quality of low order Padé approximants rather than the convergence properties for sequences of P.A. It is natural to ask whether appropriate functions $b_{Li}(x)$ and $c_{Lj}(x)$ can be found such as to ensure

$$[L+i/L]f(x) - b_{Li}(x) \leq f(x) \leq [L+j/L]f(x) + c_{Lj}(x), \quad (22)$$

where i and j are 0 or ± 1, e.g. Correction terms of this kind have been studied[20] for i= -1, j=0 and for functions of the type

$$f(x) = \int_0^\infty \frac{d\varphi(u)}{1+ux}, \quad x \geq 0, \quad (23)$$

where $\varphi(u)$ is of bounded variation and piecewise differentiable on $0 \leq u \leq \infty$. In this case the correction terms have the simple structure

$$b_{L,-1}(x) = x^{2L} b_L / [Q_{L-1/L}(x)]^2, \quad (24)$$

$$c_{L,0}(x) = x^{2L} c_L / [Q_{L/L}(x)]^2,$$

where b_L and c_L are positive constants independent of x. The polynomials Q(x) are equal to the denominators of the respective Padé approximants and hence known quantities. It is an attractive feature of these expressions that the singular behaviour of the P.A. near the poles is compensated by large correction terms.

The constants b_L and c_L depend on the nature of the function f(x). If f(x) is a Stieltjes function, both constants are zero. In other cases their values can be evaluated or estimated if certain additional properties of f(x) are known. Examples of such additional properties are given in ref. 20. It is also shown that such information on f(x) is easily available in many cases. These results can readily be extented to other values of i and j. They can thus also be used to derive upper and lower bounds to f(x) by considering one particular [K/L] Padé approximant and the corresponding correction terms.

The importance of these results for the calculation of P.A. to the effective interaction is quite obvious. It seems to me imperative to concentrate further efforts not so much on convergence properties but much more on the derivation of bounds. The main questions are whether there are bounds of a similar structure for $W(x)$ and which additional properties of $W(x)$ could be used to determine the constants b_L and c_L. The first step in this direction could be taken in the framework of the schematic model outlined in sect.2.

9. CONCLUSIONS

The divergence problem of the effective interaction has been solved completely in the schematic model given in sect.2. In this case an expansion of $W(x)$ has been found which converges to the maximum overlap eigenvalue, whenever the overlaps of the two eigenvalues with the model space are different from each other.

This procedure can be extended to model spaces with larger dimensions with the aid of Padé approximants and I hope to have convinced you that , on the basis of numerical evidence, this method is equally successful. In this context I have discussed applications of P.A. of the type $[L+1/L]$ to $W(x)$ as well as its Hermitean counterpart $W^H(x)$, its matrix elements $W_{ik}(x)$ in a transformed representation or simply to the eigenvalues $\lambda_i(x)$. Often the results obtained by several of these methods complement each other and the best success is obtained by a combination. More work is required to establish the cross-connections and relative merits of these approximations.

The convergence properties of sequences of P.A. are known for meromorphic functions and for a special class of functions with branch points. In both cases it has been shown that P.A. converge in capacity with increasing order. However, the calculation of high order P.A. is not feasible in practical applications, since only the first few terms of the corresponding perturbation series expansion are available. It is thus much more important, from a practical point of view, to search for the best possible approximation on the basis of the given information. The present status of such a search for upper and lower bounds to low order P.A. is outlined in sect.8. It is certainly this aspect which merits the highest attention of those who want to explore further the mathematical properties of Padé approximants in connection with the calculation of effective interactions in nuclei.

ACKNOWLEDGEMENTS

It is a pleasure to thank Drs. G.A.Baker, M.F.Barnsley, D.Bessis, H.M.Hofmann, J.Richert and H.A.Weidenmüller for many stimulating discussions and comments on various aspects of this paper.

REFERENCES

1) T.H.Schucan and H.A.Weidenmüller, Ann.Phys.(N.Y.)73,108 (1972)
2) T.H.Schucan and H.A.Weidenmüller, Ann.Phys.(N.Y.)76,483 (1973)
3) P.J.Ellis and E.Osnes, Phys.Lett.45B, 425 (1973)
4) H.S.Wall, Analytic Theory of Continued Fractions (D.Van Nostrand, New York,1948)
5) G.A.Baker Jr., in Advances in Theoretical Physics, vol.I, (Academic Press, New York, 1965)
6) J.Zinn-Justin, Phys.Rep. 1, 55 (1971)
7) P.Graves-Morris,ed.,Padé Approximants (The Institute of Physics, London and Bristol, 1973)
G.A.Baker Jr., The Essentials of Padé Approximants, (Academic Press, New York, 1974)
8) H.Padé, Thesis, Ann.Ecole Nor. 9, Suppl.,1 (1892)
9) H.M.Hofmann et al., Phys.Lett.45B, 421 (1973) and Ann.Phys.(N.Y.) 85, 410 (1974)
10) H.M.Hofmann, J.Richert and T.H.Schucan, Z.Phys.268, 293 (1974)
11) H.A.Weidenmüller, Intern.Conference on Nuclear Structure and Spectroscopy, Amsterdam 1974, H.P.Blok and A.E.L.Dieperink, editors, Scholars Press, Amstérdam 1974, p.1
12) J.Richert, T.H.Schucan, M.H.Simbel and H.A.Weidenmüller, to be published
13) See contributions of B.H.Brandow and H.A.Weidenmüller to this conference
14) E.M.Krenciglowa et al., Phys.Lett. 47B, 322 (1973)
15) M.R.Anastasio, T.T.S.Kuo and J.B.McGrory, preprint
16) T.-S.H.Lee and S.Pittel, Phys.Lett. 53B, 409 (1975)
17) see e.g. E.Hille, Analytic Function Theory, Vol.2 (Ginn,Boston, 1962)
18) C.Pommerenke, J.Math.Anal.Appl. 41, 775 (1973)
19) J.Nuttall, preprint 1974
20) M.F.Barnsley, J.Math.Phys. 16, 918 (1975)

+ Present address: Physics Department, University of Basel, Klingelbergstr. 82, CH-4056 Basel, Switzerland.

T.H. SCHUCAN: PADE APPROXIMATIONS

Brandow: Suppose one is given a Padé approximant, and you plot up the results as a function of x. Is it a valid rule of thumb to say that in regions where the eigenvalues are behaving in a linear manner, those eigenvalues are trustworthy?

Schucan: Yes, I would consider this a valid rule of thumb.

Vincent: Which of these theorems have been extended to matrix-valued Padé approximants? Have some of them been proved for this more complicated case?

Schucan:Neither of the two theorems have been extended yet. Both are very new and the proofs (given in the references) are very complicated even in the case of special classes of scalar functions.

Pittel: [2/1] approximants have been presented for the case of the realistic ^{18}O $J = 0^+$ matrix elements. The observed pole near x = .5 was then associated with an intruder state-model space state crossing. Since 4p-2h intruder states cannot couple weakly in third-order, how can the [2/1] approximant possibly give information on the intruder-state model space state crossing?

Schucan: Checks with and without the 4p-2h contribution to W_3 have indicated that the crossing does indeed lead to a pole of [2/1]W(x) near x = .5. However, the location of the pole is quite sensitive to the choice of the matrix elements.

Vichniac: I think you should tell us what would be the comment of Cauchy on Nuttall's theorem.

Schucan: "God gives the branch points and man draws the cuts" (Cauchy). It is the merit of Nuttall to have shown how Padé solved this problem for mankind in certain cases.

Barrett: In your one "realistic" example for the J = 0, T = 1 states of ^{18}O, your [2/1] Padé result converged to the lowest two experimental 0^+ states. I thought that the work of Ellis and Engeland indicated that the second 0^+ state in ^{18}O was the deformed, intruder state. How do you understand your result in the light of their work?

Schucan: The result is very sensitive to the matrix elements used in the calculation of the effective interaction. To decide which of the two states is the intruder state, every effort has thus to be made to improve the calculation of these matrix elements.

McCullen: Doesn't Nature tell you which is which? It shouldn't depend on a choice of matrix elements.

Schucan: To resolve this ambiguity, you have to bring Nature to tell you which matrix elements are better than the others.

Lee: Let us suppose that the effective interaction is separated into an intruder-free part and a pole part, i.e.

$$H_{eff} = \mathcal{H}_{pp}^{(Q_2)} + \mathcal{H}_{pQ_1}^{(Q_2)} \frac{1}{E - \mathcal{H}_{Q_1Q_2}^{(Q_2)}} \mathcal{H}_{Q_1P}$$

where the $\mathcal{H}_{pp}^{(Q_2)}$ is the effective interaction free from the intruder state, and hence free from divergence. Do you think that the Padé approximation will be a useful method to calculate $\mathcal{H}_{pp}^{(Q_2)}$?

Schucan: It could very well be. Even within the radius of convergence Padé approximants have been used successfully in many mathematical and physical applications in order to accelerate the convergence. The final answer to this question will be given, if the appropriate correction terms à la Barnsley can be found.

Kümmel: Do I understand you correctly that the convergence of the Padé approximants to eigenvalues with eigenfunctions with largest overlap between model and exact eigenfunctions is still a conjecture?

Schucan: I gave the proof for the schematic model. For model spaces with higher dimensions the result is a conjecture supported by numerical evidence. It remains to be seen whether Nuttall's theorem can be extended to this case.

Kümmel: It is known that this overlap in extended many body systems actually is very small. Does this fact not disturb you?

Schucan: No, it doesn't. There are still M eigenfunctions with larger model space overlap than all the others, at least in a generic sense.

Koltun: It's the only game in town.

Pittel: The problem of approximating V_{eff} in the presence of intruder states is most severe when the crossing takes place very near x = 1. Have you tested the Padé approximation scheme in any such cases?

Schucan: No, if the level crossing near x = 1 is connected with a pole of the [2/1] P.A., at least one eigenvalue has a singularity and is thus not trustworthy. Whether this is the case or not, depends on the coupling strength of the crossing Q-state.

Vichniac: I think Dr. Kümmel refers to the thermodynamic limit, where indeed the unperturbed states become orthogonal to the true states. But in finite nuclei we are still far from the thermodynamic limit.

SHELL MODEL DIAGONALIZATIONS IN AN EXPANDED SPACE

David J. Rowe
Department of Physics
University of Toronto
Toronto, Canada M5S 1A7

I am sure that many people at this conference regard heaven as the ultimate realization of the SM(shell-model) effective interaction. It is appropriate therefore, that the first publication concerning the path to glory should be found in the Old Testament, where the following account of Jacob's remarkable revelation is recorded: "And he (Jacob) dreamed, and behold a ladder set up on earth, and the top of it reached to heaven." (Genesis 28, v12)

But progress in science is slow and it was many years before Kuo and Brown [1], with the help of Brueckner-Bethe-Goldstone theory, were able to tackle the infinite number of steps and come within a step of this lofty goal.

JACOB'S LADDER

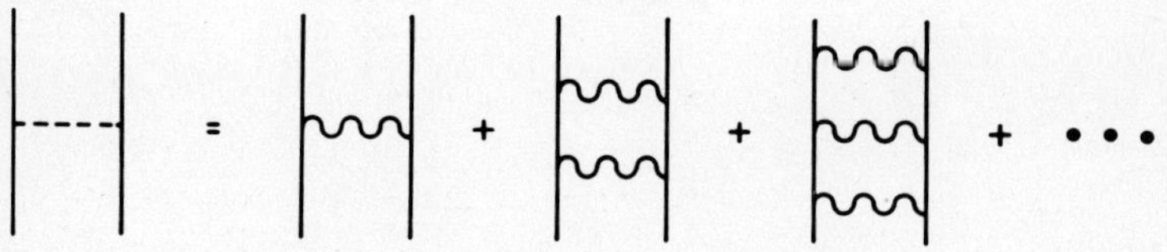

Figure 1

The Brueckner G-matrix --- a single step to heaven!

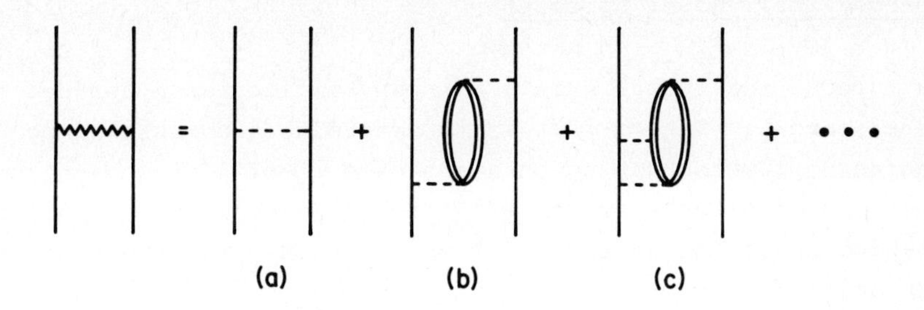

Figure 2

The final step - the SM effective interaction
(a) the bare G-matrix (b) a 2nd. order renormalization due to core excitation; cf. fig. 3 (c) one of the series of valence particle-excited core interactions.

Unfortunately the final step - the calculation of the SM effective interaction - turned out to be also one of infinite complexity; cf. fig. 2. Kuo and Brown tried to make it with a simple bubble (2nd order perturbation theory) and came very close. Since then considerable time and effort have been spent to include higher order diagrams but have come little closer to the experimental data than the first efforts.

There may be many reasons for this. One is the possibility that Kuo's G-matrix elements are unsatisfactory. Another is that many more terms in the perturbation expansion are needed. One would therefore, like some exact calculations, summing the whole series,to resolve such questions. Unfortunately only St. Peter has a computer large enough and only he really knows the correct input data - the two-nucleon interaction, if such exists. What one can do though is set up a model problem which can be solved exactly with the aid of a large computer. Such a model can then be deployed to generate pseudo-experimental data and enable one to examine the convergence of perturbative solutions to the model problem. Furthermore, by making the model as realistic as possible, subject only to the constraint that it be solvable, one can compare its predictions with experiment and hopefully learn something from its shortcomings.

The Expanded Shell Model Approach

We suppose that there exists a known Hamiltonian H defined on a Hilbert space L. The problem is to determine a corresponding Hamiltonian H_{eff} which acting on a subspace L_M of L will have the same spectrum as H for a subset of states. We shall refer to L as the expanded space, L_M as the model space and H_{eff} as the effective Hamiltonian.

H_{eff} can be defined as follows: Let $|\psi_\alpha\rangle$ be an eigenstate of H in L,

$$H\,|\psi_\alpha\rangle = E_\alpha\,|\psi_\alpha\rangle\ ,$$

and $|\phi_\alpha\rangle$ its projection onto L_M

$$|\phi_\alpha\rangle = P\,|\psi_\alpha\rangle\ .$$

If the vector space L_M is N dimensional, it is possible to select N eigenstates $|\psi_\alpha\rangle$ having non-negligible projections onto L_M and which together span L_M. These states are clearly not orthonormal. It is therefore, convenient to define biorthogonal states $\langle\tilde{\phi}_\alpha|$ such that

$$\langle\tilde{\phi}_\alpha\,|\,\phi_\beta\rangle = \delta_{\alpha\beta} \qquad \alpha,\beta = 1,N.$$

The effective Hamiltonian is then defined

$$H_{eff} = \sum_{\alpha=1}^{N} |\phi_\alpha\rangle\, E_\alpha\, \langle\tilde{\phi}_\alpha|$$

which clearly has the required property

$$H_{eff}\,|\phi_\alpha\rangle = E_\alpha\,|\phi_\alpha\rangle \qquad \alpha = 1,N.$$

If the N eigenstates $|\psi_\alpha\rangle$ are chosen to be the N lowest energy states whose projected states $|\phi_\alpha\rangle$ together span L_M then, in the limit that L is the full infinite dimensional Hilbert space of nuclear states, the above definition of H_{eff} becomes identical to that given by the folded linked cluster expansions of Brandow [2] and of Kuo-Lee-Ratcliffe [3]. As in the linked cluster expansion, the eigenstates $|\phi_\alpha>$ are not in general orthogonal and consequently H_{eff} cannot, in general, be hermitian.

Effective transition operators are defined in a parallel way.

Application to the sd-shell

For the purposes of this review we shall consider primarily a model space L_M consisting of valence particle states belonging to (2s,1d) configurations outside of an inert ^{16}O closed-shell core. The expanded space L will contain, in addition, configurations having an excited ^{16}O core. For the two-particle problem, for example, we can express this algebraically

$$L_M = (sd)^2$$

$$L = (sd)^2 \oplus \sum_\lambda (sd)^2 \lambda$$

where λ indicates a core excited state, which may be a simple $2\hbar\omega$ 1p-1h (1 particle - 1 hole) state [4] or a TDA (Tamm-Dancoff Approximation) or RPA (Random Phase Approximation) phonon [5].

Some of the diagrams coupling the two valence shell particles to a core excitation are illustrated in fig. 2 and the diagrams which may be included in the description of a core excitation are illustrated in fig. 3.

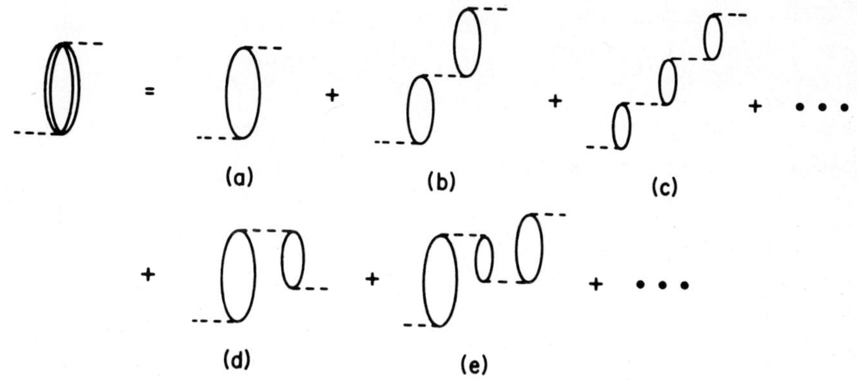

(a) (b) (c)

(d) (e)

Figure 3

Core excitation diagrams. (a) is a simple 1p-1h excitation (a bubble). The forward iterated sequence of bubbles, (a),(b),(c) ..., add up to TDA core excitations, while the complete set of forward and backward iterated sequences add up to RPA excitations.

We adopt the following notation: shell model calculations in which the basis for core excitations are the 1p-1h states will be denoted SM; those for which the core states are TDA states by TDASM; and those with RPA states RPASM. These latter calculations, in which the core excitations are determined in a preliminary calculation, are usually referred to as CPC (core-particle coupling) calculations and are discussed in detail in the papers of, for example, LRW II [5] and Goode and Siegel [6]. Since the 1p-1h states and the TDA states span the same core space, SM and TDASM results should be identical. However, they may differ in practice because, with the addition of the valence particles, the TDASM states become slightly non-orthogonal and overcomplete and, in the calculations reported here, this was not taken into account, although it could and ideally should have been. The same is true of the RPASM calculations. However, all basis states were properly anti-symmetrized, normalized and coupled to good angular momentum and isospin.

The largest dimension of L encountered in the above SM calculations is $\sim$ 900 which is about as large as we (at Toronto) have been willing to handle, although with the techniques of the Glasgow group one can now consider much larger L. For example, Watt, Cole and Whitehead [7] (WCW) included all $2\hbar\omega$ excitations in their expanded space.

By considering the 0, 1 and 2 particle systems, we obtain the first three terms in the expansion

$$H_{eff} = H^{0}_{eff} + H^{1}_{eff} + H^{2}_{eff} + \cdots$$

where H^{0}_{eff} gives the ^{16}O ground state binding energy, which we take as reference point and hence define to be zero, H^{1}_{eff} is a 1-body Hamiltonian giving the lowlying $5/2^+$, $1/2^+$ and $3/2^+$ levels in ^{17}O and ^{17}F, and H^{2}_{eff} is the effective 2-body interaction which we are primarily interested in calculating and which together with H^{1}_{eff} should give the lowlying A = 18 spectra. To obtain the 3-body component of the effective Hamiltonian we would also have to consider the 3 particle system, as indeed Barrett, Halbert and McGrory [9] have done for smaller L_M and L.

We now consider the Hamiltonian H, expressed as the sum of a 1-body part H_o and a residual interaction V

$$H = H_o + V.$$

There are four popular choices of H_o:

(i) is a simple harmonic oscillator Hamiltonian, for both valence particles and core excitations. A particle-hole core excitation then has unperturbed energy

$$\varepsilon_p - \varepsilon_h = n\hbar\omega , \qquad n \text{ integer}$$

and all valence particle states are degenerate. We shall label results calculated with this choice of H_o by the subscript ($\hbar\omega$;deg). It is the simplest and hence most popular choice for perturbative calculations but the least realistic.

(ii) is to base H_o directly on experimental energy levels. Thus H_o is chosen such that H^1_{eff} will give precisely the experimental energies for the active (*i.e.* valence) particle states. This means that the single-particle renormalizations, illustrated diagrammatically in fig. 4, must be subtracted from the A = 17 experimental energies to give the unrenormalized single-particle

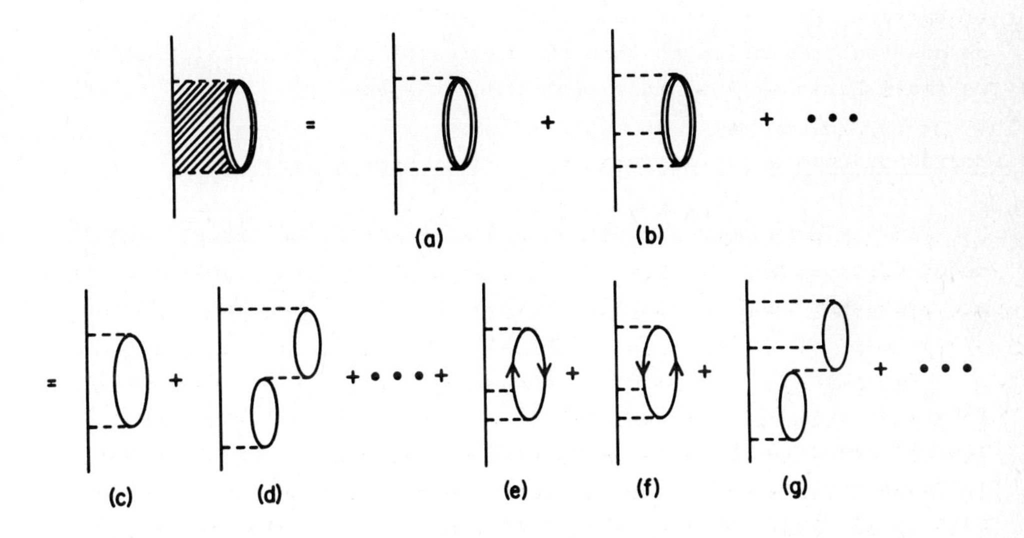

Figure 4

Diagrams which contribute to the renormalized single-particle energies.

Hamiltonian H_o appropriate for the extended space. The energies for the passive particle and hole states can be taken directly from experiment. We shall label results for this choice by the subscript (exp). This choice of H_o is the most realistic and is in accord with the folded linked-cluster theory.

(iii) is a hybrid, whereby the core excitations are described by a harmonic oscillator H_o but the valence-particle energies are derived as in (ii). Results for this choice will be labelled by the subscript ($\hbar\omega$).

(iv) is to calculate H_o from first principles by adding the interaction energy of each particle with the closed-shell ^{16}O core to its kinetic energy as in a Hartree-Fock calculation. This is the procedure followed by WCW [7] and will be labelled by their initials. It is the most fundamental choice, or would be if the single-particle basis were Hartree-Fock rather than harmonic oscillator, but it is also very demanding on the model in the sense that its success at predicting H^2_{eff} depends very much on its success at predicting H^1_{eff}, which in (ii) is taken from experiment.

Finally the matrix elements of the 2-body interaction V are equated with the G-matrix elements of Kuo . The G-matrix elements of Barrett, Hewitt and McCarthy [9] have also been used for this purpose but, in order to compare results of different calculations, we shall here consider only Kuo's matrix elements.

Correspondence with the folded-linked-cluster expansion

In making a correspondence between a so-called 'exact' shell-model diagonalization and the folded-linked-cluster expansion, there are important fundamental differences that should be kept in mind.

First of all the extended shell model space L is necessarily a finite subspace of the full infinite dimensional Hilbert space of nuclear states. The corresponding linked-cluster expansion must therefore only contain intermediate states belonging to L. This much is obvious. However, it is easy to overlook diagrams which cancel out of the linked-cluster expansion in the full Hilbert space but which are not fully cancelled for any finite subspace. The prime offenders, as first pointed out by Goode [10], are the disconnected

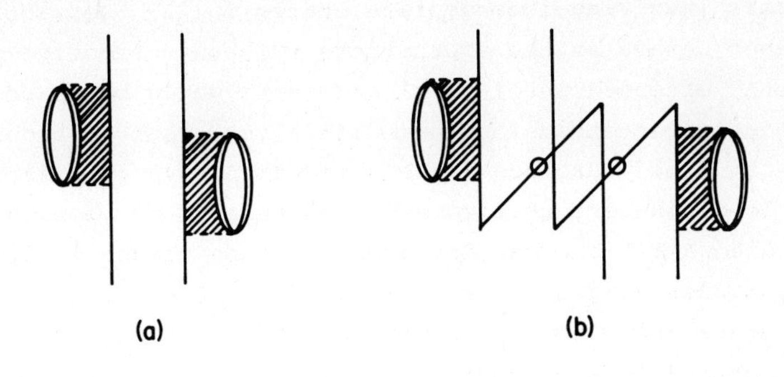

Figure 5

Disconnected diagrams which exactly cancel only in the full infinite-dimensional linked-cluster expansion.

diagrams of Fig. 5. Diagram 5(b) is a folded diagram. Such diagrams appear in the linked-cluster expansion in factoring out unlinked diagrams for which the intermediate states are in the model space. Thus diagram 5(b) is generated in the shell-model diagonalization. Now, in the full diagrammatic expansion, 5(b) is exactly cancelled by 5(a) and thus neither of these disconnected diagrams appears explicitly. However, for a finite space of intermediate states, diagram 5(a) may not be present to effect the cancellation. In particular, it is not present in the sd-shell model problem, presented above, because it contains an intermediate state in which the core is doubly excited and which is not in L. Thus the disconnected folded diagrams of the type shown in fig. 5(b) remain uncancelled in the diagrammatic expansion corresponding to this particular shell-model diagonalization.

A second point of departure concerns the meaning of an interaction line. In extended shell-model calculations we assume, because it would be very difficult to do otherwise, that V in the Hamiltonian

$$H = H_o + V$$

is a non-singular two-body interaction. However, we may go on to equate the matrix elements of V with those of a G-matrix, which is strictly a reaction matrix and not a two-body interaction; *i.e.* G is already an infinite sum of ladder diagrams. Ideally of course,

G would be designed for the extended shell-model space and would exclude ladders involving intermediate states in L. When acting on L it would then behave as the appropriate two-body effective interaction for the extended shell-model and there would be no problem. In the linked-cluster expansion there is in any event no problem because one can simply exclude ladder diagrams, such as diagram (e) of fig. 4. In a non-perturbative approach this is not possible. Thus in the diagrammatic expansion corresponding to the shell-model calculation, ladder diagrams must be included. Depending on what G-matrix elements one uses, there may therefore, be some double-counting in the model.

Certainly the expanded shell-model procedure has its limitations but it does present an exactly solvable model problem and one for which there is a well-defined diagrammatic expansion. It should also be remembered that one can always add or subtract diagrams from the shell-model subset.

Some results

Some calculated $J^\pi = 0^+$ $T = 1$ matrix elements of the 2-body effective interaction are given in table 1. Also the lowlying spectra of the mass 18 nuclei, together with the theoretical predictions are shown in figs. 6 - 9.

	ref.	(4,4)	(5,5)	(6,6)	(4,5)	(5,4)	(4,6)	(6,4)	(5,6)	(6,5)
G	1	-1.236	-2.049	-0.087	-0.626	-0.626	-3.025	-3.025	-0.526	-0.526
PT2(exp)	4	-1.725	-2.027	-0.299	-0.816	-0.816	-3.441	-3.441	-0.630	-0.630
SM(exp)	4	-1.569	-1.980	-0.373	-0.863	-0.834	-2.951	-3.033	-0.627	-0.613
PT2($\hbar\omega$)	5	-1.991	-2.004	-0.409	-0.893	-0.893	-3.606	-3.606	-0.673	-0.673
PT3($\hbar\omega$)	11	-1.720	-1.873	-0.460	-1.011	-1.003	-3.065	-2.987	-0.664	-0.669
SM($\hbar\omega$)	12	-1.541	-1.963	-0.317	-1.028	-0.817	-3.169	-2.818	-1.069	-0.608
TDASM($\hbar\omega$)	5	-1.743	-1.935	-0.154	-0.962	-0.853	-2.733	-2.888	-0.700	-0.647
RPASM($\hbar\omega$)	5	-2.010	-2.752	-0.450	-1.108	-0.725	-2.777	-2.880	-0.997	-0.593
SM($\hbar\omega$;deg)	12	-1.107	-1.352	-0.098	-0.919	-0.915	-2.975	-2.967	-0.568	-0.626

$4 = 1d_{5/2}$, $5 = 2s_{1/2}$, $6 = 1d_{5/2}$; $(a^2 J{=}0 T{=}1 | V_{eff} | b^2 J{=}0 T{=}1)$ is denoted (a,b)

TABLE 1

$J^\pi = 0^+$ $T = 1$ matrix elements of the 2-body effective interaction.

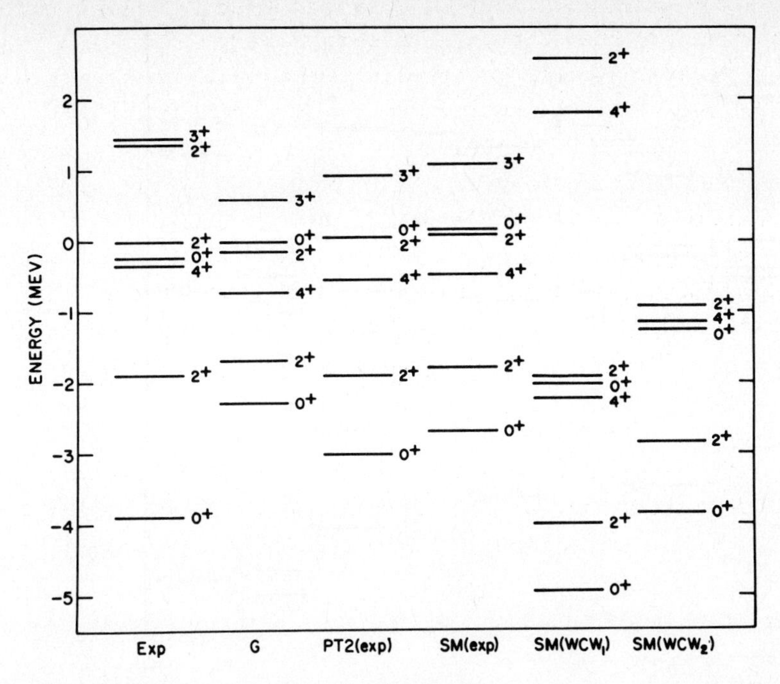

Figure 6

Experimental and calculated T=1 spectra for ^{18}O. WCW_1 and WCW_2 refer respectively to the 1p-1h and all $2\hbar\omega$ calculations of Watt Cole and Whitehead. The other labels are explained in the text.

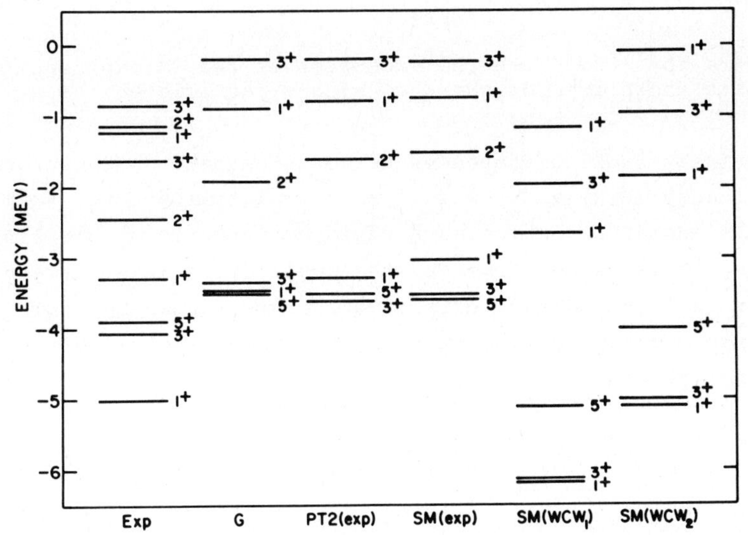

Figure 7

Experimental and calculated T=0 spectra for ^{18}F; cf. caption to fig. 6.

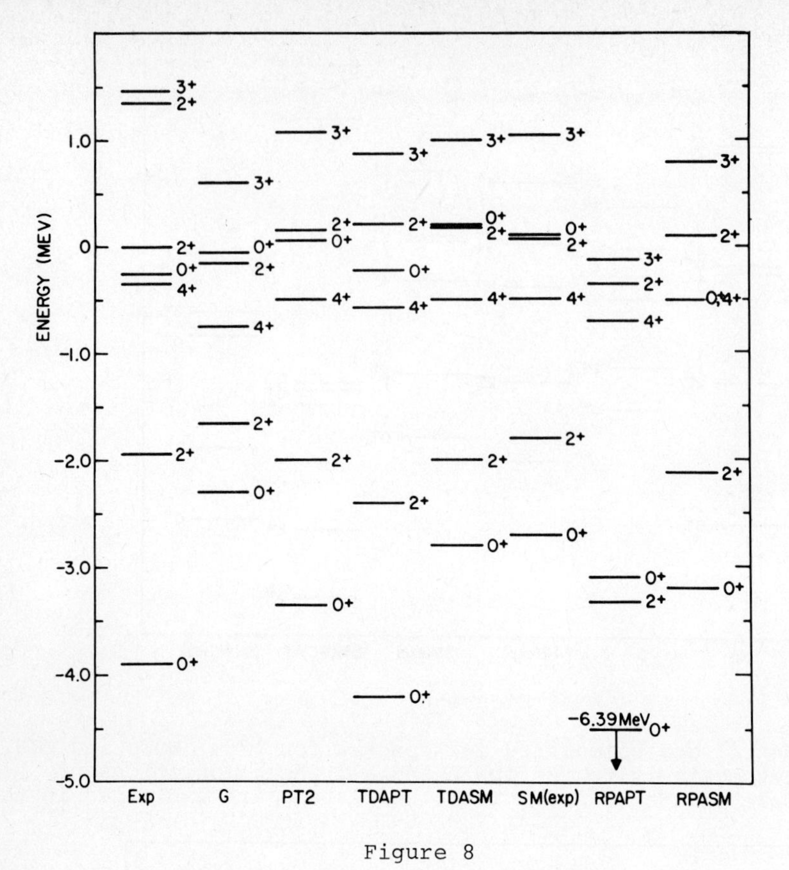

Figure 8

T=1 spectra for ^{18}O. With the exception of the SM(exp) spectrum, all results shown are for the ($\hbar\omega$) single-particle energies.

The relatively small differences between the bare G, 2nd order perturbation theory (PT2) and the full SM calculation, for the (exp) single-particle energies, indicates a rapid convergence of the linked-cluster series. This is confirmed by the small difference between the zero order (e^0) and SM(exp) effective charges, shown in table 2, and by the large overlaps of the expanded SM eigenstates with the model space (not shown).

For the ($\hbar\omega$) results we can compare the bare G (1st order), 2nd (PT2), 3rd order (PT3) and the all orders SM matrix elements, given for $J^\pi = 0^+$, T = 1 in table 1. This sequence is far less indicative of a convergent perturbative expansion than the (exp) results. In fact, it is found that the 1st, 2nd and 4th SM($\hbar\omega$) eigenstates have the largest overlap with the model space, rather than the lowest three,

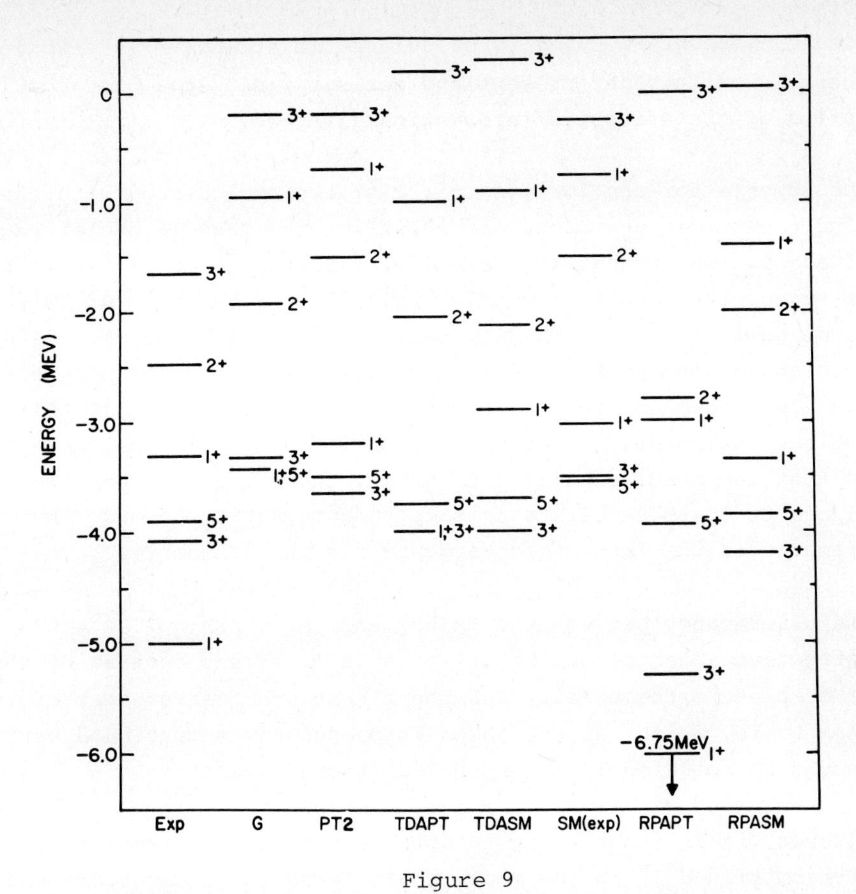

Figure 9

T=0 spectra for ^{18}O; cf. caption to fig. 8.

	e^0	PT1(ħω)	PT2(ħω)	TDAPT(ħω)	SM(ħω)	SM(exp)	exp
ref.		13	13	14	4	4	6
P	1	1.119	1.050	1.242	1.071	1.062	1.75
N	0	0.269	0.315	0.362	0.376	0.245	0.54

TABLE 2

A = 17 effective charge for the $5/2^+ \rightarrow 1/2^+$ E2 transition and which are therefore used to calculate the effective interaction. Thus there is an intruder present and not surprisingly a perturbative calculation of the effective interaction diverges.

The spectra for the lowlying J^{π}, T states, calculated with the ($\hbar\omega$) single particle energies, are shown for the mass 18 nuclei in figs. 8 and 9. Unfortunately the SM($\hbar\omega$) results are not shown in the figures, since the results are only available for the $J^{\pi} = 0^{+}$, T = 1 states, given in table 3. The nominally equivalent TDASM ($\hbar\omega$) results are shown and again exhibit the lack of convergence of the perturbative series. Also shown are the 2nd order perturbative results (TDAPT) in which the intermediate core excitation is treated in the TDA. It is seen that the TDAPT results are in many cases even further from the all orders TDASM results than simple perturbation theory. Even more dramatic is the lack of convergence for the RPA series.

The discrepancy between the SM($\hbar\omega$) and the TDASM($\hbar\omega$) $J^{\pi} = 0^{+}$, T = 1 effective interaction, shown in table 1, arises because of the lack of complete orthogonality and the slight overcompleteness of the TDASM basis, which was not taken into account as mentioned earlier. This should be remedied in future calculations.

However, it is interesting to note that a diagonalization of the non-hermitian H_{eff} in the model space leads to eigenvectors and eigenvalues for SM($\hbar\omega$) and TDASM($\hbar\omega$) which are extremely similar. The eigenvalues and minimal overlaps of eigenvectors with the SM($\hbar\omega$)

	E_1	E_2	E_3	Overlaps with corresponding SM($\hbar\omega$) eigenvectors
SM(exp)	-2.70	0.13	10.54	>0.999
SM($\hbar\omega$)	-2.75	0.22	10.61	1.000
SM($\hbar\omega$;deg)	-2.24	0.78	11.00	>0.998
TDASM($\hbar\omega$)	-2.79	0.20	10.66	>0.999
RPASM($\hbar\omega$)	-3.18	-0.51	10.38	>0.996

TABLE 3

The eigenvalues (in Mev) resulting from diagonalizing some of the $J^{\pi} = 0^{+}$ T = 1 effective interactions with experimental single-particle energies.

eigenvectors are shown, for several of the above effective interactions, in table 3. The reason for the close similarity in the mass 18 spectra is due to the dominance of H^1_{eff} over H^2_{eff} and in all cases the former was taken from experiment. Larger differences would presumably show up for A>18. Table 3 does emphasize, however, the sensitivity of the effective interaction to the details of the A = 18 spectrum and hence draws attention to the need for considerable accuracy in this part of SM or CPC calculations, *i.e.* a 10% error in the spectrum can lead to say a 40% error in the effective interaction matrix elements.

Finally, we mention the SM($\hbar\omega$;deg) results. Starkland and Kirson [12] have pointed out that neither the (exp) or ($\hbar\omega$) calculations are really directly comparable with PT2($\hbar\omega$) or PT3($\hbar\omega$) which employ pure harmonic oscillator energy denominators. Their SM($\hbar\omega$;deg) effective interaction, shown for $J^\pi = 0^+$, T = 1 in table 1 would appear to exhibit even more strongly the lack of convergence of the perturbative expansion. However in this case there is no intruder present.

The SM($\hbar\omega$;deg) calculation, although appropriate for comparisons with the presently available perturbative results, is somewhat anomalous. For no one would begin to believe the SM($\hbar\omega$;deg) spectrum, since as we have already observed, the single-particle energies of the valence particles play such a dominant role. (Note that the $J^\pi = 0^+$, T = 1 spectrum given in table 3 corresponds to using the SM($\hbar\omega$;deg) effective interaction with experimental single-particle energies and it is therefore, not the spectrum generated in the expanded SM($\hbar\omega$;deg) diagonalization.) Such anomalies would be obviated if experimental rather than harmonic oscillator single-particle energies were used in perturbative calculations.

Are the expanded shell-model calculations realistic?

In spite of the interest of the ($\hbar\omega$) calculations , as regards questions of convergence, intruder states and comparisons with perturbative calculations, they can hardly be considered realistic. For example an ($\hbar\omega$) calculation predicts the monopole 0^+ T=0 excitation of ^{16}O to be at 10 MeV and 5 MeV, respectively, in TDA and RPA. These values are much lower than the 25 MeV and 24 MeV obtained with (exp) single-particle energies and not surprisingly result in excessive core polarization in an effective interaction or effective charge calculation.

What about the (exp) calculations? Comparison of the spectra, in figs. 6-7, with experiment indicates that a more attractive effective interaction is needed. The question is then posed: "to what do we attribute the discrepancy between the SM(exp) results and experiment?"

Consider first the imperfections and misuse of the G matrix. An estimate of the effects of double counting ladder diagrams [15] indicates that the SM effective interaction may be slightly too attractive. More serious, however, is the fact that the G-matrix is calculated in a harmonic oscillator rather than a Hartree-Fock basis. As a result the monopole core polarization contributions may be substantially overestimated as demonstrated by the calculations of Ellis and Osnes [16] and Rowe [17]. These errors are of course, in addition to others due to approximations introduced in calculating the G-matrix.

It is known from CPC calculations that the core excitations which contribute most to the renormalization of the effective interactions and charges are the T = 0 0^+ and 2^+ excitations. We have already remarked that the monopole core polarization may be excessive. What about the quadrupole?

It is well-known that many major shells are needed to realistically describe nuclear deformations and rotations. The small admixtures of higher shells in, for example, unrestricted Hartree-Fock calculations may be small but they can also be coherent and build up large nuclear quadrupole moments. Thus they are expected to be important also for the dynamic core polarizations that renormalize effective interactions and charges. Certainly the failure of the SM(exp) calculations to reproduce the E2 effective charges indicates that the quadrupole core polarization has been very inadequately treated.

Unfortunately, the inclusion of higher configurations in a shell model calculation rapidly renders the dimensions prohibitive An interesting qualitative approach to the problem has recently been proposed by Harvey [18]. The idea is to exploit a knowledge of the E2 effective charges to learn about the renormalizations of the effective interaction due to quadrupole core polarization.

However, ultimately there is a need for detailed realistic microscopic calculations in which each phase of the calculation is pushed to its limits. It will require reliable and accurate Brueckner -Hartree-Fock G matrix elements and perturbative and SM calculations in still larger spaces. But certainly there is little virtue in pursuing divergent perturbation expansions to higher and higher orders.

Thus it seems that we are still stumbling over the rungs of the ladder and it will be a while yet before we can all rest in peace.

References

[1] T.T.S. Kuo and G.E. Brown, Nucl. Phys. 85 (1966)40; T.T.S. Kuo, Nucl. Phys. A103 (1967) 771.

[2] B.H. Brandow, Rev. Mod. Phys. 39 (1967) 771.

[3] T.T.S. Kuo, S.Y. Lee and K.F. Ratcliff, Nucl. Phys. A176 (1971) 172.

[4] N. Lo Iudice, D.J. Rowe and S.S.M. Wong, Nucl. Phys. A219 (1974) 171; Phys. Letts. 37B (1971) 44 (Herein referred to as LRW I).

[5] N. Lo Iudice, D.J. Rowe and S.S.M. Wong, Nucl. Phys. (in press) (Herein referred to as LRW II).

[6] P. Goode and S. Siegel, Phys. Letts. 31B (1970) 418.

[7] A. Watt, B.J. Cole and R.R. Whitehead, Phys. Letts. 51B (1974) 435.

[8] B.R. Barrett, E.C. Halbert and J.B. McGrory, "Effective three-body forces in truncated shell-model calculations" (preprint).

[9] B.R. Barrett, R.G.L. Hewitt and R.J. McCarthy, Phys. Rev. C3 (1971) 1137.

[10] P. Goode, Nucl. Phys. A241 (1975) 311.

[11] B.R. Barrett and M.W. Kirson, Nucl. Phys. A148 (1970) 145.

[12] Y. Starkland and M.W. Kirson, Phys. Letts. 55B (1975) 125.

[13] P.J. Ellis and S. Siegel, Phys. Letts. 34B (1971) 177.

[14] S. Siegel and L. Zamick, Nucl. Phys. A145 (1970) 89.

[15] M.W. Kirson, Phys. Letts. 32B (1970) 399.

[16] P.J. Ellis and E. Osnes, Phys. Letts. 41B (1969) 97.

[17] D.J. Rowe, Phys. Letts. 44B (1973) 155.

[18] M. Harvey, Independent particle description of collective motion, "Enrico Fermi" Summer School lectures, Varenna, 1974.

D.J. ROWE: SHELL MODEL DIAGONALIZATIONS IN AN EXPANDED SPACE

Pittel: To what extent might the discrepancy between your calculated effective charges and the experimental effective charges be due to your neglect of ground state correlations (i.e. 3p-2h correlations)?

Rowe: I am not sure. One needs to do the calculation. However, one does have some indication from the RPASM results, since the RPA does take account, in an approximate way, of the 2p-2h ground state correlations in core transitions. It is known that RPA tends to overestimate ground state correlation effects so an explicit inclusion of 3p-2h configurations will probably contribute less to the enhancement of the effective charge than found in RPASM.

Beck: How do you treat the spurious admixtures to the states in the expanded space shell model? Do you consider them as belonging to the "pseudo-experimental results"?

Rowe: The one-phonon spurious states do not participate in the problem because of their negative parity. The two-phonon spurious states are primarily 2p-2h states and therefore outside of our expanded space. We therefore ignored them. The justification for this has since been given by Watt, Cole and Whitehead who eliminated the small 1p-1h spurious center-of-mass components from their expanded space and found that it made negligible difference.

Manakos: I think that spurious admixtures are important for the monopole 1p-1h states of ^{16}O. If they are removed then the lowest 0^+ T = 0 state is shifted up by 4 MeV or more as far as I remember. On the other hand, RPA will make matters worse, if one would use self-consistent one-particle energies and wave functions, since according to the Thouless theorem spurious states would then appear at zero energy.

Rowe: It is not clear to me that RPA will make matters worse in this respect. After all the theorem you refer to says that the spurious center-of-mass mode should decouple exactly from the non-spurious modes. The RPA, of course, does not say anything directly about two-phonon states such as the coupling of spurious two-phonon and non-spurious one-phonon states. I am surprised by the magnitude of the shifts you mention since, as I said before, the spurious components in the space are very small.

Ellis: Could I point out that the experimental third 0^+ level at about 5 MeV in ^{18}O was omitted from your figures? I would also like to point out that there is a great deal of evidence from phenomenological calculations that it is the third 0^+ level which is mostly of a two-particle structure whereas the second 0^+ is mainly a four particle-two hole state. As far as I know the only way to shift the second calculated 0^+ up towards 5 MeV is to calculate the bare G-matrix in a Woods-Saxon or Hartree-Fock basis.

Rowe: Thank you for pointing out the omission. Both comments are clearly pertinent.

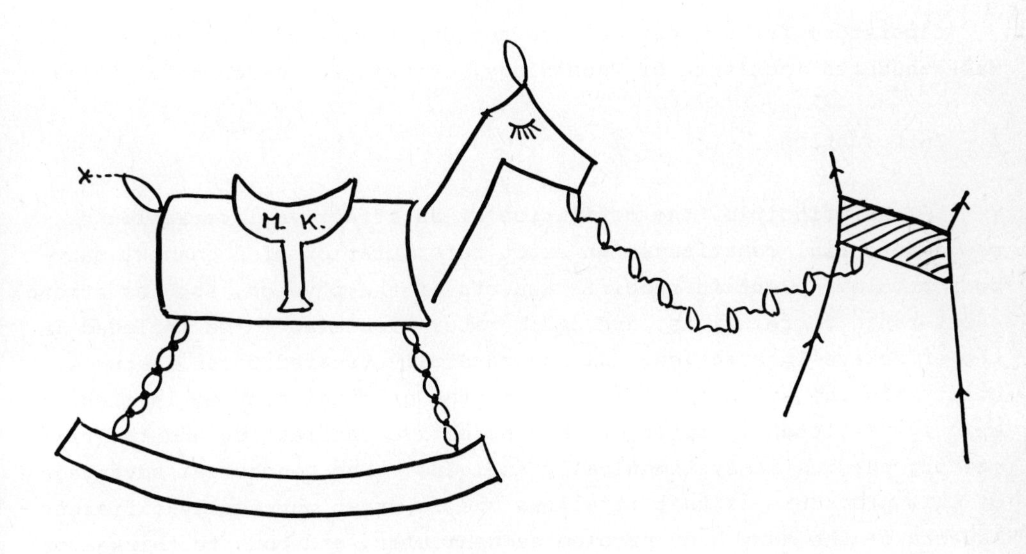

"DENSITY DEPENDENCE IS SIMPLY APPROXIMATING G"

DENSITY DEPENDENT INTERACTIONS*

J. W. Negele+

Laboratory for Nuclear Science and Department of Physics,
Massachusetts Institute of Technology, Cambridge, Massachusetts 02139

I. Introduction

In principle, the definition of an effective interaction or pseudopotential constitutes an exact reformulation of a quantal many-body problem. Certain specific aspects of the physics, such as strong short-range correlations, are deliberately isolated to be included in the effective interaction. Having carefully treated those features built into the effective interaction, the original problem is then exactly rewritten in terms of this effective interaction and the remaining physics is systematically included. The conceptual advantage of this procedure is that it allows one to treat physically different aspects of the many-body problem sequentially, and thus to think about them one at a time.

In practice, the exact partitioning of physics between the effective interaction and the rest of the problem is never carried out. Rather, almost all of the real many-body problem is thrown away in order to arrive at a computationally managable theory. Thus, one may seek to define an effective interaction such that a shell model calculation in a specified model space should approximate certain nuclear states, such that the Hartree Fock (HF) approximation should describe gross ground state properties, such that the random phase approximation (RPA) should approximate collective excited states, or such that

---

* Work supported in part through funds provided by the Energy Research and Development Administration under Contract AT (11-1)3069.

+ Alfred P. Sloan Foundation Research Fellow.

the time dependent Hartree Fock (TDHF) approximation should describe the time evolution of a certain class of wave functions.

In recent years, density dependent interactions have been both used and abused as effective interactions for such applications. Thus, the purpose of this work is to consider to what extent such interactions can be derived from microscopic theory and to specify what physics is included as well as what is omitted. The conceptual advantage of partitioning the complete problem such that one doesn't have to think about all the parts at once has the obvious danger that if an effective interaction is too convenient, one might forget to think about it at all. The Skyrme force is a good example of this danger. Although we shall show that the Skyrme force is a satisfactory approximation to a microscopic effective interaction for a restricted set of Hartree Fock ground states, we shall also demonstrate that it is not suitable without modification for use in shell-model, RPA, or TDHF calculations.

The starting point for our development is the G-matrix, which in a finite nucleus we shall define as follows:

$$G(W) = v - v \frac{Q}{QH_oQ-W} G(W) \tag{1}$$

where

$$Q = \sum_{ab} |ab\rangle \langle ab| \ .$$

Matrix elements will be understood to be antisymmetrized and occupied and unoccupied orbitals will be denoted by upper and lower case letters respectively. To fully define G, the projector onto unoccupied states, Q, and the operator QH_oQ must be defined, which, in turn, requires that the single particle potential generating the basis be defined. For the present discussion, we will assume that $QH_oQ = QTQ$, i.e., the potential energy for excited states is zero, and that the particle-hole matrix elements of the single-particle potential are generated by variation of[1]:

$$\langle H\rangle = \sum_N \langle N|T|N\rangle + \frac{1}{2}\sum_{MN} \langle MN|G(\varepsilon_M + \varepsilon_N)|MN\rangle \tag{2}$$

where

$$\varepsilon_N = T_N + \sum_D \langle ND|G(\varepsilon_N + \varepsilon_D)|ND\rangle$$

The variation of Eq. 2 is most straightforwardly effected by considering an infinitesimal unitary transformation of the form

$$\begin{aligned} |A'\rangle &= |A\rangle + \lambda_{Aa}|a\rangle \\ |a'\rangle &= |a\rangle - \lambda^*_{Aa}|A\rangle \end{aligned} \tag{3}$$

Note that a transformation between particle states or between hole states leaves $\langle H \rangle$ unchanged, so that only the particle-hole matrix elements are determined. Variation of λ^*_{Aa} yields

$$\begin{aligned} \langle a|U|A\rangle &= \sum_N \langle aN|G(\varepsilon_A+\varepsilon_N)|AN\rangle \\ &+\frac{1}{2}\sum_{MNn} \frac{\langle MN|G(\varepsilon_M+\varepsilon_N)|AN\rangle\langle an|G(\varepsilon_M+\varepsilon_N)|MN\rangle}{\varepsilon_a+\varepsilon_n-\varepsilon_M-\varepsilon_N} \\ &-\frac{1}{2}\sum_{\substack{MN\\mn}} \frac{|\langle MN|G(\varepsilon_M+\varepsilon_N)|MN\rangle|^2}{\varepsilon_m+\varepsilon_n-\varepsilon_M-\varepsilon_N}\langle aM|G(\varepsilon_A+\varepsilon_M)|AM\rangle + \ldots\ldots \end{aligned} \tag{4}$$

plus a specified class of higher order terms. The three terms above correspond to the diagrams in Figure 1.

This choice of single particle potential suffers from two deficiencies and should ultimately be improved. The kinetic spectrum for particles introduces an unphysical gap at the Fermi surface, motivated primarily by the asymmetry between particle and hole line insertions in a theory in which only upward ladders are summed. This could be remedied either by the prescription introduced by Mahaux[2] or by starting with a completely symmetrical theory. Furthermore, the variation of Eq. 2 yields single particle propagators which lead to overcounting when substituted back into the expression for $\langle H \rangle$, so that unless one is willing to explicitly subtract off correction terms, a more complicated expression should be varied. The subse-

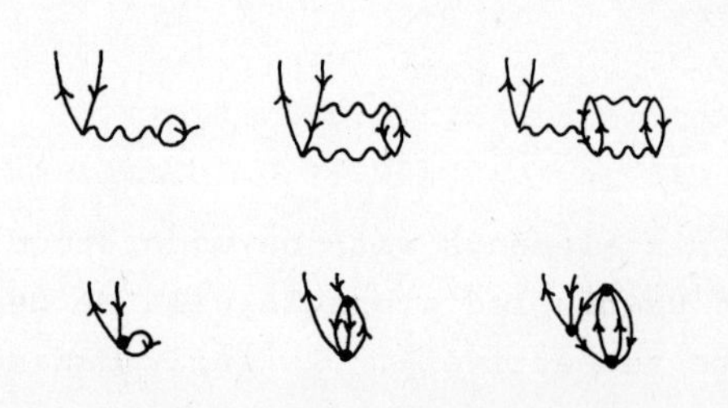

Fig.1 Diagrams for the single particle potential defined in Eq. 4. Each diagram is drawn two ways, with the G-matrix denoted by a conventional wavy line or by a heavy dot to denote both direct and exchange contributions.

quent development could be generalized to include the complications arising from remedying either of these deficiencies, so these limitations will be inessential to the main point of this work.

The variational procedure introduced above to define the particle-hole matrix elements of the single particle potential is easily extended to obtain the finite nucleus counterpart of the particle-hole interaction in Landau Fermi liquid theory. The counterpart to the second derivative of the energy with respect to occupation number $\frac{\partial^2 \langle H\rangle}{\partial n(k)\partial n(k')}$ is $\frac{\partial^2\langle H\rangle}{\partial\lambda_{Bb}\partial\lambda^*_{Aa}}$. The first derivative with respect to λ^*_{Aa} creates a final state with particle a and hole A and the second derivative with respect to λ_{Bb} creates an initial state with particle b and hole B. As before, only particle-hole transformations can change $\langle H\rangle$, so the emergence of a particle-hole interaction is automatic. Variation of the three terms in Eq. 4 with respect to λ_{Bb} yields the diagrams shown in Fig. 2 as well as a number of more complicated contributions. Even with the over-simplified choice of $\langle H\rangle$ and QH_0Q made above, systematic analysis and interpretation of all the contributions becomes very complicated, and is beyond the scope of this work. The lowest order contribution enumerated in Fig. 2 will be sufficient for the subsequent considerations in Section 5.

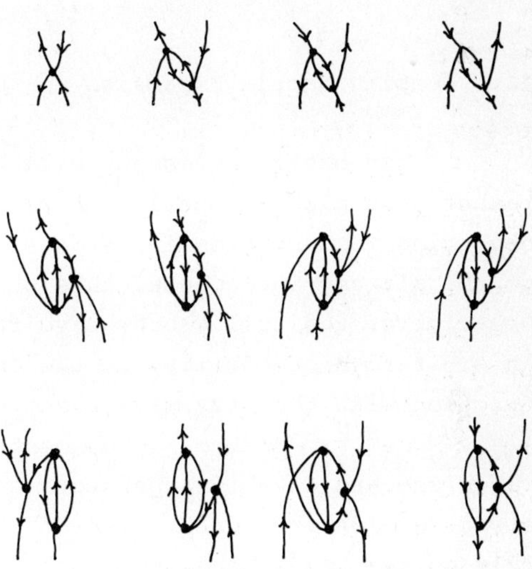

Fig. 2 Diagrams in the effective particle-hole interaction obtained by differentiating with respect to λ_{Bb} and λ^*_{Aa}.

The reaction matrix, G(W), in Eq. 1 is fully specified by the Slater determinant of occupied states which, in turn, is specified by the single particle density matrix.[+] Hence, G(W) is manifestly a density-

[+]For the case $QH_0Q = QTQ$, this follows from the fact that Q can be written in terms of $\rho(x,x')$ as in Eq. (6). If H_0 includes some potential energy due to interactions with particles in the Fermi sea, QH_0Q becomes a correspondingly more complicated functional of $\rho(x,x')$.

matrix dependent interaction. The main question, then, is how this density-matrix dependence of G(W) can be reduced to a density dependence, and whether this is just some crude ansatz or if it is the first step in a completely general systematic approximation procedure. In Section 2, we shall show that the essential feature is the local behavior of the density matrix, and thus shall expand the density matrix about the center of mass of two interacting particles. This then leads to the justification of the local density approximation presented in Section 3. In Section 4, we shall show that the density-dependent effective interaction derived for a restricted set of determinantal wave functions is very similar to the phenomenological Skyrme force. Finally, in Section 5, the serious limitations of this interaction are emphasized and the problems involved in generalizing it for other applications are discussed.

II. Density Matrix Expansion

Our basic philosophy will be to concentrate on the local, rather than global behavior of wave functions and operators. In beginning with a hole-line expansion for $\langle H \rangle$ in Eq. 2, we have, of course already implemented this philosophy. We never seek the global behavior of the full N-body wave function, for if each particle has an excitation probability ε, the overlap of the unperturbed wave function with the true wave function is exponentially small, of order $(1-\varepsilon)^N \sim e^{-\varepsilon N}$. Rather, we concentrate on expectation values of finite range one and two-body operators such as ρ, T, and v so that the errors are only of order $(1-\varepsilon)^2 \sim 1-2\varepsilon$. Thus, we are clearly focussing only on the local behavior of the many-body wave function within the range of the two-body potential and ignoring all the hopelessly complicated global behavior of the wave function.

Turning now to the lowest order term in the hole-line expansion for $\langle H \rangle$, Eq. 2, we observe that this is fully specified by the single particle density matrix

$$\rho(x,x') \equiv \sum_{M=1}^{A} \psi_M(x)\, \psi_M^*(x') \tag{5}$$

The G-matrix in Eq. 1 depends upon $\rho(x,x')$ in two non-trivial ways.

The Pauli projection operator onto normally unoccupied intermediate states may be written in terms of the density matrix as follows:

$$\langle xy|Q|x'y'\rangle = (\delta(x-x')-\rho(x,x'))(\delta(y-y')-\rho(y,y')) \qquad (6)$$

In addition, the available energy parameter W is defined in terms of single particle energies which depend upon $\rho(x,x')$.

Although $\rho(x,x')$ is much simpler than the full many-body wave function, it still has very complicated global behavior. For example, it obeys the global relation $\int \rho(x,x'')\rho(x'',x')d\,x'' = \rho(x,x')$. Notice, however, that $\rho(x,x')$ only contributes to $\langle H\rangle$ for $|x-x'|$ within the range of the two-body potential, since the range of G is the range of v and expansion of the integral equation for G shows that Q is always surrounded by v's. Thus, we have the familiar situation of a rather complicated global quantity which only contributes locally, and it makes sense to design an approximation which describes the local behavior very accurately without wasting effort on global behavior which is never physically utilized.

Our basic tool to study the local behavior of the density matrix is the Density Matrix Expansion (DME).[4] To illustrate the basic idea as simply as possible, we first consider a one-dimensional system of Fermions. For one-dimensional "nuclear matter" comprised of plane wave states occupied between $-k_F$ and k_F, the exact density matrix is $\rho(r_1,r_2) = j_o(k_F|r_1-r_2|)\rho$. The reason this nuclear matter density matrix goes to zero for $|r_1-r_2|>\pi/k_F$ follows from the fact that products of wave functions are being summed over the Fermi sea, and since the wave functions near the top of the sea change sign within a half wave length π/k_F, the sum $\sum_k \psi_k(r_1)\psi_k^*(r_2)$ becomes incoherent at separations comparable to this distance. In an arbitrary one-dimensional potential, we expect the local wavelength of the wave functions at the top of the Fermi sea to be similar to that of nuclear matter at the same density, so the nuclear matter density matrix should yield a good first approximation to the fall-off of the exact density matrix with $|r_1-r_2|$. Hence, we seek an expansion of the exact density matrix such that the first term reproduces the nuclear matter result. Changing variables to $R = \frac{r_1+r_2}{2}$ and $s = r_1-r_2$, we write a formal expansion

$$\rho(R+\tfrac{s}{2},\ R-\tfrac{s}{2}) = e^{\frac{s}{2}(\partial_1-\partial_2)}\sum_A \phi_A(R_1)\phi_A^*(R_2)\Big|_{R_1=R_2=R} \ , \qquad (7)$$

where ∂_1 acts on R_1 and ∂_2 acts on R_2. Using the identity

$$e^{ik_F s(\frac{\partial_1-\partial_2}{2k_F i})} = \sum(2\ell+1)i^\ell j_\ell(k_F s)P_\ell(\frac{\partial_1-\partial_2}{2k_F i}) \quad , \tag{8}$$

we obtain

$$\begin{aligned}(R+\frac{s}{2},R-\frac{s}{2}) &= j_o(k_F s)\rho(R) + i\ 3s\ \frac{j_1(k_F s)}{k_F s}\ J(R) \\ &+s^2\frac{15}{2}\ \frac{j_2(k_F s)}{(k_F s)^2}\ \left[\tfrac{1}{4}\rho''(R)-\tau(R)+\tfrac{1}{3}k_F{}^2\rho(R)\right] + \ldots.\end{aligned} \tag{9}$$

where

$$\rho(R) = \sum_A|\phi_A(R)|^2, \quad \tau(R) = \sum_A|\frac{\partial}{\partial R}\phi_A(R)|^2$$

and

$$J(R) = \frac{1}{2i}\ \sum_A(\frac{\partial\phi_A(R)}{\partial R}\ \phi_A^*\ (R) - \phi_A(R)\frac{\partial\phi_A^*(R)}{\partial R}\ )$$

Thus, we have arrived at a completely general expansion for the density matrix in powers of the relative coordinate s, which has been arranged such that its first term is exact in nuclear matter, and each higher order term vanishes identically in nuclear matter. Since we are in effect expanding the difference between the exact and nuclear matter results in powers of s and will eventually calculate matrix elements involving products of short range forces or operators times the density matrix, we expect to be able to truncate the expansion at reasonably low order.

The three-dimensional case is more cumbersome, and we shall restrict our attention to the case of expanding either the proton or neutron density matrix subject to the assumption that time-reversed states are filled pairwise. By virtue of this time-reversal assumption, the current term vanishes and the square of the angle-average is identical to the angle average of $|\rho(x,x')|^2$ through terms quadratic in s. Hence, we may use the same technique as above by expanding the angle-average of the density matrix about $R = \frac{r_1+r_2}{2}$ in terms of $s = r_1-r_2$ as follows

$$\begin{aligned}\int \frac{d\Omega_s}{4\pi}\rho(\vec{R}+\frac{\vec{s}}{2},\vec{R}-\frac{\vec{s}}{2}) &= \int \frac{d\Omega_s}{4\pi}\sum\psi_M(\vec{R}+\frac{\vec{s}}{2})\psi_M^*(\vec{R}-\frac{\vec{s}}{2}) \\ &= \int\frac{d\Omega_s}{4\pi}\quad e^{\vec{s}\cdot\frac{\vec{\nabla}_1-\vec{\nabla}_2}{2}}\sum_M\psi_M(R_1)\ \psi_M^*(R_2)\Big|_{R_1=R_2=R} \\ &= \frac{3j_1(k_F s)}{k_F s}\ \rho(\vec{R}) + \frac{35s^2 j_3(k_F s)}{2(k_F s)^3}\left[\tfrac{1}{4}\nabla^2\rho(\vec{R})-\tau(\vec{R})+\tfrac{3}{5}k_F{}^2\rho(\vec{R})\right] +\ldots.\end{aligned} \tag{10}$$

where now $\tau(\vec{R}) = \sum_M |\vec{\nabla}\psi_M(\vec{R})|^2$.

The structure of the three-dimensional expansion is analogous to the one-dimensional case. The first term is the Slater approximation, which is exact for nuclear matter, and the second term introduces corrections which are quadratic in s. The $\frac{3}{5}k_F^{\ 2}\ \rho(R)$ in the correction term simply subtracts off the coefficient of s^2 in the Slater approximation, so that the actual quantity governing the leading off-diagonal behavior of the density matrix is the local value of $\frac{1}{4}\nabla^2\rho(R)-\tau(R)$ which may also be expressed as $\frac{1}{2}(\sum_N \psi_N\nabla^2\psi_N-\tau(R))$.

The accuracy of this expansion is demonstrated in Figure 3

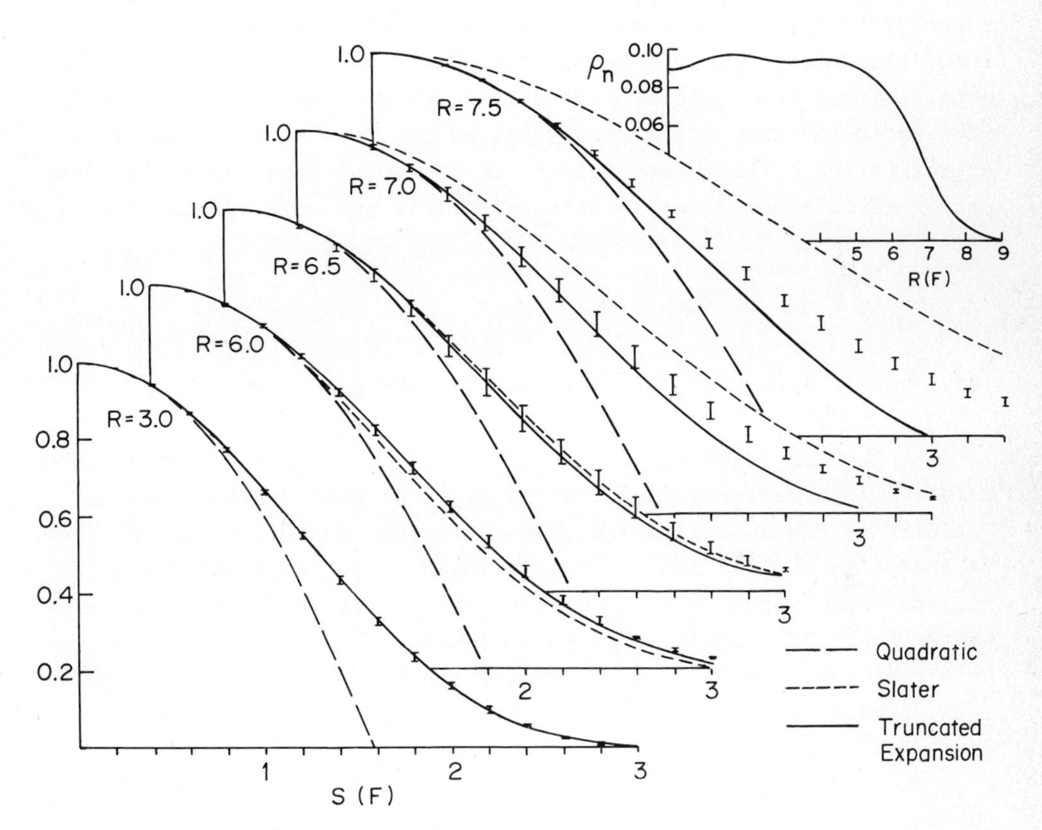

Fig. 3 Comparison of the square of the exact neutron density matrix in ^{208}Pb with the DME expansion. The neutron density distribution is shown for reference, and for five values of c.m. coordinate R, the square of the density matrix is plotted as a function of the relative coordinate s. The error bars denote the extrema of the exact density matrix in the radial direction and perpendicular to the radial direction. The long dashes denote the result when only the quadratic terms in a Taylor expansion are retained, the short dashes indicate the Slater approximation, and the solid curves correspond to Eq. 10.

which shows the square of the exact neutron density matrix in ^{208}Pb compared with several approximations. In the interior, at R=3fm, the extremal values of the exact density matrix in the radial direction and perpendicular to the radial direction, denoted by the error bars, are indistinguishable from the Slater and DME results, confirming our intuitive expectation that the nuclear matter approximation is very accurate in the interior of a large nucleus. In the surface, the Slater approximation does not yield serious errors until the density reaches about 1/3 of the central density, and even in the extreme surface the DME yields an excellent approximation within the range of the two-body nuclear potential. Thus, we conclude that knowing $\rho(R)$, we may obtain a reasonable approximation to the density matrix within the range of the nuclear force of the position R and knowing in addition $\nabla^2\rho(R)$ and $\tau(R)$, we may construct an excellent approximation. This understanding of the local behavior of the density matrix gives rise to two important applications to be discussed below.

III. Justification of the Local Density Approximation

The LDA is often described in terms of using a nuclear matter G-matrix in a finite nucleus, which gives no real feeling for the validity of the approximation. An equivalent definition of the LDA is that Q in the definition of G(W) in Eq. 1 is replaced by the nuclear matter projector Q_{NM}. The available energy W is just a parameter, and thus requires no approximation. From Eq. 6, however, we note that replacing Q by Q_{NM} is equivalent to replacing ρ by ρ_{NM}. Thus, recalling that Q is only utilized for values of the relative coordinate within the range of the nuclear force, and returning to Fig. 3, we conclude the LDA is extremely accurate throughout most of the nuclear interior and reasonably accurate through most of the nuclear surface.

It is worthwhile to contrast the LDA approximation to Q with the harmonic oscillator (HO) approximation to Q. By the same argument as above, the HO approximation may be expressed by replacing $\rho(r,r')$ by $\rho_{HO}(r,r')$. But in heavy nuclei, the one-body HO density, $\rho_{HO}(R)$, differs tremendeously from $\rho(R)$ so that for the case of ^{208}Pb sketched in Fig. 4, the normalization of $\rho^2_{HO}(R,s)$ is very seriously in error. Thus, although the oscillator approximation might superficially seem appealing because it corresponds exactly to a finite Slater determinant and satisfies the global relation $\rho^2 = \rho$, it is in fact far less suitable for evaluating G than the LDA which sacrifices irrelevant global behavior in lieu of an accurate description of the relevant local behavior.

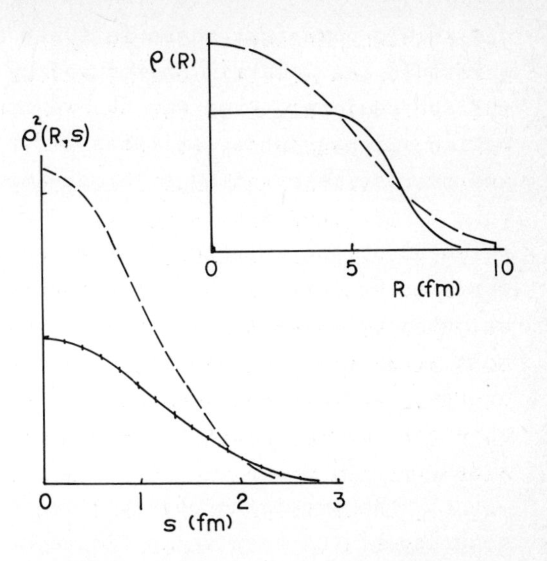

Fig. 4 Sketch of the harmonic oscillator approximation to the density matrix, The exact and harmonic oscillator one-body densities as a function of R are shown by the solid and dashed lines in the upper right hand graph, and the DME and oscillator density matrices at R=0 as a function of s are denoted by the solid and dashed lines in the main graph.

There have been several numerical investigations relating to the validity of the LDA, which I do not have time to review.[6,7] The results of one recent comparison[1] of similar oscillator space and LDA calculations of ^{40}Ca with the Reid potential and no phenomenological adjustments are encouraging, although not definitive, since as usual, the comparison is hampered by annoying technical differences in the calculations. After correction for three differences for which quantitative estimates were feasible, the oscillator space Brueckner-Hartree Fock result of Ref. 8 and the LDA result of Ref. 9 yielded charge densities which agreed reasonably well. The remaining discrepancy is certainly an overestimate of the inaccuracy of the LDA since two technical differences concerning the non-self-consistency of the oscillator Q and the parameterization of the LDA effective interaction at densities beyond nuclear matter density are of the proper sign to help resolve this discrepancy even though quantitative evaluation has proved impractical. The binding energies in the two calculations after correction for the three calculable technical differences agree to better than 0.1 MeV per particle. Although this agreement may be partly fortuitous given the two additional effects for which numerical corrections proved impossible, one should note that there is substantial cancellation of the small errors in the G-matrix introduced by the LDA in the near and far surface regions. From Fig. 3, we observe that the Slater approximation slightly underestimates $\rho(r,r')$ in the near surface (6fm in Pb) and overestimates it by a larger amount in the extreme surface (beyond 7 fm). The small error in G of one sign in the near surface weighted by the interior density substantially compensates the somewhat larger error in G of the opposite sign in the extreme surface weighted by a small fraction of the interior density. Thus, the most stringent test of the LDA is the density distribution, not the energy, and future investigations of the LDA should concentrate on an accurate comparison of technically comparable density calculations.

One promising possibility for significantly improving the accuracy of LDA results in finite nuclei, and thus assessing the accuracy of the approximation, is to include the quadratic DME correction terms in Eq. 5. Defining $\Delta\rho(x,x') \equiv \rho(x,x') - \rho_{NM}(x,x')$ and using Eq. 6, Q may be written exactly as follows:

$$\langle xy|Q|x'y'\rangle = \langle xy|Q_{NM}|x'y'\rangle + \langle xy|\,\Delta Q|x'y'\rangle \tag{11}$$

where

$$\begin{aligned}\langle xy|\,\Delta Q\,|x'y'\rangle &= -\Delta\rho(x,x')\,|\delta(y-y')-\rho_{NM}(y,y')| \\ &\quad -|\delta(x-x')-\rho_{NM}(x,x')|\;\Delta\rho(y,y') \\ &\quad +\Delta\rho(x,x')\;\Delta\rho(y,y')\end{aligned}$$

Approximation of $\Delta\rho$ by $\frac{35}{2sk_F^3}\, j_3(sk_F)\,|\frac{1}{4}\nabla^2\rho(R)-\tau(R)+\frac{3}{5}k_F^2\rho(R)|$ from Eq. 10 then yields a simple expression for ΔQ which should be extremely accurate within the relevant range of s. Using Eq. 1, which accounts for the fact that Q and T are not simultaneously diagonal in a finite nucleus and maintains the orthogonality of occupied and unoccupied states, it is straightforward to calculate the correction to G arising from ΔQ. Thus, one obtains a theory in terms of $G(\rho,\frac{1}{4}\nabla^2\rho-\tau+\frac{3}{5}k_F^2\rho,W)$ which can still be formulated variationally and should be extremely accurate. Siemens[11] has proposed a related program, and has further suggested an efficient technique for eliminating the usual angle-average approximation in Q.

Thus far, we have described a systematic, computationally convenient procedure for approximating the reaction matrix, G(W) in any nucleus, which should be applicable to a wide variety of nuclear structure problems. In the next section, we introduce further approximations to obtain an effective Hamiltonian of much greater simplicity, but with a much more limited range of applicability.

IV. Derivation of an Effective Hamiltonian

The preceding argument concentrated on the fact that G in Eq. 1 depends only on the local behavior of the density matrix. We now turn our attention to the fact that given G, $\langle H \rangle$ in Eq. 2 also depends primarily upon the local behavior of the density matrix. Schematically, assuming local direct and exchange effective interactions have been constructed and lumping the energy dependence of G, into a correction term H_W to be discussed subsequently $\langle H \rangle$ may be written:

$$\langle H \rangle = \langle T \rangle + \int \rho(r) v_{DIR}(r-r')\rho(r')d^3rd^3r' + \int |\rho(r,r')|^2 v_{EXCH}(r-r')d^3rd^3r' + \langle H_{COUL} \rangle + \langle H_W \rangle \tag{12}$$

Using the expansion, Eq. 10, for $\rho(r,r')$ and similarly expanding the direct term yields the approximate Hamiltonian density

$$\begin{aligned}\mathcal{H}(R) &= \frac{\hbar^2}{2m}(\tau_n+\tau_p) + A(\rho_n,\rho_p) \\ &+B(\rho_p,\rho_n)\tau_p + B(\rho_n,\rho_p)\tau_n + C(\rho_n,\rho_p)|\vec{\nabla}\rho_n|^2 \\ &+C(\rho_p,\rho_n)|\nabla\rho_p|^2 + D(\rho_n,\rho_p)\,\vec{\nabla}\rho_n\cdot\vec{\nabla}\rho_p\end{aligned} \tag{13}$$

where τ_n, τ_p, ρ_n and ρ_p are, of course, functions of R and

$$\langle H \rangle = \int d^3R\,\mathcal{H}(R) + \langle H_{COUL} \rangle + \langle H_W \rangle \quad .$$

The functions A,B,C, and D appearing in Eq. 13 are specified integrals of v_{DIR} and v_{EXCH} times the functions of relative coordinate appearing in Eq. 10. The density dependence of these functions is of two origins, arising from both the density dependence of G and from the k_F-dependent weighting functions in Eq. 10.

A variety of phenomenological effective interactions currently used in Hartree-Fock calculations may be expressed in the form given in Eq. 13, and are reviewed in detail in Section 2.1 of Ref. 12. The widely used Skyrme interaction[13,14] for example, may be expressed:

$$\begin{aligned}
A(\rho_n,\rho_p) &= \frac{t_0}{4}(1-x_0)(\rho_n^2+\rho_p^2) + t_0(1+\frac{x_0}{2})\rho_n\rho_p \\
&\quad +\frac{t_3}{4}\rho_n\rho_p(\rho_n+\rho_p) \\
B(\rho_n,\rho_p) &= (\tfrac{3}{8}t_2+\tfrac{1}{8}t_1)\rho_n+\tfrac{1}{4}(t_1+t_2)\rho_p \qquad (14) \\
C(\rho_n,\rho_p) &= \tfrac{3}{32}(t_1-t_2) \\
D(\rho_n,\rho_p) &= \tfrac{1}{8}(3t_1-t_2)
\end{aligned}$$

Numerical values for the Skyrme parameters obtained by linearizing the functions A,B,C, and D about the proton and neutron densities occuring in the interior of Lead are compared with two sets of phenomenologically determined Skyrme parameters in Table 1.

Table 1

Interaction	t_0 (MeV fm^5)	t_1 (MeV fm^5)	t_2 (MeV fm^5)	t_3 (MeV fm^6)	X_0
Skyrme II	-1169.9	585.6	-27.1	9331	0.34
Skyrme III	-1128.8	395.0	-95.	14000	0.45
DME	-1259.	481.1	-45.1	14960	0.50

The overall qualitative agreement is quite good, indicating that only the very simple averages of the two body effective interaction which are built into the DME functional $\mathcal{H}(R)$ are relevant to the gross structure of the ground states of nuclei with time-reversed states filled pairwise. In this sense, we believe that we have "derived" the Skyrme parameters from the effective interaction of Ref. 9.

In spite of the similarity between the DME and Skyrme forces, there are definite advantages to using the Hamiltonian density $\mathcal{H}(R)$ in Eq. (13) instead of the Skyrme interaction. The strongest argument for using it, rather than a phenomenological interaction fitted to the mass table, is that it embodies most of the physics contained in the realistic G-matrix while maintaining the computational simplicity of the Skyrme force. For example, the range of the nonlocality associated with the effective mass

$$m^*_n(r) = m\left[1+\frac{2m}{\hbar^2}B(\rho_n,\rho_p)\right]^{-1}$$

arises directly from the off-diagonal range of the density matrix and the long range part of the G-matrix, which is essentially the long range component of the bare nucleon-nucleon potential. From Fig. 3 and the fact that the long range part of the potential is well determined from scattering phase shifts, we conclude that the range of non-locality is theoretically well determined. The two phenomenological parameters which must be added to the interaction to obtain the proper binding energy and saturation in some heavy nucleus, say ^{208}Pb, are at least included in a physically motivated way. Since higher order terms such as three-body forces and three body clusters require at least two nucleons to be close together and yield their largest contributions in triplet-even states, these processes are parameterized by a density-dependent zero-range force in the triplet even channel. Thus, when one calculates nuclear properties which have not been directly built into the effective interaction, one may have considerable confidence that these properties are being determined by the physics of the nuclear interaction and not by the limitations of a restrictive ansatz in parameterizing a phenomenological force or by a parameter in a phenomenological force which has not been properly constrained by fitting pertinent data.

V. Limitations of the Effective Hamiltonian

The Hamiltonian density $\mathcal{H}(R)$ in Eq. 13 should be understood as a generalized Skyrme interaction in which each of the coefficients has density and isospin dependence specified by a realistic G-matrix. Since it also agrees substantially with phenomenological Skyrme parameters, any theoretical limitations imposed on it are also applicable to the Skyrme interaction.

A strong hint that the Skyrme parameterization is subject to serious limitations occurs when one attempts to derive it from a short range expansion of a relaistic effective interaction, truncated after second derivative terms. If such a short range expansion were valid, one would indeed be justified in using the Skyrme force in any application for which the realistic effective interaction were appropriate. To investigate this short range expansion interpretation, Dr. Richard Sharp[15] has recently performed the following illuminating calculation. A very simple density-independent effective interaction is defined, comprised of a short-range and a long-range Gaussian for both even and odd state forces. The strengths of the Gaussians are adjusted to yield nuclear matter saturation at k_F= 1.36 fm^{-1} with BE/A = 15.7 MeV and a binding energy maximum of 140MeV in ^{16}O at the observed rms radius for a harmonic oscillator trial function. The even state effective interaction looks quite realistic, with a zero at 0.6 fm, a maximum repulsion of 450 MeV at the origin and a maximum attraction of 115 MeV at 0.95 fm, whereas the odd state potential has somewhat larger attractive and repulsive components than expected. Presumably, a density dependent effective interaction would accommodate a more realistic odd state component. Using $\mathcal{H}(R)$ from Eq. 13, with A,B,C and D calculated from this Gaussian effective interaction, yields a nuclear matter binding energy per particle of 15.7 MeV, as it should by construction, and a binding energy of 167 MeV in ^{16}O. In contrast, performing a Taylor series expansion in momentum space of the Gaussian effective interaction and truncating after the quadratic terms yields a binding energy per particle in nuclear matter of -9.3 MeV, i.e. nuclear matter is unbound by 9.3 MeV, and a binding energy in ^{16}O of only 41 MeV. Qualitatively similar results occur when one attempts a short range expansion of the G-matrix used in Ref. 9. The reason for these unphysical results with the short range expansion is obvious from examination of Fig. 3. The long dashed lines in that figure denote a quadratic expansion of the density matrix about the diagonal, and correspond to the approximation made to the density matrix in the momentum space Taylor series expansion. Evidently, beyond 1 fm, the long range attraction is weighted much less heavily than it should be, and beyond 1.5 fm it is even weighted with the wrong sign, thus explaining the severe loss of attraction manifested in the results. By virtue of the unrealistic odd state force, Sharp's model somewhat overestimates the inaccuracies in both the DME and the short range expansion, but I believe the lesson is still very clear. The Skyrme force must be understood as a very special average of the effective interaction over the occupied states of the Fermi sea, and the values and density dependence of the parameters contain a combination of information about both the true effective interaction and the Fermi Sea over which it is averaged. Hence, the Skyrme parameters in no way describe the interactions between valence particles, and one must return to the underlying effective interaction to learn about the interactions between valence nucleons.

Given that we expect $\mathcal{H}(R)$ and Skyrme forces to be of much more limited validity than the original effective two-body interaction G(W), we now proceed to briefly catalog several salient limitations.

A. Finite Range Effects

In Eq. 12, the reduction of the finite range Hartree term to an expansion in terms of $\rho(R)$ and $\nabla^2\rho(R)$ was an inessential simplification performed only to establish contact with the Skyrme parameterization. However, as explained in Refs. 4 and 12, such a reduction gives rise to unphysical enhancement or damping of the shell fluctuations in Hartree-Fock density distributions. In Fig. 5 results for a finite range effective interaction are compared with results using the effective Hamiltonian $\mathcal{H}(R)$ derived from the same interaction and one observes that the differences can be quite significant. When the finite range direct term is retained in

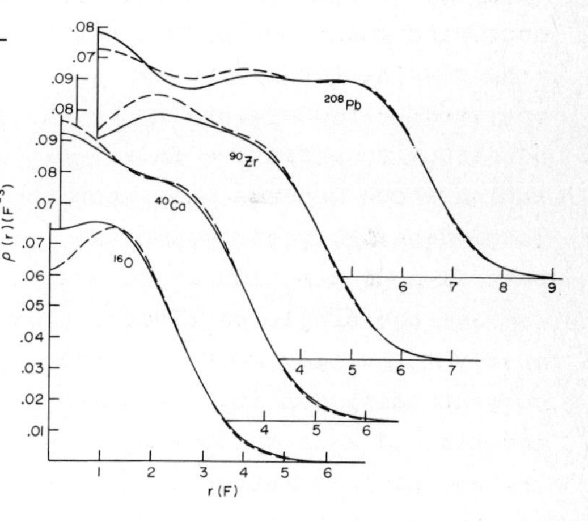

Fig. 5 Point proton density distributions calculated with the DME (solid curves) and with the finite-range effective interaction (dashed curves).

Eq. 12 and only the exchange term is expanded, the fluctuations become virtually indistinguishable from those obtained in the full finite range calculations. Thus, we conclude that a realistic, finite range direct interaction should definitely be used and note in passing that this is an insignificant computational complication since the finite range direct Coulomb force is always included in any event.

B. Energy Dependence

The expression for <H> in Eq. 2 depends upon the density matrix both through Q and through the single particle energies E_M and E_N. Thus far, we have emphasized the dependence of Q on the local density and thereby obtained a density-dependent G-matrix, G(W), in which the energy still appears explicitly as a parameter. Since the single particle energies E_M and E_N are not determined locally, their density matrix dependence cannot be transformed into a density dependence with the same quantitative precision as for the Pauli operator. If one wants to construct an effective interaction which depends only upon the local density, then the most natural approximation is to use the average single particle energy in nuclear matter at the corresponding density. In the interior of a large nucleus, the average nuclear matter energy differs negligibly from the average of the true single particle energies, but in the surface, the discrepancy is quite significant, as shown in Fig. 6. Here, the dashed lines denote the local average of the single par-

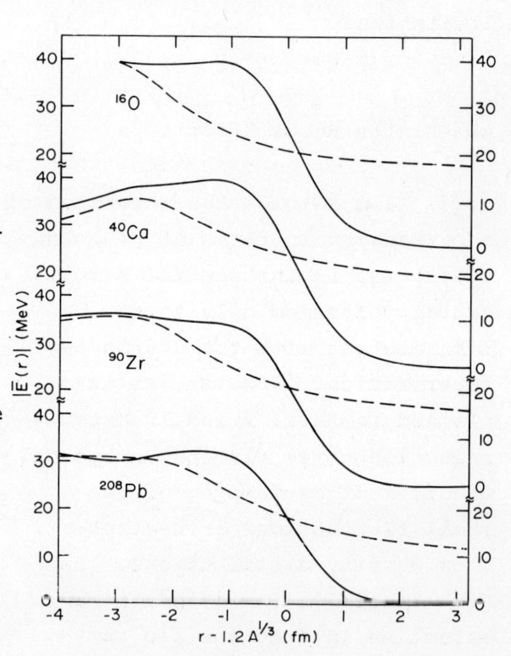

Fig. 6 Comparison of the magnitude of the average single particle energy for neutrons in nuclear matter at the local density (solid curves) with the exact average single particle energy for neutrons (dashed curves) in four closed shell nuclei.

ticle energies in four finite nuclei as a function of radius, which clearly approach the removal energy of the last particle for large r. The solid lines indicate the average single particle energy in nuclear matter at the local density and systematically exceed the exact average in the near surface and underestimate the exact average in the extreme surface. From knowledge of $\frac{\partial G}{\partial W}$, which may be obtained from the correlated two-body wave function when G is calculated, it is straightforward to derive the perturbative correction term H_W in Eq. 12, and analogous corrections for the single particle eigenvalues. From Fig. 6, it is evident that the correction is localized in the surface, being attractive in the near surface (since the nuclear matter energy denominators are larger than the finite nucleus denominators, yielding too little attraction) and repulsive in the extreme surface. The near surface is weighted more heavily, and the net correction yields a significant contribution to the surface energy, reducing its magnitude from -21.61 MeV to -19.06 MeV and yielding reasonable agreement with the semiempirical value of -18.56 MeV.

The present treatment of energy dependence is necessarily crude because of the fact that the single particle energies are essentially global quantities, and only certain averages of single particle energies are easily defined locally. Further effort in treating the energy dependence is clearly warranted, but until some breakthrough occurs, it seems prudent to regard a purely density dependent force having no energy dependence with a generous dose of skepticism.

C. Spin Dependence

The simplest context in which to discuss the spin dependence of the effective interaction is to consider the Landau interaction in uniform nuclear matter, which may be written:

$$\langle \vec{k}_1\vec{k}_2|v|\vec{k}_1\vec{k}_2\rangle = \frac{\hbar^2\pi^2}{2k_F m^*}\left(F(\theta)+F'(\theta)\sigma_1\cdot\sigma_2+G(\theta)\tau_1\cdot\tau_2+G'(\theta)\sigma_1\cdot\sigma_2\tau_1\cdot\tau_2\right) \tag{15}$$

where $\cos\theta = \frac{\vec{k}_1\cdot\vec{k}_2}{|k_1||k_2|}$. Superficially, the Hamiltonian $\mathcal{H}(R)$ in Eq. 13 appears to be an ideal starting point, since it is exact for nuclear matter. However, upon further consideration, it is evident that the spin and isospin content originally present in the singlet even (SE), triplet even (TE), singlet odd (SO) and triplet odd (TO) components of the effective interactions cannot be fully recovered from $\mathcal{H}(R)$ by double differentiation. Although the neutron and proton densities appear separately and thus may be differentiated independently, the assumption that time-reversed states are filled pairwise has irrevocably locked the occupation of the spin-up Fermi sea to the occupation of the spin-down Fermi sea. Hence the resulting effective interaction only represents the projection of Eq. 15 onto the subspace of spin-saturated wave functions and thus only $F(\theta)$ and $G(\theta)$ are actually determined.

The real difficulty concerning spin dependence arises when $\mathcal{H}(R)$ is rewritten in terms of the Skyrme parameters in Eq. 14. For spin-saturated systems, $\mathcal{H}(R)$ is then essentially equivalent to the Skyrme force,

$$\langle\vec{k}|v_{12}|\vec{k}'\rangle = t_0(1+X_0P_\sigma)+\tfrac{1}{2}t_1(k^2+k'^2)+t_2\vec{k}\cdot\vec{k}' +\tfrac{1}{6}t_3(1+P_\sigma)\rho \qquad (16)$$

where P_σ is the spin exchange operator. This parameterization, however, specifies all four terms in Eq. 15, and thus makes definite predictions about the spin as well as isospin dependence of the interaction. Since the Skyrme parameters are essentially determined by fitting observable properties of spin-saturated nuclei, or because they may be derived from $\mathcal{H}(R)$ in which the real spin-dependence is manifestly averaged, it is evident that the apparent spin dependence in Eq. 16 is purely illusory. Given that the spin dependence in Eq. 16 is ficticious, it is not surprising that it leads to spin collapse as shown by Bäckmann et al.[16] To derive the actual spin dependence from the effective interaction, it is easy to simply recalculate $\mathcal{H}(R)$ for the SE, TE, SO, and TO components of the two-body effective interaction separately, and then construct, F, F', G, and G' by taking the appropriate linear combinations.

D. Angular Dependence

Closely related to the ficticious spin dependence in Eq. 16 is the ficticious angular dependence. $\mathcal{H}(R)$, arising from the SE, TE, SO, and TO components of the two-body effective interaction, clearly contains contributions from all partial waves. The Skyrme force in Eq. 16, however, appears to contain only s-waves (t_0, t_1 and t_3) and p-waves (t_2). When the nuclear matter energy is expressed in terms of $\mathcal{H}(R)$, the directions of all momenta have been integrated over, and only the magnitude of k_F remains. Hence, double differentiation clearly cannot recover the angular dependence of the Landau parameters in Eq. 15, and thus only the lowest Legendre coefficients F_o, F'_o, etc. are determined. Again, the t_2 term gives apparent θ dependence, since $\vec{k}\cdot\vec{k}' = \frac{1}{2}k_F\,(1-\cos\theta)$, but this is just as illusory as the spin dependence.

E. Wave Function Weighting

At the beginning of this section, we emphasized that the Skyrme force must be understood as a particular average of the two-body effective interaction over the occupied states of the Fermi sea. Thus, in any application, one must ask if the particular average built into $\mathcal{H}(R)$ is relevant to the specific application.

In considering a shell-model calculation for two particles outside a closed shell in state $[n,\ell,j]$ coupled to J=0, R. Sharp[15] has shown that the interaction between the two particles may be written:

$$V = \int d^3r_1 d^3r_2\; v(|r_1-r_2|)\; \{\, |\rho^{\ell}(r_1,r_2)|^2 + |\vec{\rho}^{\,\ell}(r_1,r_2)|^2 \} \tag{17}$$

where the scalar and vector density matrices, ρ^{ℓ} and $\vec{\rho}^{\,\ell}$ are proportional to $\sum_{m=-\ell}^{\ell} R_{n\ell j}(r_1)R_{n\ell j}(r_2)\; Y_{\ell m}(1)Y^*_{\ell m}(2)$ and $\sum_{m=-\ell}^{\ell} R_{n\ell j}(r_1)R_{n\ell j}(r_2)\; mY_{\ell m}(1)Y^*_{\ell m}(2)$ respectively. For the case of two $d_{5/2}$ particles outside an ^{16}O core, these density matrices are shown in Fig. 7. Note that although the scalar part is of essentially the same structure as the density matrix of the core, and thus well represented by the wave-function weighting built into the DME, the vector part is completely different and will involve completely different moments of the two-body interaction. This is thus another strong argument against using the Skyrme interaction for shell-model applications. The dashed lines in Fig. 7 represent a first attempt to generalize the DME to the case of valence interactions. Although the scalar piece is well approximated, as might be expected from its similarity to the core density matrix, the vector part is not at all well represented by a simple expansion. The basic problem is that we have not found a counterpart of the Slater density about which to expand, as in the scalar case.

A similar wave function weighting problem arises when one attempts to deal with time-odd Slater determinants in collective motion problems. In Eq. 9, for the case of one-dimension, we explicitly wrote out the time-odd current term J as well as the time even terms involving ρ, ρ'', and τ. When Eq. 9 is squared, multiplied by the two-body effective interaction, and integrated over the relative coordinate, one obtains a Hamiltonian density involving ρ^2, $\rho[\frac{1}{4}\rho'' - \tau + \frac{1}{3}k_F{}^2\rho]$, J^2, and higher order terms. One difficulty is that, as in the case of $\vec{\rho}^{\,\ell}$, we have not found a relevant quantity about which to expand the current term, and we have no assurance that the odd terms in Eq. 9 are in any sense optimal. A second problem is posed by Galilean invariance. Since a Galilean transformation conserves the quantity $\rho\tau - J^2$, the coefficient of $-\rho\tau$ in the above expansion should be equal to the coefficient of J^2, a property which is not satisfied by Eq. 9. Thus, substantial difficulties remain in generalizing the effective Hamiltonian to collective motion applications, and until they are resolved, one should not really trust Skyrme forces for any but the most schematic calculations.

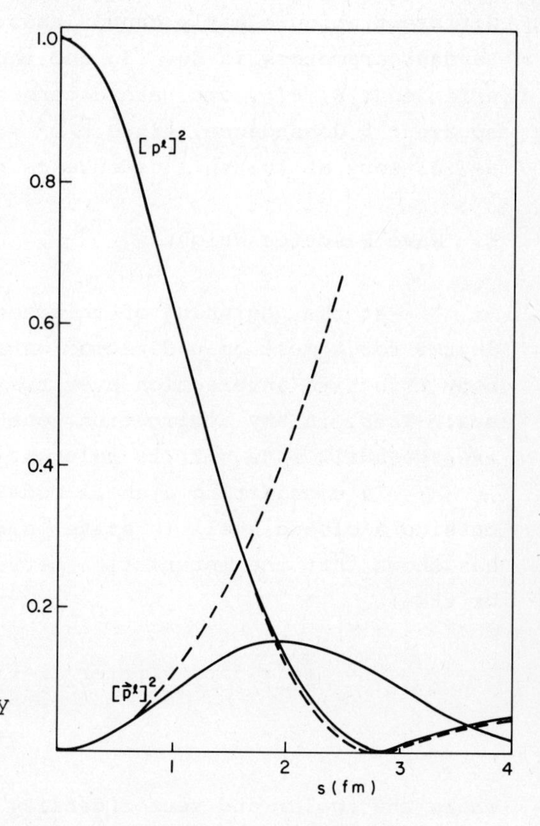

Fig. 7 Angle-averaged scalar and vector density matrices as a function of s for $d_{5/2}$ nucleons in Oxygen at R=2.5fm. The solid lines are exact and the dashed lines correspond to the DME.

VI. Conclusion

In this work, density dependent interactions have been treated on two different levels. The most fundamental level deals with G(W) and involves a systematic expansion of the density matrix about the center of mass of the two interacting particles. It has been shown that only the local behavior of the density matrix is essential. Thus a systematic expansion for G(W) is obtained such that the first term depends only on the local density and the next term involves the local kinetic energy density. These ideas generalize straightforwardly to approximating the three-body Faddeev equations in finite nuclei, and offer the most computationally practical scheme I am aware of for use in heavy nuclei.

The second level of approximation is to obtain an approximate Hamiltonian density which offers the computational simplicity of Skyrme-like forces. Although such a reduction offers tremendous technical simplification in calculations, it suffers from a serious lack of generality. Thus, $\mathcal{H}(R)$ is not in fact valid for many of the applications for which the Skyrme force is presently being used and serious effort should be devoted to generalizing it for each of these applications.

Acknowledgment

The author is grateful to Richard Sharp for communicating results prior to publication.

References

1. K.T.R. Davies, R.J. McCarthy, J.W. Negele, and P.U. Sauer, Phys. Rev. C10, 2607 (1974).

2. C. Mahaux, Proceedings of "Hartree Fock and Self-Consistent Field Theories in Nuclei", Trieste, Italy, North Holland (1975).

3. C. Bloch, Studies in Statistical Mechanics, Vol. III, ed. J. De Boer, North Holland (1965).

4. B.H. Brandow, Ann. Phys. (N.Y.) 57, 214 (1970).

5. R.W. Jones, F. Mohling and R.L. Becker, Nucl. Phys. A220, (1974) 45.

6. C.W. Wong, Nucl. Phys. A91 (1967) 399.

7. C.W. Wong and S.A. Moszkowski, Nucl. Phys., to be published.

8. K.T.R. Davies and R.J. McCarthy, Phys. Rev. C4 (1971) 81.

9. J.W. Negele, Phys. Rev. C1 (1970) 1260.

10. J.W. Negele and D. Vautherin, Phys. Rev. C11 (1975) 1031.

11. P.J. Siemens, private communication.

12. J.L. Friar and J.W. Negele, Advances in Nuclear Physics, ed. M. Baranger and E. Vogt, to be published (1975).

13. D. Vautherin and D.M. Brink, Phys. Rev. C5 (1972) 626.

14. M. Beiner, H. Flocard, Nguyen Van Giai, and P. Quentin, Nucl. Phys., to be published.

15. R.W. Sharp, Jr., to be published.

16. S.O. Bäckmann, A.D. Jackson, and J. Speth, Phys. Lett., to be published.

J.W. NEGELE: DENSITY-DEPENDENT INTERACTION

Harvey: Is the Skyrme interaction appropriate in the calculation of fission barriers in view of your claim that it is not appropriate for such collective calculations as TDA and RPA?

Negele: There are certainly problems with penetrabilities arising from uncertainties in the mass parameter, but the barrier height should be as believable as any ground-state energy, given that one is dealing with a nucleus with time-reversed states filled pairwise.

Sauer: You compared the charge densities of ^{40}Ca and ^{208}Pb arising from renormalized Brueckner-Hartree-Fock (RBHF) and density-dependent Hartree-Fock (DDHF). You stressed the fact that the differences are due to the particle-hole insertion, second-order in the reaction matrix, included in DDHF, but omitted in RBHF. If one takes your plots literally, this particular particle-hole insertion appears especially operative in ^{208}Pb. I doubt this. The enhanced discrepancy in ^{208}Pb, enhanced as compared to ^{40}Ca, is most likely due to the poor non-selfconsistent Pauli operator used in RBHF.

Negele: This is quite likely. I have never plotted the oscillator density corresponding to the $\hbar\Omega$ used in the RBHF calculations, so I have no valid feeling for the magnitude or even the sign of the effect, but I certainly should not have given the impression that the figure accurately represents the quantitative effect of each contribution.

Sauer: In order to reproduce the reaction matrix as faithfully as possible by a density-dependent interaction, you approximate the Pauli operator very accurately and keep the energy-dependence. However, you also replace the intrinsically nonlocal, i.e. momentum-dependent reaction matrix by a local, i.e., momentum-independent, potential. Have you also checked this approximation?

Negele: Phil Siemens and I each independently checked matrix elements from 0 to $2k_F$ and found the accuracy to be at the few percent level for individual matrix elements and better on the average. The tensor force may have been somewhat less accurate for individual matrix elements, but again was well reproduced on the average.

Becker: Doesn't your adjustment of the interaction give a larger change in the radius than do the second- and third-order terms in the single-particle potential?

Negele: I do not have any table of numbers with me, but I believe not. Especially in light nuclei, the added binding due to the adjustment tends to significantly decrease the radius while the added saturation tends to increase the radius. Thus the net shift in rms radius due to adjustment is rather small in light nuclei and probably even vanishes for some particular case.

Becker: Does the variational prescription you have used give an on-energy-shell hole-hole matrix element of the second-order term in the single-particle potential? If so, is it included in your calculations?

Negele: The variation I have described in this talk is based on an exact variation with respect to the density matrix dependence and does not give any information concerning hole-hole matrix elements. Rather the definition of ε_N is given by eq. 2. This is different from the variation in my original calculations which did include an approximate hole-hole potential. Perhaps I should emphasize how fundamentally different a finite nucleus is from nuclear matter. In nuclear matter momentum conservation hands you a basis, and all you have to worry about are single-particle energies, i.e. the hole-hole potential. In a finite nucleus, although the hole-hole potential problem remains, a new and much more serious problem is the particle-hole potential which determines the basis and thus the spatial extent of the system. That is why I emphasized the particle-hole potential so strongly - it is of crucial importance for any numerical applications, and we have no guidance from nuclear matter concerning its definition.

Kallio: Since I seem to be the only European defending the Skyrme force, I would like to make the following remarks concerning your criticism of the density dependent Skyrme force. Clearly the quality of the Hartee-Fock calculation is quite high compared with the simplicity of the force. I have always thought it to be a rather good representation of S = 0 and S = 1, ℓ = 0,1 part of the G-matrix. Clearly it does not have the starting energy dependence, but I do not see any great difficulty in modifying it, so as to have this dependence, since it is nearly linear. Finally, if one would like to use your modified interaction for a configuration mixing calculation, it, like the Skyrme force, must be readjusted for the tensor force.

Negele: I think we will have to agree to disagree about basic philosophy. Certainly, by adding enough parameters, you can fit anything you want. My reservation is whether you have accomplished anything of physical significance. Regarding starting energy dependence, I believe Bruce Barrett and Dick Becker would not be too happy with a linear approximation to the very pronounced energy dependence they have shown for relevant G-matrix elements.

Sprung: If you want to use Skyrme to do RPA you should go back to the Reid G-matrix and pick up things that were averaged out when the LDA local effective interaction was derived. Only the central force was used because you were mainly interested in the binding energy and other bulk properties. (I have given the LS and S_{12} forces for G-0, my effective interaction, in a paper in Nucl. Phys.)

CALCULATION OF OTHER EFFECTIVE OPERATORS+

Paul J. Ellis
School of Physics and Astronomy
University of Minnesota
Minneapolis, Minnesota

1. INTRODUCTION

We have heard a good deal about the calculation of effective interactions for use in shell model calculations. Of course this is not the whole story since all the other operators one needs should be designed for the configuration space chosen. In this brief review I would like to focus mainly on the effective quadrupole operator required for the calculation of E2 transitions and moments (see ref. 1 for more details). We should begin by recalling the basic perturbation theory framework which underlies the actual calculations.

2. FORMALISM

Since the formalism for the effective interaction has already been extensively discussed, it should be sufficient to outline briefly the approach of Brandow[2] oversimplifying a little as we go. As usual the Hamiltonian is split into unperturbed and perturbing parts $H = H_0 + V$ and we wish to approximate the solution of the many-body Schrödinger equation $H\Psi = E\Psi$. Using operators P and Q which project into and out of the desired shell model (or valence) space we can write the true wave function as

$$\begin{aligned} \Psi &= (P+Q)\Psi \\ &= \Psi_D + \frac{Q}{E-H_0} V\Psi = \Omega\Psi_D \ . \end{aligned}$$

Here Ψ_D is the component ofthe true wave function in the model space and we have defined a model operator Ω which when operating on the model wavefunction Ψ_D yields the true wave function. Clearly Ω satisfies the equation

$$\Omega = 1 + \frac{Q}{E-H_0} V\Omega = 1 + \frac{Q}{E-H_0} V^{EFF}$$

and in the last equality we have pointed out that contact can be made here with the effective interaction, V^{EFF}. Now we wish to separate out those parts of Ω which refer to the non-degenerate core in the absence of valence particles (Ω_c) and those parts which operate on the valence particles themselves (Ω_v). This can

+Work supported in part by ERDA Contract No. AT(11-1)1764.

be done using expansions of the type

$$(\varepsilon + \Delta - H_o)^{-1} = (\varepsilon - H_o)^{-1} \left(1 - \frac{\Delta}{\varepsilon - H_o} + \left(\frac{\Delta}{\varepsilon - H_o} \right)^2 - + \cdots \right),$$

together with the factorization theorem, so that we obtain

$$\Omega = \Omega_v \Omega_c \quad , \text{ with}$$

$$\Omega_v = 1 + \frac{Q}{E_v - H_o} V \Omega_v \quad \text{and } E = E_c + E_v \; .$$

Here by splitting the total energy into core and valence parts and expanding out the core parts we can remove all "vacuum fluctuation" diagrams which contribute to the core energy E_c. Thus the diagram of Fig. 1A is removed. In addition by suitably expanding out the valence parts all reference to the valence particles is removed from Ω_c. We can now calculate the matrix elements of a one-body operator O.

$$\langle f|O|i\rangle = \langle \Psi_f|\Psi_f\rangle^{-\frac{1}{2}} \langle \Psi_f|O|\Psi_i\rangle \langle \Psi_i|\Psi_i\rangle^{-\frac{1}{2}}$$

$$= \frac{\langle \Psi_{Df}|\Omega^{+}_{vf}\, O\, \Omega_{vi}|\Psi_{Di}\rangle}{\langle \Psi_{Df}|\Omega^{+}_{vf}\, \Omega_{vf}|\Psi_{Df}\rangle^{\frac{1}{2}} \langle \Psi_{Di}|\Omega^{+}_{vi}\, \Omega_{vi}|\Psi_{Di}\rangle^{\frac{1}{2}}} + \delta_{if} \langle O\rangle_{core}$$

We have oversimplified here somewhat because one may have pieces from Ω_v and Ω_c in the bra and ket which join together in such a way that a linked diagram is obtained. These cross terms are to be included in the valence contribution—they cause no problem and occur naturally when one writes out the diagrams which can contribute.

The last stage of the analysis is to write $E_v = \varepsilon + \Delta E_v$, where ε is the unperturbed valence energy and to expand out ΔE_v in both the numerator and denominator of the above expression thus producing folded diagrams. The result is most easily stated using the Q box notation.[3] We recall from Ratcliff's talk that $V^{EFF} = Q(\varepsilon) + \frac{dQ(\varepsilon)}{d\varepsilon} Q(\varepsilon) + \cdots$, where the Q box sums all non-folded diagrams (with Rayleigh-Schrödinger denominators) and the second term puts in all once folded diagrams. The norm we require is essentially just the derivative of V^{EFF} so we have

$$\langle \Psi_{Di}|\Omega^{+}_{vi}\, \Omega_{vi}|\Psi_{Di}\rangle = 1 - \frac{dQ(\varepsilon_i)}{d\varepsilon_i} - \frac{d^2 Q(\varepsilon_i)}{d\varepsilon_i^2} Q(\varepsilon_i) \cdots .$$

The $\frac{dQ(\varepsilon_i)}{d\varepsilon_i}$ term is sometimes called a divided Q box, since a squared denominator is present at some point which divides the diagram into two halves. We can also write

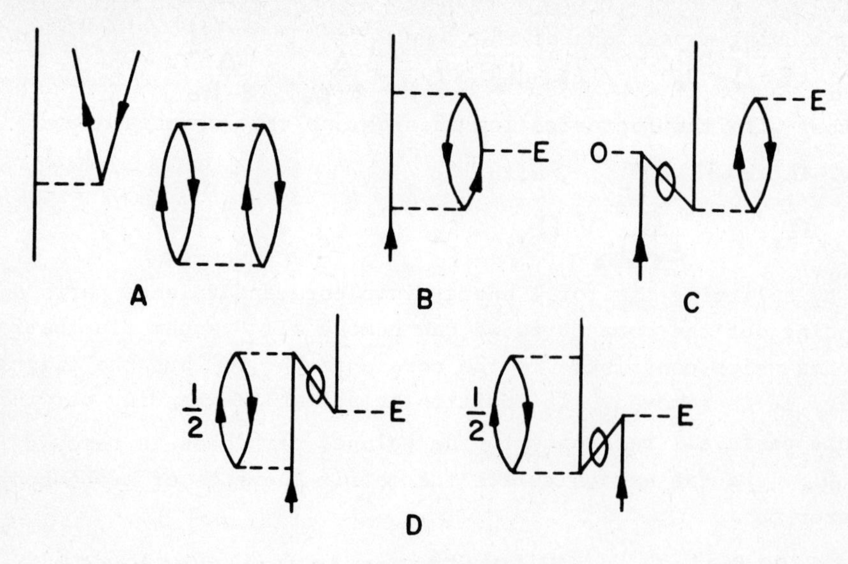

Figure 1

$$\langle \Psi_{Df} | \Omega^{+}_{vf} O \, \Omega_{vi} | \Psi_{Di} \rangle = X(\varepsilon_f, \varepsilon_i) + Q(\varepsilon_f) \frac{dX}{d\varepsilon_f}(\varepsilon_f, \varepsilon_i) + \frac{dX}{d\varepsilon_i}(\varepsilon_f, \varepsilon_i) \, Q(\varepsilon_i) \; \ldots .$$

where the X box now sums all unfolded diagrams containing the operator O. Finally we have

$$\langle f | O | i \rangle = \left(1 + \frac{1}{2} \frac{dQ(\varepsilon_f)}{d\varepsilon_f} \ldots\right) \left\{ X(\varepsilon_f, \varepsilon_i) + Q(\varepsilon_f) \frac{dX}{d\varepsilon_f}(\varepsilon_f, \varepsilon_i) + \frac{dX}{d\varepsilon_i}(\varepsilon_f, \varepsilon_i) \, Q(\varepsilon_i) + \ldots \right\} \left(1 + \frac{1}{?} \frac{dQ(\varepsilon_i)}{d\varepsilon_i} \ldots \right) + \delta_{if} \langle O \rangle_{CORE}$$

Although the derivation seems unavoidably complicated the result is simple and examples of the valence diagrams which occur are given in Figure 1, where the one-body operator is represented by the letter E. Thus Figure 1B shows an unfolded diagram arising from X, Figure 1C shows a folded diagram arising from the $\frac{dX}{d\varepsilon_i}$ term. Figure 1D shows folded diagrams arising from the norms of the initial and final states producted with the lowest order contribution to X — the bare operator. The extra factor of 1/2 arises from the

binomial expansion of $(1-x)^{-1/2} = 1 + \frac{1}{2}x + \ldots$.
We now have an expansion for effective operators which involves Rayleigh-Schrödinger energy denominators and which can be shown to be completely linked in the general case of several valence particles (note however that it is necessary to use orthogonal model space wave functions). This is the desired result and we can now go on to discuss the calculations which have been performed.

3. ORDER-BY-ORDER CALCULATIONS

Much of the work has centered on the calculation of the effective quadrupole operator in the sd shell region and, as I remarked, I want to spend most of my time on this. The operator is of course $Q_{2m} = \sum_i e_i r_i^2 Y_{2m}(\Omega_i)$ and it is customary to normalize the results by dividing by the proton matrix element evaluated in an unperturbed harmonic oscillator basis so as to express the results in terms of an effective charge, i.e.

$$e_{fi}^{EFF} = \frac{\langle f | Q_{2m} | i \rangle}{\langle \phi_f | Q_{2m} | \phi_i \rangle}$$

Note that the core contribution to the numerator is zero, since $J = 0$ for the core and Q is a tensor of rank two. As regards experimental data in mass 17, the odd neutron data yield values of about 1/2. For the odd proton the $1/2^+$ to $5/2^+$ γ-decay yields a value of 1.75. This large value is surely due to the fact that the $1/2^+$ level is bound by only 0.1 MeV and if one applies a simple correction for this effect using wave functions calculated in a Woods-Saxon well one obtains a value somewhat less than 1.5. So an additional effective charge of about 1/2 on neutrons and protons is indicated, which is consistent with shell model studies in this region.

Let us start by examining the results obtained order-by-order in the G matrix (I don't need to remind you that the strong repulsion in the nucleon-nucleon interaction at short distances requires the summation of ladder diagrams so that V is replaced by G). There is presumably no formal bar to obtaining convergence here since there are no intruder states in mass 17. However, one might worry about low lying particle-hole states which would be hard to describe in perturbation theory. They do produce some effect for the $1/2^+$ to $5/2^+$ γ-decay, but not for the ground state quadrupole moment.[4] This could lead to convergence problems.

		$d_{5/2} - d_{5/2}$ N	P	$d_{3/2} - s_{1/2}$ N	P	
ZERO ORDER		0	1	0	1	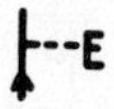
FIRST ORDER	NON HF	0.28	0.11	0.20	0.06	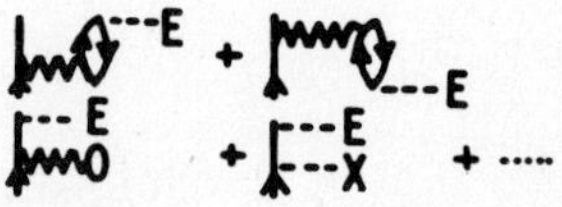
	HF	0	0.08	0	0.34	
	TOTAL	0.28	0.19	0.20	0.40	
SECOND ORDER	NON HF	0.06	-0.06	0.01	-0.16	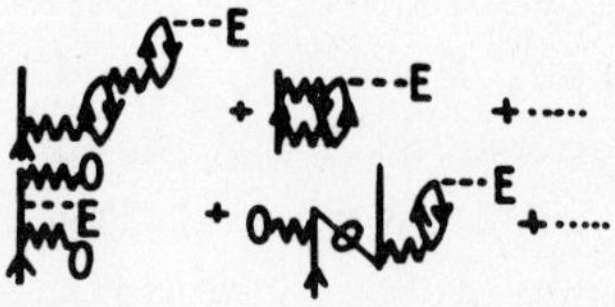
	HF	-0.18	-0.07	-0.08	0.11	
	TOTAL	-0.12	-0.13	-0.07	-0.05	

Table 1. Order by order results for the effective charge.

Working for us, however is the presence of r^2 in the operator which effectively cuts off the intermediate state summations for the core polarization diagram in contrast to the case of the effective interaction. Some selected results[5] are given in Table 1; a harmonic oscillator unperturbed Hamiltonian has been used and the intermediate states are restricted (roughly) to be of $2\hbar\omega$ excitation energy ($\hbar\omega\sim 14$ MeV). The notation $d_{3/2} - s_{1/2}$ indicates the effective charge for a transition between the $d_{3/2}$ and $s_{1/2}$ states and N(P) refer to neutron (proton). Two different sets of G-matrix elements have been used in the computations, but the indications are strong that this will not change the qualitative conclusions. If one simply looks at the non-Hartree-Fock results labelled NON HF implicitly assuming that the HF and single particle insertions cancel, e.g. the diagrams shown in the first order HF row, there is no evidence of convergence for protons. We may remark that the second order TDA and vertex correction diagrams shown are large, but cancel less strongly than for the effective interaction. Of course Table 1 indicates that the HF effects are important. Sizeable corrections are seen already in first order for protons as might be expected since we are evaluating matrix elements of r^2 for weakly bound valence particles. The second order totals are seen to be smaller than first order and of opposite sign. The ratio is very roughly -1/3 and if one simply assumes a geometric series this implies that the full result is 3/4 of the first order value. This is too small so one must conclude that there is no smooth convergence to the desired value order-by-order in G.

There are a couple of more detailed points which emerge with greater clarity here than in the effective interaction calculations. The implicit assumption is being made that similar qualitative ideas apply in the two cases—this seems to have some support from the strong correlation in sign between corresponding diagrams. Firstly it seems unwise to rely on the cancellation of the number conserving sets. An example is given in Figure 2. If E were the number

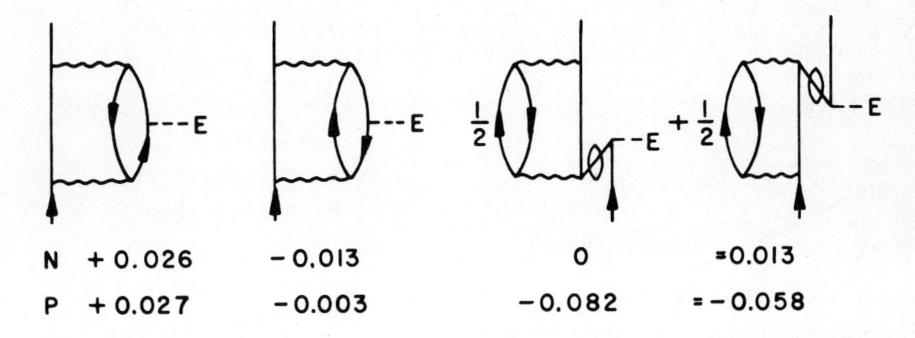

Figure 2. Contributions from a number-conserving set to the $d_{5/2}$ - $d_{5/2}$ effective charge.

operator exact cancellation would ensue, but clearly for the quadrupole operator the cancellation is poor. Secondly the diagrams of Figure 3 were not included since this would involve double counting. Would convergence be improved by doing a double partitioned calculation and including this pair? The answer is probably no—while they give a significant contribution for neutrons of about 0.06, the figure for protons is very small, 0.01. (Other diagrams would also enter but the indications are that they are small).

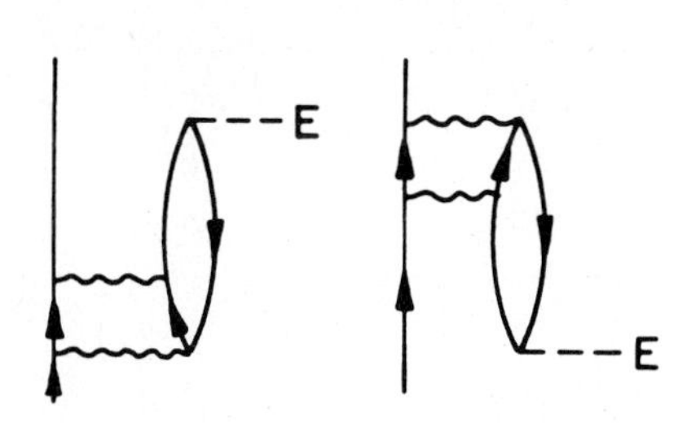

Figure 3.

$$\delta\mu = 0.379\ \delta\langle s\rangle - 4.705\ \delta\langle \tau s\rangle - 0.500\ \delta\langle \tau \ell\rangle$$

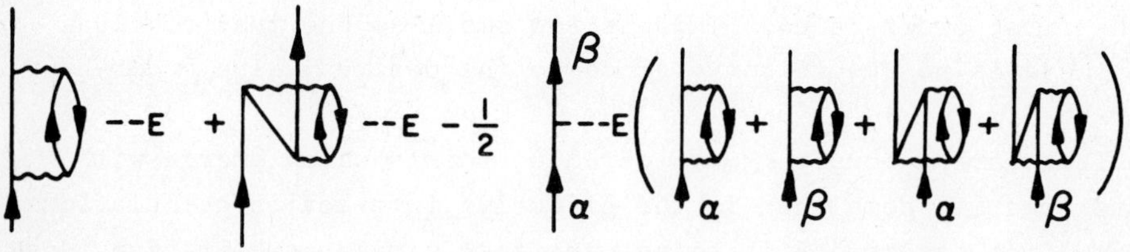

Figure 4.

We remarked that the form of the quadrupole operator is helpful in restricting the intermediate state summation for the core polarization diagram. However, for other diagrams this will not be the case and it is useful here to study the magnetic moment work of Shimizu, Ichimura and Arima.[6] The deviation of the magnetic moment from the single particle Schmidt value can be written in the form shown in Figure 4. Note that $\delta\langle \ell\rangle$ can be removed since $\delta\langle \ell\rangle + \delta\langle s\rangle = 0$ as j is a good quantum number. For a core with closed L-S shells there is no first order effect and the second order diagrams are shown in Figure 4. In the first two diagrams the M1 operator (labelled E) can act on any of the intermediate lines at the level indicated. The third term arises from the normalization of the wave function and the diagrams are more properly written in folded form. Notice

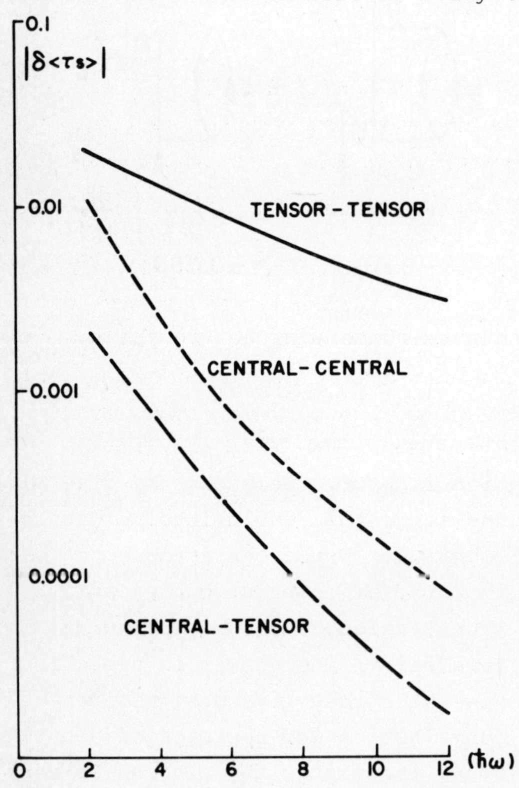

Figure 5. Contributions to $|\delta\langle \tau s\rangle|$ from central and tensor forces as a function of intermediate state excitation energy.

	$\delta\langle s\rangle$	$\delta\langle \tau s\rangle$	$\delta\langle \tau \ell\rangle$
G - MATRIX $(2\hbar\omega)$	-0.013	-0.019	-0.239
TENSOR FORCE $(4-12\hbar\omega)$	-0.078	-0.034	-0.230
TOTAL	-0.091	-0.053	-0.469

Table 2. Second order corrections to the magnetic moment of the $d_{5/2}$ orbital in mass 17.

that the diagrams of Figure 4 comprise two number-conserving sets. In addition to studying $2\hbar\omega$ intermediate states, Shimizu et.al. considered excitations up to $12\hbar\omega$ using the Hamada-Johnston tensor force together with a short range correlation function to cut out the hard core. The full line inFigure 5 shows that the pure tensor contribution to the diagrams of Figure 4 drops off slowly as a function of excitation energy. Curves are also plotted for a phenomenological Yukawa central potential and it is seen that the pure central and central-tensor cross terms drop off quite rapidly. Table 2 also shows that the tensor force couples in the highly excited intermediate states strongly so that they contribute at least as much as the states of $2\hbar\omega$excitation energy. Hartree-Fock effects were not included in these calculations, but Habbal and Mavromatis[7] have pointed out that such effects are negligible for closed LS shells. This is to be expected since only two independent diagrams contribute. As regards comparison with experiment exchange current corrections should of course be taken into account and if this is done the results do not look unreasonable. What would the inclusion of these highly excited intermediate states imply for the effective charge ? The calculations have not been done, but if the dominant effect is to enhance the number conserving sets by a factor of 2 or 3 convergence would look distinctly worse. Of course it has been tacitly assumed in most calculations that the low and high lying states should be treated differently, with the hole line expansion for the latter. This may still be valid.

It should be mentioned that effective operators will in general have n body parts. For instance in mass 18

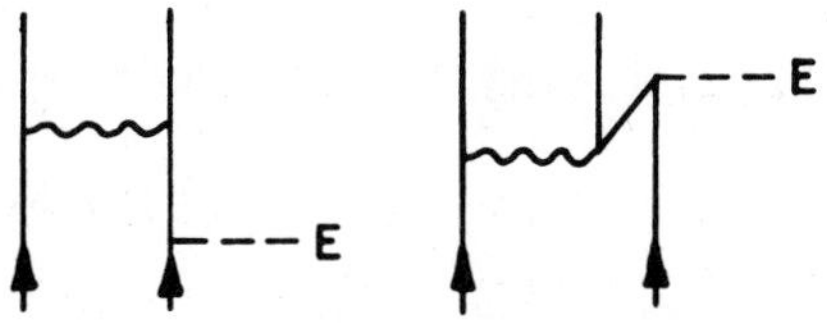

E

E

Figure 6.

the lowest order two-body effective operator is given by the diagrams of Figure 6. Rather little work has been done on such effects but Harvey and Khanna[8] have made an estimate for mass 18 based on a phenomenological quadrupole-quadrupole interaction. They find that the matrix element of the two-body operator varies from about 5 to 20% of the corresponding one-body matrix element. This is really rather large, particularly when one reflects that the number of pairs increases rapidly as particles are added so that the two-body pieces are more strongly weighted. However Lo Iudice, Rowe and Wong[9] have used a diagonalization approach and obtained a much smaller figure, 1% or less. The situation is thus unclear, but one hopes that such effects are small—empirically this appears to be the case.

4. PARTIAL SUMMATIONS

It seems that the order-by-order approach does not work very well for the effective charge, so that it is natural to ask what is the result of summing selected sets of diagrams? Perhaps the most basic summation is that carried out by Siegel and Zamick[10] many moons ago. As shown in Figure 7 they generalized the core polarization diagram A by allowing the particle and hole to interact, thus generating the TDA series B, which can be written as a particle-hole ladder summation. Including in addition the "backward-going" vertices the RPA diagrams C are also generated. As indicated this simply corresponds to different descriptions of the particle-hole pair—unperturbed states for A, TDA phonons for B and

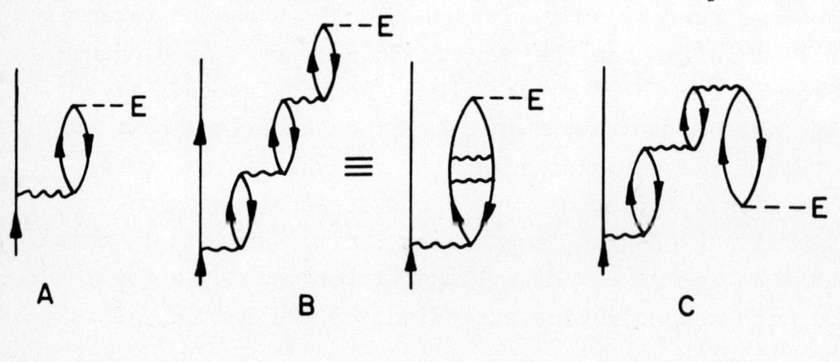

$$\sum \frac{\langle (sd)\,|\,G\,|\,(sd)\,2^+\ \text{PHONON}\rangle\langle 2^+\ \text{PHONON}\,|\,Q\,|\,0\rangle}{-E_{\text{PHONON}}}$$

$$\frac{1}{2\hbar\omega} \longrightarrow \frac{1}{2\hbar\omega + A} \longrightarrow \frac{1}{2\hbar\omega + A + B}$$

Figure 7.

RPA phonons for C. The result is simple if we only have one state and results in the propagator replacements shown in Figure 7. Thus if the particle-hole matrix element A is attractive an enhancement will result in TDA. A further enhancement is given in RPA if the "backward-going" vertex B is attractive. In actual calculations both the correlation of the wave function and the energy shift are equally important. Some results obtained for mass 17 are given in Table 3 and we see a large enhancement for the TDA and an even larger effect in RPA. This, ofcourse, is correlated with the position of the isoscalar quadrupole phonon which is also shown. It appears to be somewhat low here, in fact in ^{40}Ca it is near to collapse at 6.8 MeV in the RPA, compared to current giant resonance work which would place it at around 25 MeV in ^{16}O and 20 MeV in ^{40}Ca.

The next effect to be incorporated is the screening of the particle-hole interaction--a sample diagram is labelled SCREEN in Table 3. The results of Kuo and Osnes[11] show that screening

	$d_{5/2}-d_{5/2}$		$d_{3/2}-S_{1/2}$		$E_x(2^+ T=0)$
	N	P	N	P	
1ST ORDER	0.33	1.10	0.24	1.05	28.0
TDA	0.50	1.32	0.32	1.17	15.3
RPA	0.66	1.48	0.38	1.23	14.5
SCREENED RPA	0.44	1.25	0.30	1.14	20.0
SCCE	0.32	1.13			
SHELL MODEL	0.43	1.14	0.29	0.94	
0TH ORDER HF	0	1.10	0	1.63	
HFRPA	0.16	1.19	0.14	1.70	29.5

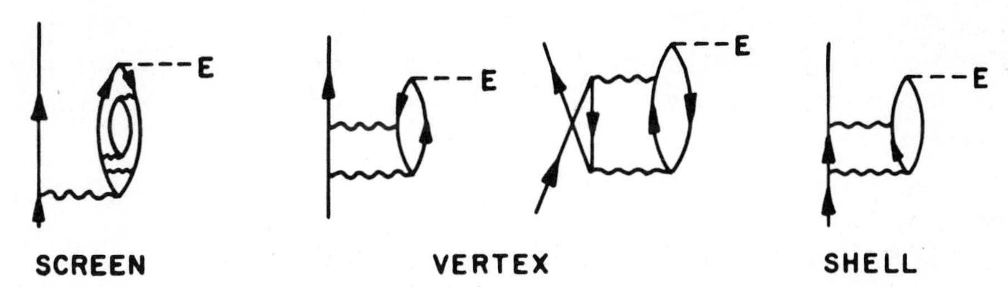

Table 3. The effective charge calculated in various approximations. Also shown is the position of the isoscalar quadrupole phonon.

reduces the RPA result to the TDA one. It also serves to push up the quadrupole phonon to a more reasonable energy. We might also note here the obvious point that the results obtained depend on the energy denominators used; smaller enhancements are found if empirical single particle energies are used rather than oscillator ones. Kirson[12] goes further than this by including in addition vertex renormalization effects such as illustrated in Table 3. He includes vertex and self-screening corrections self-consistently to all orders to obtain the self-consistent coupled equation (SCCE) results of Table 3. The final outcome of summing this vast set of diagrams is quite close to first order which is disappointing since this is too small. The effect is, in fact, less dramatic than for the effective interaction since the vertex corrections are smaller here.

Finally we can contrast this explicit diagram summation approach with the shell model approach of Lo Iudice, Rowe and Wong[9]. They take the one particle state, along with all two particle-one hole configurations of $2\hbar\omega$ excitation energy and diagonalize the matrix. Such a procedure includes the TDA diagrams and many other diagrams as well. We see from Table 3 that the neutron results are close to TDA whereas the proton results are close to first order. This seems to be understandable since they include some, but not all, of the vertex corrections of Kirson, and in addition they include ladder diagrams of the type labelled SHELL in Table 3 which push things in the opposite direction. (There is clearly some double counting here and much argument has been generated about how much!) As we have remarked the ladder diagram is much larger for neutrons than protons so it is not surprising that the neutron shell model result is relatively the larger. One might also expect that the normalization of the wave function in the shell model calculations is significant particularly for the proton case, a feature which has been emphasised by Goode, West and Siegel[13] in a somewhat different context.

Thus far we have discussed effects due to the coupling of quadrupole phonons treated in various approximations without worrying about monopole or Hartree-Fock corrections which we have previously seen to be large. What is the effect of using an unperturbed Hartree-Fock Hamiltonian instead of a pure oscillator one? One effect is to make the calculation much more difficult! Other effects are shown in Table 3. We see that the bare effective charge for protons, which just involves the ratio of the matrix element of r^2 in the HF basis to that in an oscillator basis, is enchanced.[5] Similar results are obtained using Woods-Saxon wave functions for these weakly bound valence states and surely such an effect is needed to get reasonable proton charges. Also shown in the table are the results of an RPA calculation using a HF basis.[14] The drastic reduction in the RPA result arises primarily from the pulling in of the occupied state wave functions and the pushing out of the unoccupied ones. The calculations were done with the Sussex matrix elements[15] which may not be entirely reliable, but I believe the qualitative trend. Indeed Köhler[16] has pointed out that since core polarization effects correspond physically to disturbing the ^{16}O core, they will be smaller if the core is initally in a state of equilibrium. By carrying out a HF calculation we have minimized the total energy, which corresponds to an equilibrium condition.

We have seen that both screening and HF effects serve to stabilize the quadrupole phonon. This is even more dramatic for the monopole case where the isoscalar phonon is close to collapse in RPA. Even so attempts to calculate the monopole core polarization have not been encouraging--the sign of the isotope shift is incorrect unless one puts in an explicit density dependence.[17]

So what seems to be emerging from all this is the not very surprising result that it is rather important to get the monopole and quadrupole fields about right, as has been strongly emphasised by the Chalk River group. In this connection it is useful to discuss briefly the model calculations of Harvey.[18] What he does is to take a Hamiltonian consisting of just a harmonic oscillator potential and a quadrupole-quadrupole interaction and carry out a deformed Hartree-Fock calculation. This results in a HF single particle Hamiltonian containing an oscillator Hamiltonian H_0 (frequency ω_0) plus a quadrupole deformation term. For axial symmetry this takes the form $H_\delta = H_0 - \hbar\omega_0 \delta Q_{20}$, the eigensolutions of which are easily obtained in a Cartesian representation. The self-consistency condition is $-\delta = x \langle \chi_\delta | Q_{20} | \chi_\delta \rangle \equiv x \langle Q_{20} \rangle_\delta$, where x is the strength of the interaction. Harvey considers the case of a prolate wave function in ^{20}Ne

$$\chi_\delta = \left| (000)^4 \; (001)^4 \; (010)^4 \; (100)^4 \; (002)^4 \right|$$

where (n_x, n_y, n_z) gives the number of quanta in the x, y and z directions. Then the quadrupole moment $\langle Q_{20} \rangle_\delta$ can be evaluated with this wave function and self-consistency obtained. Harvey then tries to reproduce the quadrupole moment $\langle Q_{20}\rangle_\delta$ starting with a spherical basis (δ=0) and interpret the results in terms of perturbation theory. This is non-degenerate perturbation theory since the state X_δ has a single component X_0 in the chosen model space of $(sd)^4$ outside an ^{16}O core. He shows that $\langle Q_{20}\rangle_\delta = \langle Q_{20}\rangle_0 - 2(2S_3+S_1)x\varepsilon\langle Q_{20}\rangle_0 + 3(4S_3-S_1)x^2\varepsilon^2\langle Q_{20}\rangle_0^2 + \ldots$ where the "effective charge" $\varepsilon = \langle Q_{20}\rangle_\delta / \langle Q_{20}\rangle_0$ so that this embodies the HF self-consistency. (For ^{20}Ne S_1 = 14 and S_3 = 22). The first two terms solved self-consistently yield the RPA approximation shown in Figure 8. Note that the unperturbed vacuum state does not have good angular momentum, hence the diagrams shown are non-zero. Further terms in the equation yield diagrams such as those labelled bb-bd and bb-bd-IBG and these must be included self-consistently thus building up very complicated diagrams. The results are given by the intersection of the 45° line in Figure 8 with the curves and we see that only the bb-bd-IBG approximation is close to the exact result labelled UDHF. Although the comparison to other work is not entirely straightforward because of the simplicity of the interaction and of the many-body contributions which are implicitly included, we may roughly identify bb-bd with vertex renormalized RPA. The enhancement seen here appears to be due to the particle-particle interactions (which should be included in this calculation). The remaining diagrams labelled IBG will be recognized as none other than our old friends the number conserving sets. They seem to be important, although interestingly enough less so for the effective interaction

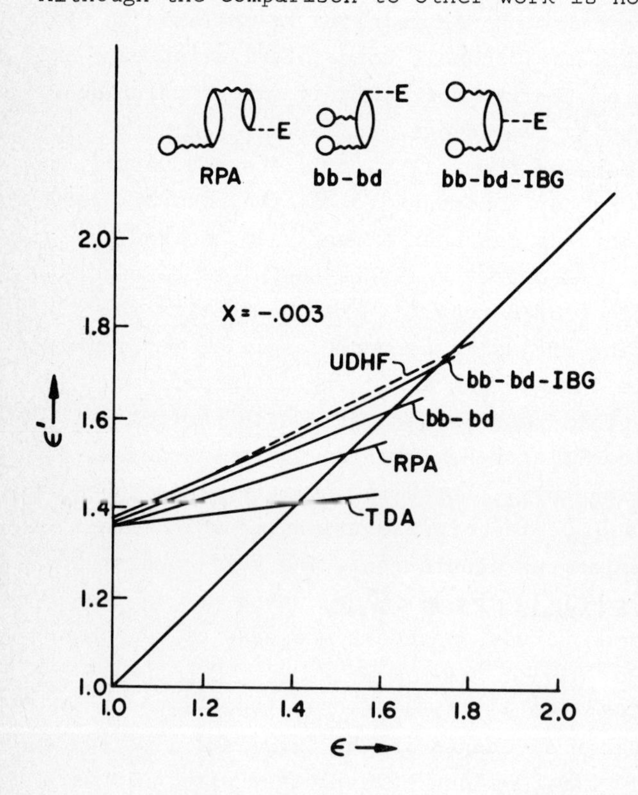

Figure 8.

(the total energy in this connection),

5. CONCLUDING REMARKS

What have we learnt, apart from the fact that the problem we have set ourselves is a difficult one? Firstly, order-by-order calculations don't seem to work and there seems to be no reason why they should. Secondly, it is important to get the monopole and quadrupole fields about right, but we do not yet know how to do this. I simply remark that if one calculated the bare contribution in a Hartree-Fock basis and took the oscillator TDA renormalization correction one would get reasonable orders of magnitude for both the effective charge and effective interaction. Unfortunately I don't see any justification for doing simply this. One feature that is missing from current treatments is the number-conserving sets. Hopefully it is not necessary to go to high excitation energy, although Shimizu et.al.[6] say it is. Perhaps one should try to construct a full TDA theory, including diagrams of the form indicated in Figure 9, together with appropriate folded diagrams. One has the picture here that the valence particle travels along interacting with the core and producing correlated TDA particle-hole excitations. Finally of course these particle-hole pairs must be destroyed so that one ends up in a single particle valence state. There are many other effects one might wish to build in, but I don't have anything to report on this as yet, so on that speculative note I will close.

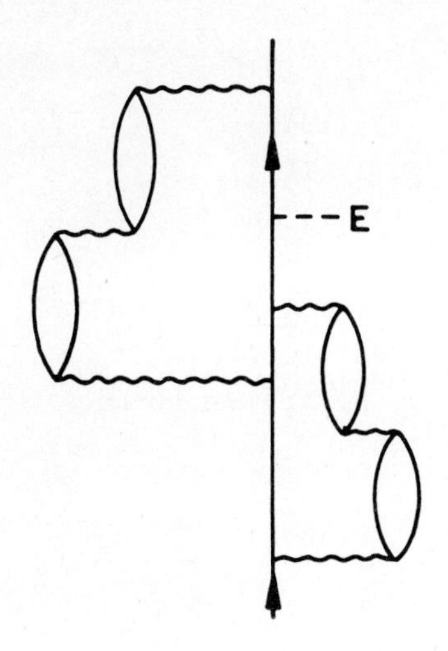

Figure 9.

REFERENCES

1. Barrett, B. R., Kirson, M. W., in Advances in Nuclear Physics, (M. Baranger and E. Vogt, eds.) Vol. VI (Plenum Press, N. Y., 1973), p. 219.
2. Brandow , B. H., Revs. Mod. Phys. 39, 771 (1967).
3. Kuo, T. T. S., Lee, S. Y., Ratcliff, K. F., Nucl. Phys. A176, 65 (1971); Krenciglowa, E. M., Kuo, T. T. S., preprint.
4. Engeland, T., Ellis, P. J., Nucl. Phys. A181, 368 (1972); Goode, P., Nucl. Phys. A172, 66 (1971).
5. Ellis, P. J., Siegel, S., Phys. Lett. 34B, 177 (1971); Ellis, P. J., Mavromatis, H. A., Nucl. Phys. A175, 309 (1971).
6. Shimizu, K., Ichimura, M., Arima, A., Nucl. Phys. A226, 282 (1974).
7. Habbal, S. R., Mavromatis, H. A., Nucl. Phys. A223, 174 (1974).
8. Harvey, M., Khanna, F. C., Nucl. Phys. A155, 337 (1970).
9. Lo Iudice, N., Rowe, D. J., Wong, S. S. M., Nucl. Phys. A219, 171 (1974).
10. Siegel, S., Zamick, L., Nucl. Phys. A145, 89 (1970).
11. Kuo, T. T. S., Osnes, E., Nucl. Phys. A205, 1 (1973).
12. Kirson, M. W., Ann. Phys. (N. Y.) 66, 624 (1971) and 82, 345 (1974).
13. Goode, P., West, B. J., Siegel, S., Nucl. Phys. A187, 249 (1972).
14. Ellis, P. J., Osnes, E., Phys. Lett. 42B, 335 (1972).
15. Elliott, J. P., Jackson, A. D., Mavromatis, H. A., Sanderson, E. A., Singh, B., Nucl. Phys. A121, 241 (1968).
16. Köhler, H. S., preprint.
17. Ellis, P. J., Osnes, E., Phys. Lett. 49B, 23 (1974).
18. Harvey, M., preprint.

P.J. ELLIS: CALCULATION OF OTHER EFFECTIVE OPERATORS

Barrett: What interaction did you use in your Hartree-Fock calculation?

Ellis: The original Sussex matrix elements.

Barrett: Since the original Sussex interaction does not saturate, how would your Hartree-Fock results be changed, if you used the new Sussex interaction which does saturate?

Ellis: It is true that ^{16}O is underbound and too small. Recent results of Malta and Sanderson which include the effect of an arbitrary hard core adjusted to give the correct radius for ^{16}O show smaller effects. However, I do not fully understand these calculations as one still needs a much larger oscillator parameter for the occupied than for the unoccupied states if one uses oscillator wave functions which give good overlap with the Hartree-Fock results. Calculations with simple forces indicate that this will reduce the matrix elements which enter. Possibly there is a counter balancing effect from the energy denominators which become smaller with the hard core. At any rate I feel that the qualitative features of the results are to be stressed, and, as we heard from Michael Kirson, other interactions show similar trends.

Towner: In one of the tables, there was a row labelled zeroth-order Hartree-Fock, and the entry for $d_{3/2}$-$s_{1/2}$ proton transition was 1.63. This value is just the ratio of the expectation value of $\langle r^2 \rangle$ in a Hartree-Fock basis relative to that in a harmonic oscillator basis. A ratio of 1.63 seems to me to be very large - is this value understood?

Ellis: It is indeed very large. The Hartree-Fock effects are much larger for the $d_{3/2}$ and $s_{1/2}$ wave functions than the for $d_{5/2}$ case. This seems reasonable, since the $d_{3/2}$ state is the least strongly bound (in fact, it is unbound) and since the $s_{1/2}$ wave function has an extra node. Rather similar results are obtained using Woods-Saxon wave functions which gives me some confidence in the numbers.

Harvey: A comment on the question of the two-body effective E2 operator! The experimental data in ^{18}Ne compared with those in ^{18}O do, I believe, indicate an effective E2 operator of the magnitude that Khanna and I indicated. Also there is a difficulty in the understanding of observed E2 transitions between oscillator states in

^{18}F - a problem originally pointed out by Benson and Flowers. The data would seem to suggest an effective E2 operator, but we have not been able to understand this in the phenomenological approaches. Finally, I should emphasize that in the determination of the effective E2 operator by Khanna and myself, we chose a Q·Q interaction that would first fit the observed one-body effective operator - an essential feature if one is to say anything about the two-body part.

Ellis: In the weak coupling calculations that Engeland and I carried out, we obtained rather reasonable agreement for the mass-18 data, using just a one-body effective operator. It is true that transitions between states of a different particle-hole structure are sensitive to small components in the wave function, but the need for a two-body E2 operator is not evident looking at the two-body part.

Kallio: You and many other people have been calculating contributions to effective charges from very high-lying oscillator shells, especially in the case of the tensor force. Do you have any idea what would happen if you made use of some other spectrum? You don't find it unreasonable using only oscillator states?

Ellis: The calculations I quoted were those of Shimieu, Arima and Ichimura. I do not find it very appealing to use pure oscillator states at such high excitation energy, but I do not know what would happen if one used some other representation. I doubt that the large tensor force effects would go away.

THEORY VERSUS THEORY AS A TEST OF THE EFFECTIVE INTERACTION*

L. ZAMICK
Department of Physics
Rutgers University, New Brunswick, New Jersey 08903

Invited Talk: Tucson Conference on Effective Interactions and Operators, University of Arizona, June 2-6, 1975

ABSTRACT: Using density dependent interactions of the Skyrme type, we calculate the energies of the giant isoscalar monopole and quadrupole states in T.D.A. and R.P.A. Then we use a collective model involving the Inglis cranking formula. Very close agreement is obtained. The simplicity of using oscillator wave functions with Skyrme forces, especially for constructing the equation of state, is noted. Then the isoscalar effective charges are calculated in the R.P.A. and compared with the Hartree Fock calculations. There are significant differences, and attempts to make the results converge, by making the deformation parameters for the valence nucleon different from those of the core, are discussed. Most of the calculations were done by Michaela Golin.

In this talk I will show that by comparing one theory against another one can learn a great deal about the effective interaction between nucleons in a nucleus. I will use the giant monopole and giant quadrupole states, as well as the effective charge, as examples. The theories involved are the T.D.A., R.P.A., adiabatic cranking model and the Hartree-Fock.

I feel that in order to properly test an effective interaction one should use it to calculate everything - the binding energy, the single-particle energies, the particle-particle and particle-hole matrix elements. Hybrid calculations, in which for example single-particle energies are taken from experiment , or alternately, the ad-hoc prescriptions like making single-particle energy differences equal to some multiple of $\hbar\omega$ (this is neither theory nor experiment), do not really test the interaction and often lead to confusing results.

But to undertake such a program one needs at the very least an interaction which leads to nuclear saturation, preferably at the right energy and the right radius. The Skyrme interactions, which

*Supported in part by the National Science Foundation.

were reviewed by Vautherin and Brink, do just this, and so we will use them. A somewhat truncated version of the interaction is:

$$V = -\alpha\delta(\vec{r}_1-\vec{r}_2) + \Delta\delta(\vec{r}_1-\vec{r}_3)\delta(\vec{r}_2-\vec{r}_3) + \frac{\beta}{2}[k^2\,\delta(\vec{r}_1-\vec{r}_2) + \delta(\vec{r}_1-\vec{r}_2)k^2]$$

where $\vec{k} = (\vec{k}_1-\vec{k}_2)/2$.

There is an attractive two-body delta term to hold the nucleus together. There is a repulsive three-body delta term to prevent the nucleus from collapsing to a point. In binding energy calculations for nuclei such as ^{16}O and ^{40}Ca this interaction is equivalent to a linear density-dependent interaction $(\Delta/6)\rho((\vec{r}_1+\vec{r}_2)/2)\delta(\vec{r}_1-\vec{r}_2)$. In single particle energies one uses $\Delta/4$ and for particle-particle and particle-hole interactions $\Delta/2$ is used.

The β term is called a finite range or velocity-dependent interaction. It is also repulsive and can by itself lead to saturation. One really doesn't need two kinds of repulsion to achieve saturation. But the two repulsive terms are required to get other properties such as the single particle spectrum. For example, the smaller β is (whilst readjusting α and Δ so as to still fit the binding energy and radius) the more compressed is the single particle spectrum.

We shall use three versions of this interaction.

a) Zero Range Force $\beta = 0$

	α MeV $[fm]^3$	Δ MeV $[fm]^6$	$\hbar\omega$
^{16}O	1085.132	20196.307	13.36
^{40}Ca	1107.294	19563.729	10.60

b) Finite Range Force $\Delta = 0$

	α MeV $[fm]^3$	β MeV $[fm]^5$	$\hbar\omega$
^{16}O	1285.310	1100.328	13.36
^{40}Ca	1314.24	1199.11	10.60

c) The VBII' Interaction

α = 1191.667 MeV $[fm]^3$, β = 585.6 MeV $[fm]^5$, Δ = 9447.725 MeV $[fm]^6$ for ^{16}O and α = 1208.346 MeV $[fm]^3$, β = 585.6 MeV $[fm]^5$, Δ = 10011.116 MeV for ^{40}Ca.

In the above, the parameters α, Δ and β were adjusted so that when trial Slater determinants involving harmonic oscillator wave functions were used, equilibrium was reached at a value $\hbar\omega = 13.36$ MeV for ^{16}O and $\hbar\omega = 10.60$ MeV for ^{40}Ca. ($b = 1.76$ fm, $b = 1.98$ fm, respectively.) The binding energies used were $E_B = 139.5$ MeV, $E_B = 410.8$ MeV, respectively.

For the zero range interaction, all the repulsion comes from the linear density-dependent term; for the finite range interaction all the repulsion comes from the velocity-dependent term. These can be regarded as two extremes of the Vautherin-Brink series. The third interaction is very close to the Vautherin-Brink II interaction.

Equations of State

It turns out that Skyrme forces and oscillator Slater determinants are made for each other. This is because of the scaling properties. If $\Psi(b,1\ldots A)$ is a Slater determinant involving only the oscillator length parameter b ($\hbar\omega = \frac{\hbar^2}{mb^2}$) then the trial energy with the Skyrme force will depend only on b.

$$E(b) = \langle\Psi|H|\Psi\rangle$$

$$= \frac{\Sigma'}{2b^2} - \frac{A(\alpha)}{b^3} + \frac{B(\beta)}{b^5} + \frac{C(\Delta)}{b^6}$$

where A, B, C are independent of the oscillator length parameter b. The condition for a minimum is simply $\frac{dE(b)}{db} = 0$. The first term above is the kinetic energy $\Sigma' = \frac{\hbar^2}{m}\Sigma$

$$\Sigma = \sum_{\substack{\text{occupied}\\ \text{states}}} (2n+\ell+3/2) \qquad n = 0,1,\ldots$$

If we use a deformed oscillator trial wave function in which there are 2 length parameters $b_x = b_y$, b_z, then it is convenient to introduce $b_o = (b_x b_y b_z)^{1/3}$, and the equation of state now becomes

$$E(b_o,b_z) = \frac{\Sigma_{x+y}' b_z}{2b_o^3} + \frac{\Sigma_z'}{2b_z^2} + \left[\frac{2b_z}{3b_o^3} + \frac{1}{3b_o^2}\right]\frac{1}{b_o^3}B(\beta)$$
$$- \frac{A(\alpha)}{b_o^3} + \frac{C(\Delta)}{b_o^6}$$

The minimum conditions are: $\left(\frac{\partial E}{\partial b_z}\right)_{b_o} = 0$ and $\left(\frac{\partial E}{\partial b_o}\right)_{b_z} = 0$. In the above

$\Sigma_{x+y}' = \frac{\hbar^2}{m}\Sigma(N_x+\frac{1}{2})+(N_y+\frac{1}{2})$ $\Sigma_z' = \frac{\hbar^2}{m}\Sigma(N_z+\frac{1}{2})$, where N_x, N_y, N_z are the number of quanta in the x, y, and z directions.

Note that of all the potential energy terms only the finite range term depends on b_z.

The Random Phase Approximation Versus the Adiabatic Cranking Model

I now address myself to the calculation of the mean energies of the giant isoscalar monopole (breathing mode) state and giant isoscalar quadrupole states in ^{16}O and ^{40}Ca.

The E2 state is defined as $|E2\rangle = \sum_i r^2(i) Y_{2,\mu}(i)|0\rangle$ where $|0\rangle$ is the ground state. Likewise $|E0\rangle = \sum_i r^2(i)|0\rangle$, except that one only keeps the part which is orthogonal to the ground state.

The mean energies of these states are

$$\bar{E}^2 = \frac{\langle EL|H|EL\rangle}{\langle EL|EL\rangle}$$

The EL operator acting on the shell model ground state produces a one particle-one hole state. For example

$$|E0\ {}^{16}O\rangle = \sqrt{\frac{3}{18}}|1s_{1/2}\ 0s_{1/2}{}^{-1}\rangle + \sqrt{\frac{10}{18}}|1p_{3/2}\ 0p_{3/2}{}^{-1}\rangle + \sqrt{\frac{5}{18}}|1p_{1/2}\ 0p_{1/2}{}^{-1}\rangle .$$

There are many E2 particle-hole components, the most important of which are $0f_{7/2}0p_{3/2}{}^{-1}$, $0f_{5/2}0p_{1/2}{}^{-1}$, and $0d_{5/2}0s_{1/2}{}^{-1}$.

The T.D.A. and R.P.A. methods are very familiar so I won't bother to discuss them except to remind you that we do not use $2\hbar\omega$ for the energy difference of the single particle and single hole - we calculate the difference.

But I will take some time to discuss the adiabatic cranking model. Note that we are applying this model to the average energy of the state.

If q is a collective variable, then the collective Hamiltonian is $\frac{1}{2} B \dot{q}^2 + \frac{1}{2} Cq^2$ and the energy of the first excited state is $\hbar\Omega = \sqrt{C/B}$.

The problem is always what one should choose for a collective variable. For the monopole state we can choose b itself, for the quadrupole state b_z. The mass parameters are then given by the Inglis cranking formula:

$$\text{MONOPOLE} \qquad B_M = 2\hbar^2 \sum_n |\langle n \tfrac{\partial}{\partial b}\, 0\rangle|^2/(E_n - E_o)$$

QUADRUPOLE $\quad B_Q = 2\hbar^2 \sum_n |\langle n \frac{\partial}{\partial b_z} 0\rangle|^2/(E_n-E_o)$

We take $E_n-E_o = 2\hbar\omega$. Let us consider the quadrupole case in more detail. Since the wave function has the structure $f(x\sqrt{b_z}/b_o^{3/2}, y\sqrt{b_z}/b_o^{3/2}, z/b_z)$ we can replace $\partial/\partial b_z$ by

$$\frac{1}{2b_o} [x \frac{\partial}{\partial x} + y \frac{\partial}{\partial y} - 2z \frac{\partial}{\partial z}]$$

which in turn can be replaced by the commutator

$$\frac{1}{2b_o} (\frac{-2m}{\hbar^2}) [H_o,(x^2+y^2-2z^2)]$$

so that

$$B_Q = \frac{m^2}{8\hbar^2 b_o^2} \sum_n (E_n-E_o)|\langle n(x^2+y^2-2z^2)0\rangle|^2 .$$

But the sum is nothing more than the energy weighted sum rule for quadrupole transitions: $\sum_n (E_n-E_o)|\langle n|Q_o|0\rangle|^2 = 4\hbar^2/m \langle r^2\rangle$.

Thus $B_Q = m\,\Sigma/2 \qquad (\Sigma = \mathrm{Sum}(2n+\ell+3/2))$.

This result has been previously derived by Araujo and is in Bohr-Mottelson, Vol. II.

In the monopole case it turns out somewhat more convenient to use as a parameter the root mean square radius $R_{r.m.s.} = (b^2\,\Sigma/A)^{1/2}$, in which case $B_{r.m.s.} = mA$, the mass of the nucleons. This is identical to a classical result in which we evaluate the kinetic energy of a sphere in which the velocity of a point is proportional to its distance from the center.

To get the parameter C is easy because we already have the equation of state and we expand about the equilibrium. In the monopole case

$$E(b) = E(b_{eq}) + \frac{1}{2} C(b-b_{eq})^2.$$

Alternately,

$$E(b) = E(b_{eq}) + \frac{1}{2} \frac{(b-b_{eq})^2 AK}{b_{eq}^2}$$

where K, the nuclear compressibility is defined as $\frac{b^2}{A} \frac{\partial^2 E}{\partial b^2}$. We find then that the frequency of the vibration is

$$\hbar\Omega = \sqrt{\frac{\hbar^2 K}{m}} \Big/ (r.m.s.)$$

where (r.m.s.) is the root mean square radius. This is identical to the classical result. In the oscillator model (r.m.s.) = $(\Sigma/A)^{1/2}b$.

We need to know the compressibility. I previously derived it for a zero range interaction $-\alpha\delta(\vec{r}_1-\vec{r}_2) + \gamma\rho^{\sigma}(\frac{\vec{r}_1+\vec{r}_2}{2})\rho(\vec{r}_1-\vec{r}_2)$, where σ is the power of the density. Since this force has only 2 parameters and we are using it to fit 2 parameters - namely the binding energy E_B and the oscillator parameter b, then we should be able to express K in terms of these. I choose to express it in terms of the mean kinetic energy and total binding energy. $AK = 9E_B + \langle T\rangle + \sigma(3\langle T\rangle + 9E_B)$. This tells us that K is linear in σ, the power of the density. Hence the energy of the breathing mode state will depend on the nature of the force in a critical manner. Here it depends on the power of the density.

By the way, if we use a schematic interaction $\lambda r^2(1)r^2(2)$, and choose λ so that the system saturates at the proper value of $\hbar\omega$, we find the energy of the breathing mode state is $\sqrt{6}\,\hbar\omega$.

The C parameter for the quadrupole state is very easy to obtain in the case of a zero range force. In that case all the dependence on b_z is contained in the kinetic energy. One obtains the simple result, previously obtained by Mottelson, Hamamoto, and Suzuki,

$$\hbar\Omega_{2+} = \sqrt{2}\hbar\omega \ .$$

The above authors did not explicitly use short range interactions, in fact, Suzuki uses the long range quadrupole-quadrupole force. But they were able to anticipate what short range forces would do.

If we do include the finite range, the generalization is not too difficult. The finite range energy goes as k^2, somewhat reminiscent of the kinetic energy. We get

$$E_{2+} = \sqrt{2}\hbar\omega \left[1 + \frac{\text{Finite range energy}}{\text{Kinetic energy}}\right]^{1/2}$$

Since the finite range energy is chosen to be positive, the energy of the quadrupole state will be raised by increasing this energy. Since the single particle-single hole splitting is increased by having more finite range energy, this result is perhaps not surprising.

However, the breathing mode state energy does <u>not</u> follow the single particle energy. For the zero range force $\sigma = 1$, but for the finite range force, for which the finite range term goes as $\frac{1}{b^5}$, the value of σ is 2/3. Hence we get a slightly <u>lower</u> energy with the finite range force than with the zero range force despite the fact that the single particle-single hole splitting goes up. The fact that a finite range force and a $\rho^{2/3}$ density-dependent force scale the same way has been noted by Köhler.

The next four tables show a comparison of the collective formulas above with usual particle-hole calculations. The calculations for the monopole state of ^{16}O were done by Richard Sharp and the remaining calculations by Michaela Golin.

<u>Quadrupole State</u>

^{16}O ($2\hbar\omega$ = 26.72 MeV)

Interaction	$\overline{\varepsilon_P-\varepsilon_H}$	T.D.A.	R.P.A.	Collective
Zero-Range	27.06	19.78	18.39	18.89
VBII'	35.46	24.60	24.42	24.76
Finite-Range	42.85	28.84	28.75	28.93
^{40}Ca ($2\hbar\omega$ = 21.20 MeV)				
Zero-Range	22.75	15.84	14.20	14.99
VBII'	30.36	19.62	19.34	19.63
Finite-Range	38.33	23.58	23.45	23.53

<u>Monopole State</u>

^{16}O				
Zero-Range	28.36	37.29	35.47	34.87
VBII'	37.18	34.84	34.32	33.01
Finite-Range	44.94	32.69	32.48	32.30
^{40}Ca				
Zero-Range	25.28	31.57	30.24	29.63
VBII'	32.92	29.52	29.06	28.24
Finite-Range	40.92	27.36	27.24	26.71

In the above the first column $\overline{\varepsilon_P-\varepsilon_H}$ is the average single particle-single hole splitting. This is to be compared with $2\hbar\omega$. We see that it is close to $2\hbar\omega$ for the zero-range interaction, but much larger for the finite range interaction.

We see that the agreement between the R.P.A. and the adiabatic cranking model is excellent in all cases. The condition for the validity of the adiabatic model is that the velocity of the collective motion is small compared with that of the Fermi motion, as pointed out by Engel *et al*. and by Schiff. The velocity is the amplitude x frequency. The amplitude for the monopole vibration is estimated to be one half the square radius of a nucleus in which a particle has been excited through 2 major shells and the square radius in the ground state.

The difference in square radius is $\frac{2b^2}{A}$. The Fermi momentum is universally 1.36 fm^2. Let us write the collective frequency as some factor times the oscillator frequency.

$$\hbar\Omega = f\hbar\omega \ .$$

In the above example f ranges from about $\sqrt{2}$ to 3. Hence the ratio

$$\frac{v_{collective}^{\ 2}}{v_{Fermi}^{\ 2}} = \{ \frac{b^2}{A} (f\hbar\omega)^2 \} / \{ \frac{\hbar^2}{m} 1.36 \}^2$$

Using $b \approx 1.0\ A^{1/6}$ and $\hbar\omega = \frac{\hbar^2}{m} A^{-1/3}$ we get

$$\frac{v_{collective}}{v_{Fermi}} = \frac{f}{1.36\ A^{5/6}} \ .$$

In ^{40}Ca if we take $f = 3$ $A^{5/6} = 21.67$, the ratio is 0.102. The above should only be regarded as a crude estimate.

What does all this have to do with convergence and divergence, which is of interest in this conference? For one thing, it suggests that greater care be done in calculating particle-hole energy denominators. Depending on the interaction, there can be a large deviation from $n\hbar\omega$. Also, working near self-consistency is important. Whereas we get the breathing mode state near 30 MeV, if one casually does the calculation with a non-saturating interaction, then quite often the breathing mode state comes below the ground state (imaginary in R.P.A.).

The Effective Charge: R.P.A. Versus Hartree-Fock

In the previous section we showed a highly controlled situation, in which the R.P.A. and collective theories gave nearly identical results.

Once one has done the R.P.A. calculations for the core it is easy to consider what happens when one adds a nucleon to this core. The nucleus becomes deformed and one can express this by assigning an effective charge to the valence nucleon. The R.P.A. calculation of the effective charge is a core polarization calculation in which one lets the valence nucleon excite a particle-hole from the core. Then one allows the particle-hole pair to either scatter or excite other particle-hole pairs. This theory has been described many times and I won't go into it here.

Just to get our definitions straight, from now on I am going to talk about the E2 isoscalar effective charge correction. With the popular prescription 1/2 for the proton, 1/2 for the neutron, this correction would be one. I.e., the isoscalar charge correction is the sum of the neutron effective charge and the proton effective charge correction.

The next table lists the isoscalar correction for ^{16}O and ^{40}Ca for various interactions.

Oxygen 16			δe		
Interaction	J_i	J_F	First	T.D.A.	R.P.A.
Zero-Range	$0d_{5/2}$	$0d_{5/2}$	0.53	0.73	1.16
VBII'	$0d_{5/2}$	$0d_{5/2}$	0.37	0.54	0.60
Finite-Range	$0d_{5/2}$	$0d_{5/2}$	0.28	0.42	0.41
Calcium 40					
Zero-Range	$0f_{7/2}$	$0f_{7/2}$	0.60	0.88	1.71
VBII'	$0f_{7/2}$	$0f_{7/2}$	0.45	0.70	0.84
Finite-Range	$0f_{7/2}$	$0f_{7/2}$	0.35	0.58	0.57
Zero-Range	$1p_{3/2}$	$1p_{3/2}$	0.41	0.64	1.29
VBII'	$1p_{3/2}$	$1p_{3/2}$	0.31	0.50	0.62
Finite-Range	$1p_{3/2}$	$1p_{3/2}$	0.25	0.42	0.42

Examining the above table we see a nice systematic, the shorter the range of the interaction, the larger the effective charge.

Let us now concentrate on the zero range force $(-\alpha\delta(\vec{r}) + \frac{\Delta}{6}\rho\delta)$. Note that the R.P.A. effective charge for the $0f_{7/2}$ orbit is 1.71 and for the $p_{3/2}$ orbit is 1.29.

I found these results disturbing, because I recalled that in the Hartree-Fock theory with zero-range interactions the isoscalar charges should be unity. If we take the equation of state and set $\partial E(b_z,b_o)/\partial b_z = 0$, we get

$$\frac{b_z^{\ 3}}{b_o^{\ 3}} = \frac{2\Sigma_z}{\Sigma_{x+y}} ,$$

This is the same as the Mottelson conditions in the 1958 Les Hauches lectures, which he obtained from self-consistency arguments.

If N_x, N_y, N_z are number of quanta in the x,y,z directions for the valence nucleon, the single particle quadrupole moment for the Hartree-Fock intrinsic state is

$$Q_K = 2(N_z + \frac{1}{2})\, b_z^{\,2} - (N_x + N_y)\, b_x^{\,2}$$

$$= \frac{(2N_z+1)b_z^{\,3} - (N_x+N_y+1)b_o^{\,3}}{b_z}$$

One builds up an oblate solution by setting $N_z = 0$. Then $N_x + N_y \equiv 2n+\ell$ (the spherical quantum numbers) and $Q_K = -(2n+\ell)b_o^{\,2}$. For orientation purposes we consider the $0f_{7/2}$ orbit for which $2n+\ell = 3$. Note that if we set $b_z = b_o$ the intrinsic moment is $Q_K = -3b^2$. The shell model moment is $-(2j-1)/(2j+2)\langle r^2\rangle$. This is also $-3b^2$.

However, if we identify this state as a J = 7/2 member of a K = 7/2 band, then the rotor model in the strong coupling implies that there should be a reduction factor

$$Q = [3K^2 - J(J+1)]/[(J+1)(2J+3)]Q_K = 21/45\ (-3b^2)\ .$$

This is about 1/2 the shell model value. Only in the limit $K \to \infty$ will Q be equal to Q_K.

But Bohr and Mottelson show that in the weak coupling limit, which is probably more applicable here, $Q \to Q_K$.

To truly sort out this problem one should do a projected Hartree-Fock calculation, as M. Harvey and F. C. Khanna have done. I have not done this. In lieu of that I take the attitude that in what follows, the intrinsic moment should be compared with the R.P.A. theory, consistent with the weak coupling limit.

To show that the effective charge is indeed unity, we define Σ for the closed shell such that $\Sigma_x = \Sigma_y = \Sigma_z = \Sigma/3$. Then

$$\frac{b_z^{\,3}}{b_o^{\,3}} = 1 + \frac{b_z^{\,3} - b_o^{\,3}}{b_o^{\,3}}\ .$$

This is equal to

$$\frac{\frac{2\Sigma}{3} + (2N_z+1)}{\frac{2\Sigma}{3} + (N_x+N_y+1)} \approx 1 + \frac{[2N_z+1 - (N_x+N_y+1)]b_o^{\,2}}{\frac{2\Sigma}{3}\, b_o^{\,2}}\ .$$

Hence,

$$\left(\frac{b_z^{\,3} - b_o^{\,3}}{b_o}\right) \frac{2\Sigma}{3} \approx [2N_z - (N_x+N_y)]b_o^{\,2} \ .$$

The left-hand side is the quadrupole moment of the core and the right-hand side of the valence nucleon. Hence the ratio is unity.

I will now discuss an attempt, not completely finished, of finding the discrepancy between 1 and 1.7, the Hartree-Fock and R.P.A. values for the effective charge of the $0f_{7/2}$ orbit.

My first thought was that when one does the R.P.A. the valence nucleon always remains an $f_{7/2}$ nucleon. On the contrary, in the above restricted Hartree-Fock calculation all the nucleons have the same values of b_o and b_z, hence the scaling property.

Perhaps to better simulate the R.P.A. calculation we should use a different trial wave function as follows: 1) All the core nucleons are assigned the same oscillator length parameters b_z, b_o. 2) The valence nucleon is assigned different values $b_{zv}; b_o$. The energy now is

$$E(b_{zv}, b_z, b_o) = \frac{2}{3}\,\frac{\Sigma' b_z}{2b_o^{\,3}} + \frac{1}{3}\,\frac{\Sigma'}{2b_z^{\,2}} + \frac{(N_x+N_y+1)'}{2b_o^{\,3}}\, b_{zv}$$

$$+ (N_z + \tfrac{1}{2})'/2b_{zv}^2 + \text{POTENTIAL ENERGY OF THE CORE}$$

$$+ \text{SINGLE PARTICLE POTENTIAL ENERGY.}$$

As before, the potential energy of the core is independent of b_z but the single particle potential energy does depend on b_z. The expression is

$$\text{s.p.e.} = \int |\psi_{7/2}(b_{zv}, b_o)|^2 \ V[\rho(b_z, b_o)] d^3\vec{r}$$

where the single particle potential is: $V = -\frac{3}{4}\alpha\rho + \frac{9}{8}(\Delta/6)\rho^2$ and ρ is the density.

The problem is now harder to carry out so I will spare you the details.

The results will be presented as follows: I will pick a value of b_{zv} and look for the value of b_z which gives the lowest energy. I will indicate the binding energy gain relative to the case where all nucleons-valence and core - are constrained to have the same value of b_z.

For ^{40}Ca we use b_o = 1.9500 fm and the value of Σ is 120. The precise definition of the effective charge correction that we use here is:

$$\delta e = \{2\Sigma/3 \times (b_z^3 - b_o^3)/b_z\}/\{-(2j-1)/(2j+2) \times (2n+\ell+3/2)b_o^2\}$$

i.e. the ratio of the intrinsic moment of the core to the shell model value for a valence nucleon.

We consider first the case where all nucleons must have the same value of b_z. We find that at equilibrium

b_z = 1.9265
δe = 0.964
BINDING ENERGY GAIN = 0 (BY DEFINITION)

Note that δe is not quite unity because of second order effects.

Next we consider the case where b_{zv} is different from b_z.

b_{zv}	b_z (MINIMUM)	δe	BINDING ENERGY GAIN keV
1.91	1.919	1.272	- 36.44
1.93	1.919	1.272	+ 19.44
*1.95	1.920	1.231	+ 67.79
1.97	1.920	1.231	+108.79
1.99	1.912	1.190	+132.42
2.01	1.921	1.190	+168.84
2.03	1.922	1.149	+188.05
2.05	1.922	1.149	+200.16
2.07	1.923	1.108	+205.23
2.09	1.923	1.108	+203.29

Note that b_{zv} = 1.95 corresponds to $b_{zv} = b_o$.

A more careful analysis shows that the absolute minimum comes at b_{zv} = 2.074 b_z = 1.9230 δe = 1.1078 and an energy gain of 205.41 keV.

We see that the quadrupole moment of the core does indeed increase if we let b_{zv} be different from b_z. A point of interest is the case $b_{zv} = 1.95 = b_o$. We see that there is an energy gain of 68 keV. That is, the trial wave function in which the valence nucleon is not allowed to deform has a lower energy than the trial wave function in which the valence nucleon has the same deformation parameters as the core. The effective charge increases by a factor of about 1.3, lending some credence to the idea that the R.P.A. corresponds to core polarization with an 'undeformed' valence nucleon.

But we get a lower energy still, by making $b_{zv} > b_o$ whilst $b_z < b_o$. That is, the deformation parameters for the valence nucleon go towards a prolate state whilst those for the core towards an oblate state. One should not overstate this - the quadrupole moment of the valence nucleon $(b_z^{\,3} - (N_x+N_y)b_o^{\,3})/b_z$, is still negative, but it is now less negative. For this state of lowest energy, the deformation of the core increases by a factor of 1.15 from the $b_{zv} = b_z$ case.

To summarize this part, the large difference between the R.P.A. and Hartree-Fock effective charges has been partially, but only partially, resolved by allowing the valence nucleon to have different deformation parameters from the core.

Afterthoughts

I see I have a few pages left so let me ramble a bit.

You may recall that Sharp and I showed that the Vautherin-Brink interactions had a crazy spin dependence and so could not be applied to structure calculations such as the interaction between two particles or a particle and a hole. This was also shown by Speth, Krewald, and Jackson - they showed that the 2^- states came below the ground states for a large variety of VB interactions. This was also shown by the Montreal Group for the VBI interaction. What right then do we have to consider the collective electromagnetic states, which after all are particle-hole states?

The isoscalar giant resonances are somewhat of an exception in the sense that they have the same spin dependence as enters in the binding energy of the ground state. To illustrate this consider a simple delta interaction $-A_T\delta(\vec{r}_1-\vec{r}_2)$, where T stands for isospin. For T = 0 S = 1 and for T = 1 S = 0. The potential energy of a closed shell like ^{16}O or ^{40}Ca is proportional to $\frac{(A_0+A_1)}{2}$, i.e. the average of the singlet and triplet strength. The particle-hole interaction that enters into the mean energy of the isoscalar states also goes as $(A_o+A_1)/2$. Hence, once one has chosen a force to fit the binding energy, one does not have to commit oneself to a spin dependence to get the mean energy of the giant isoscalar states.

For the isovector giant states it's a different story. If the particle-hole interaction for the isoscalar state is $-|a|\ (A_0+A_1)$, then for the isovector state it is $+|a|\ (A_0-A_1/3)$. In order to get the isovector states as high as they are it is necessary to make A_1 the opposite sign of A_0, i.e., $A_0 > 0$ (attractive), $A_1 < 0$ (repulsive). But a repulsive T = 1 interaction leads to antipairing and other horrible things. One could, however, argue that at least by using the VB interaction one gets a correlation between the energy of the isovector states and the symmetry energy.

Is There a Long Range Monopole Effective Interaction in Nuclei?

The difficulties encountered with the Skyrme interaction (crazy spin dependence) may be solved by introducing long range components in the nuclear interaction.

The simplest long range potential between 2 particles is a constant $v(12) = -a\ (a > 0)$. Suppose we do a Hartree-Fock calculation with this interaction in ^{56}Ni, and ask for the splitting between the single particle states $1p_{3/2}$ and $0f_{7/2}$, i.e., the first unoccupied and last occupied orbits. Consider the interaction with the filled $f_{7/2}$ shell. The $p_{3/2}$ interaction will be -16a since there are 16 nucleons in this shell. The interaction of an $f_{7/2}$ nucleon will only be -15a since the nucleon cannot interact with itself. So we will get the unoccupied $p_{3/2}$ orbit below the $f_{7/2}$ orbit by a MeV.

But a constant potential corresponds to zero force and should therefore have no consequences. Suppose we do a particle hole calculation for the configuration $p_{3/2}\ f_{7/2}^{-1}$ with the same interaction. The particle hole interaction will be +a (as shown by Bansal and French). Hence, the energy of the particle hole state is -a + a = 0, consistent with the idea that zero force has zero consequences.

The next interaction we discuss is the Bansal-French interaction $-a + b\vec{t}_1\cdot\vec{t}_2$. The particle hole transform is $a + b\vec{t}_1\cdot\vec{t}_2$. That is, it has a value a-3/4b for T = 0 and a + b/4 for T = 1.

I want to use this interaction in a different way than it was used historically. Let us consider this as a crude corrective interaction, to be added to the realistic interactions such as Kuo and Brown, in order to better fit the experimental data.

$$V_{EMPIRICAL} = V_{KUO-BROWN} + a + b\vec{t}_1\cdot\vec{t}_2\ .$$

Let us look at the isobaric analog state of ^{208}Pb in ^{208}Bi.

This is a J = 0 T = 22 state at an energy of 15.15 MeV above the J = 5 T = 21 ground state. With the Kuo-Brown interaction this state comes about 4 MeV too low.

At first one might be discouraged in using the corrective force above since it would require $\frac{b}{4}$ = 4 MeV or b = 16 MeV. However, a more careful examination shows that for the analog state the expression must be modified. This is because the particle and hole are in the same shell. We find, as shown by Golin *et al.*, that the proper expression is

$$V_{PH} = a + b/4 + N_{ex}\frac{b}{2}$$

where N_{ex} is the neutron excess, (44 in the case of ^{208}Pb). So one only needs a value of b equal to +0.18 MeV.

This is but one of many examples where long range interactions are needed. Michaela Golin and I have considered many other cases but I won't go into them now. Also, note that the empirical Schiffer interaction has long range components in it. Perhaps the direction of things in the near future will be to incorporate these long range interactions in Hartree-Fock calculations, and to find a theoretical justification for them.

References

1. The collected works of B. R. Barrett.
2. J. M. Araujo, Vibrations of Spherical Nuclei in Nuclear Reactions 2, P. M. Endt and P. B. Smith, Ed. (North-Holland, 1962).
3. A. Bohr and B. R. Mottelson, Dan. Mat. Fys. Medd. 27 #16, (1953); A. Bohr and B. R. Mottelson, Nuclear Theory 11, to be published; B. R. Mottelson, The Many Body Problem, Les Hauches (John Wiley and Sons, Inc., New York, 1958).
4. D. R. Inglis, Phys. Rev. 96, 1059 (1954); D. R. Inglis, Phys. Rev. 97, 701 (1955).
5. T. H. R. Skyrme, Phil. Mag. 1, 1043 and Nucl. Phys. 9, 615 (1959); D. Vautherin and D. M. Brink, Phys. Lett. 32B, 149 (1970); D. Vautherin and D. M. Brink, Phys. Rev. C5, 626 (1972).
6. S. A. Moszkowski, Phys. Rev. C2, 402 (1970); J. W. Ehlers and S. A. Moszkowski, Phys. Rev. C6, 217 (1972).
7. R. W. Sharp and L. Zamick, Nucl. Phys. A208, 130 (1973); R. W. Sharp and L. Zamick, Nucl. Phys. A223, 333 (1974).
8. G. F. Bertsch and S. F. Tsai, to be published in Physics Reports.
9. S. Krewald and J. Speth, Phys. Lett. 52B, 295 (1974).
10. I. Hamamoto, Proc. Conf. on Nuclear Structure Studies Using Electron Scattering, Tohoku University, Sendai, Japan (1972) p.205.
11. T. Suzuki, Nucl. Phys. A217, 182 (1973).
12. L. Zamick, Phys. Lett. 45B, 313 (1973).

13. M. Golin and L. Zamick, Collective Models of Giant States with Density-Dependent Interactions, to be published.
14. H. Flocard and D. Vautherin, Phys. Lett. 55B, 259 (1975).
15. Y. M. Engel, D. M. Brink, K. Goeke, S. J. Krieger, and D. Vautherin, preprint.
16. G. Bertsch, Nuclear Hydrodynamics, to be published.
17. M. Baranger, European Conference on Nuclear Physics, Aix-en-Provence, 1972. Journal de Physique 33, C6-61 (1972).
18. B. Giraud and B. Grammaticos, Microscopic Analysis of Collective Motion, preprint.
19. S. Siegel and L. Zamick, Nucl. Phys. A145, 89 (1970).
20. M. Harvey and F. C. Khanna, Nuclear Spectroscopy and Reactions, Part D, J. Cerney, Ed. (Academic Press, 1975).
21. G. E. Brown, Facets of Physics, D. A. Bromley and V. Hughes, Ed. (Academic Press, New York, 1970) p. 141.

Additional comments:

Chun Wa Wong pointed out to me that at least in infinite nuclear matter the effective mass is given by $m^*/m = 1+$ (finite range energy/ kinetic energy). Hence the mean energy of the isoscalar quadrupole state can be expressed as

$$E_{2^+} = \sqrt{2}\,\hbar\omega/(m^*/m)^{1/2}.$$

From Sauer's talk and remarks by Negele we learn that the valence nucleon tends to have a larger value of b, the oscillator length parameter, than the core. This might explain why I am getting the valence value of b_{zv} to be greater than b_o (prolate). I had constrained the valence value of b_o to be the same as the core value. The only way then that the valence orbit could get larger is through b_{zv}, and so it did. What I plan to do is let both b_{ov} and b_{zv} vary. What might happen, although this should be checked, is that both b_{ov} and b_{zv} get larger, but the difference remains the same. In that case the isoscalar effective charge would still be unity in the Hartree-Fock.

L. ZAMICK: OTHER TESTS OF EFFECTIVE INTERACTIONS AND OPERATORS

Negele: What sort of space do you use for RPA?

Zamick: For the RPA we use $2\hbar\omega$ excitations. With harmonic oscillators that is all there is because both the quadrupole and monopole operators connect only through $2\hbar\omega$.

Negele: Why do you use cranking instead of Thouless-Valatin (i.e. self-consistent cranking)?

Zamick: There is a subtle point there. In doing the cranking calculation we use $2\hbar\omega$ energy denominators. In doing the RPA we use calculated single-particle energies (which in the case of the pure finite range interaction lead to a much larger spacing than $2\hbar\omega$). Let's ask Rowe. He uses the Inglis model for the pushing model and gets the right answer.

Rowe: Inglis cranking gives the correct answer for the translational mass if used with a local potential well, like the harmonic oscillator. However, if one uses a non-local single-particle well, like the Hartree-Fock potential, one needs the self-consistent Thouless-Valatin cranking model to get the correct answer.

SUMMARY TALK: WHERE DO WE STAND AT THE PRESENT TIME REGARDING THE MICROSCOPIC THEORY OF EFFECTIVE INTERACTIONS AND OPERATORS?

Michael W. Kirson
Weizmann Institute of Science, Rehovot, Israel

The fundamental aim of microscopic effective interaction theory is to bridge the gap between the nucleon-nucleon interaction and the properties of nuclei. It was not many years ago that highly-respected physicists were claiming that such a program was impossible to realize, that the forces involved were simply too strong. That we are now capable of starting with a nucleon-nucleon potential which fits the two-nucleon scattering and bound-state data and computing with some quantitative reliability the low-energy spectra of many nuclei should therefore be recognized as something of a triumph. This major achievement should not be lost sight of when assessing where we stand today. It remains true, however, that the basic program is difficult to realize with any great precision, and that the broad qualitative success of microscopic calculations tends to falter when pressed - detailed agreement with experiment and solid theoretical justification for our calculations remain elusive.

In considering the basic theory of effective interactions, it seems clear that perturbation theory still reigns supreme. In addition to the technical virtues listed by some of the invited speakers, the perturbation approach has two major advantages, in my eyes - it allows the properties of neighbouring nuclei to be related to one another (it does not require that every nucleus be calculated afresh, from the beginning), and it permits a neat pictorial vizualization of the physical processes responsible for the effects calculated. The use of diagrams aids in identifying these important physical processes and in guiding the development of appropriate theoretical constructs. Such a procedure is perhaps most conspicuous in the extensive treatment of particle-hole collectivity in core polarization, as described by Sprung in his talk.

Three rather different approaches to perturbation theory were described here, all of them having the property that the resulting formal framework is well defined and apparently complete. The methods described by Brandow and by Ratcliff, though expressed in rather different language and lending themselves to somewhat different techniques of calculation, apparently lead to the same final theory, in the sense that they produce the same set of diagrams, and the same contribution from any given diagram. The differences between the two approaches boil down to a question of grouping of terms, the eternal problem of perturbation theory. Ratcliff, for instance, advocates the computation of an energy-dependent Q-box, including unlinked valence terms, to some suitable order of perturbation theory, with folded diagrams then being introduced through suitable use of energy derivatives of the Q-box. Brandow, on the other hand, argues that folded diagrams should be treated on a par with other diagrams, in each order, thus also benefitting from the mutual cancellation of folded and non-folded unlinked valence terms. The elimination of unlinked valence terms is both physically appealing and computationally attractive, though the simplicity of the energy-derivative technique makes the latter point a matter of balancing the inclusion of energy-dependence and unlinked terms against the avoidance of the explicit construction of folded diagrams. There is clearly room for much insight in the interplay of and transition between the two methods, and it might well prove instructive to examine with more care the effect on the energy-derivative method of the starting-energy dependence of the reaction matrix, as suggested by Vary.

The third approach to perturbation theory, as described by Johnson, is much more flexible but less systematic than the other methods. It is also less familiar to the bulk of practitioners in the field, which makes it somewhat forbidding. I have no doubt at all that we could all benefit considerably from becoming much more familiar with this formalism, and developing the kind of intuition necessary to make full use of its built-in flexibility. The diagrams it produces can be connected with those of the other approaches, though it tends to have fewer diagrams, with analytical expressions differing a little from those of the other methods - unlinked diagrams never occur, factorization is immediate, and hermiticity can be directly achieved (unlike the other methods, which require additional calculations to achieve hermiticity). The price of these advantages is a more varied starting-energy dependence, possibly requiring a wider range of reaction matrix elements, and the need to develop the kind of finger-tip sensitivity which can detect the efficient choice of time-base for any diagram (and here again the pictorial quality of perturbation theory is important).

In considering effective operators, one is struck by the greater apparent differences between the three basic perturbation theories. Brandow defines an effective operator, in terms of a linked expansion involving special combinatorial factors in the folded diagrams (arising from the binomial expansion of inverse-square-root normalization factors). However, this operator must be used in an orthogonal basis which is obtainable from the eigenstates of the effective interaction only by a further calculation. This is thus a two-step prescription. The same is true of the method described by Ratcliff - one computes first a numerator (no special combinatorial factors, but containing unlinked terms), then a denominator (arising from the normalization of the eigenstates of the effective interaction), and then takes the ratio. This method, which again can make use of energy-derivatives to include folded terms, is in fact a calculation of effective transition matrix elements, rather than of an effective operator. This becomes clearer when one realizes that a different formula is used for diagonal matrix elements. In contrast with these methods, Johnson's is a one-step technique directly defining that effective operator appropriate to the eigenstates of the effective interaction used, and requiring no special combinatorial factors, no explicit normalization factors and no unlinked diagrams. These advantages would seem amply to justify the effort involved in gaining greater familiarity with the method.

So much for theory. We clearly have a well-established theoretical framework in which to calculate, and we are all well aware that our problems really begin with the calculations. We have "known" for twenty years that the basic ingredient of these calculations must be the Brueckner reaction matrix, for how else can we deal with near-singular short-range correlations? But the real justification for such an approach lies in partitioning the problem into high-energy and low-energy parts, associated roughly with short-range (cluster-type) and long-range (configuration-mixing) correlations. We believe, on the basis of computations in infinite nuclear matter, that the two-body cluster term, the reaction matrix, takes adequate care of the short-range correlations, and that the remaining configuration-mixing effects can be included through perturbation theory, with relatively low-lying virtually-excited intermediate states. As abundantly demonstrated by Becker, we now have available several essentially exact methods for computing the reaction matrix, given a nucleon-nucleon interaction and a single-particle basis and spectrum, and there is no more room for the kind of double-counting controversy that has accompanied the widespread use of more approximate reaction-matrix elements. The remaining major uncertainty is in the choice of single-particle basis and spectrum. Kümmel has told us that three-body cluster calculations in nuclei call for a weak attractive single-particle potential above the fermi surface, in agreement with Rajaraman's decade-old argument that this potential is determined by the average of the long-range part of the nucleon-nucleon interaction. We also know from Ellis that results are considerably changed by using single-particle wavefunctions closer to self-consistency. But the single-particle potential is simply an auxiliary theoretical construct, introduced to simplify calculations and make them more efficient. Mahaux showed some years ago that calculations in infinite nuclear matter have been taken to the point where the results are insensitive to wide variations in this auxiliary construct. I, personally, will feel much happier when our calculations in finite nuclei reach the point where the results do not change substantially when the single-particle basis and spectrum are quite broadly altered. Until then, all calculations will continue to be done under a giant question-mark.

A question associated with the single-particle aspects of the problem arises when one considers the demonstrated importance of occupation-probability factors in computing the bulk properties of closed-shell nuclei. The blanket inclusion of such factors on all lines in all perturbation-theory diagrams would drastically damp all higher-order effects, but this could well be a spurious effect - it might just require calculations to much higher orders to restore the old results. The occupation-probability factors take into account the possibility that a given particle (hole) state may be only partly filled (empty), due to virtual excitation of more complicated configurations. But the system continues to interact while in these more complicated configurations, and this "shift in strength" should be considered together with the occupation-probability factors. This is a prime motivation for the concept of number-conserving sets. In figure 1, the first (folded) diagram produces an occupation-probability factor on the outgoing valence line, while the remaining diagrams include the contributions of the corresponding virtually-excited configurations. There is a partial

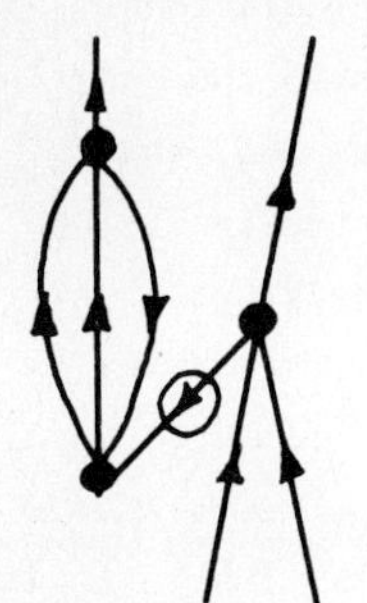
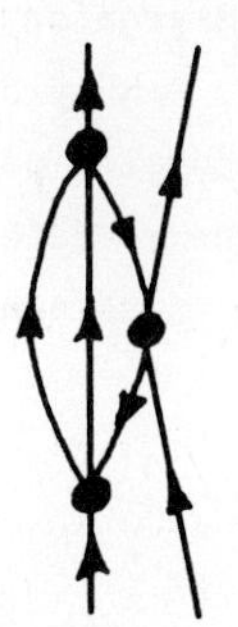
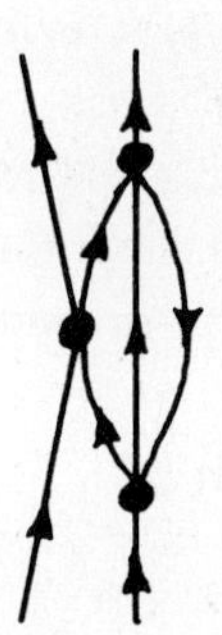

Fig. 1 - Number-conserving set

cancellation among these diagrams (one of Brandow's arguments for including folded and unfolded diagrams on the same footing, order by order), which tends to weaken the argument for including occupation-probability factors. However, it may well be justified to include that part of the occupation-probability factors which is due to the short-range correlations and hence associated with the cluster part of the calculation. This would involve a factor like .9 on every hole line in a diagram, only a mild damping effect.

Of course, as soon as one starts using perturbation theory, the question of convergence becomes a legitimate concern. There are two distinct problems here - the formal problem of mathematical convergence and the practical problem of numerical convergence. The former, as carefully explained by Weidenmüller, arises as soon as there are intruder states, and possibly with near-intruder states strongly coupled to the model space. But Schucan showed that [n+1,n] Padé approximants may well be the mathematically appropriate form of analytic continuation, formally justified (perhaps) everywhere except on a set of singular arcs, and converging to the set of states having maximum overlap with the model space. The mathematical problem could then be regarded as solved. However, the practical problem is very much with us - how many orders of perturbation theory are required before precise results are obtained, by summing the perturbation series if it converges, or by computing appropriate Padé approximants when it diverges? It is certainly encouraging, in this connection, that weakly-coupled intruders can be safely ignored. As argued some time ago by Vincent and Pittel, such states will have little effect on the final spectrum and will contribute negligibly to low orders of perturbation theory. For them, the convergence problems can simply be forgotten. However, strongly-coupled "potential intruders" (associated with branch cuts outside, but close to, the unit circle in the complex coupling-parameter plane) are likely to give rise to large effects and to slow convergence of the perturbation series.

There is little information on this problem in full-scale calculations. In the prototype mass-18 system there are weakly-coupled intruders in the observed spectra, and probably strongly-coupled potential intruders too, though it is not clear that the calculations done to date include such intruders. The complex coupling-parameter plane then contains numerous singular arcs.

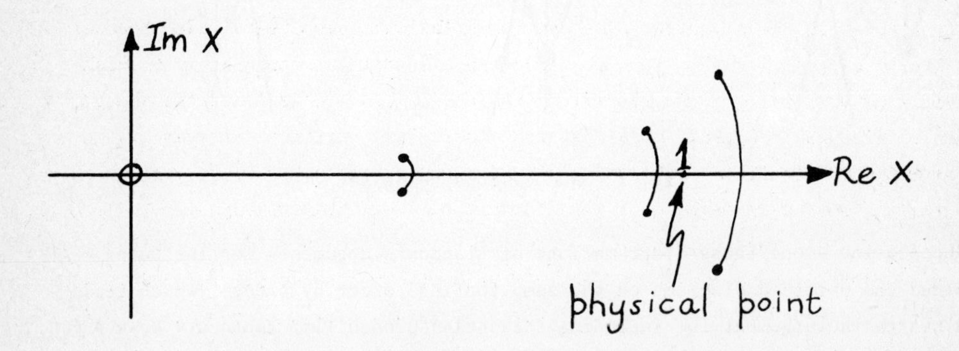

Fig. 2 - Intruder arcs for $J^{\pi}T = 0^{+}1$ in mass 18 (schematic).

Goode has shown us poor numerical convergence of averaged matrix elements through fourth order, apparently not associated with intruders, and this is consistent with the slow convergence seen in complete calculations through third order. Numerical models, based on perturbation expansions of large-matrix-diagonalization calculations, indicate that even convergent series converge slowly enough, as do Padé approximants to convergent or divergent series, that seventh or higher order results are needed to achieve even 100 keV precision. The inescapable conclusion appears to be that even though convergence is no longer a mathematical problem, attainable orders in perturbation calculations will leave an unavoidable residue of numerical uncertainty, so that precision of better than a few hundred keV can not be achieved.

The techniques of infinite partial summation and large-matrix diagonalization cannot lead us out of this numerical impasse. The former allows the selective tracing of specific physical effects through high orders, helping to pin down collective effects and suggesting where strong corrections must be carefully included, while the latter permits us to check the effect of specific sets of intermediate states on rapidity of convergence. But since both are inherently selective, and hence partial, probes, they cannot be used to pin down the final results with high precision, though they are of course very useful in establishing the degree of uncertainty in low-order calculations.

All the above supposes that the calculations to a given order can be done quite exactly. There are uncertainties associated with the choice of single-particle basis (one should include self-consistency corrections systematically) and with the dependence of reaction-matrix elements on the single-particle spectrum. Even more serious, however, is the Vary-Sauer-Wong effect, requiring sums over intermediate states to high excitation energy in order to include correctly the contribution of the tensor force. Not only does such a requirement play havoc with the perturbation theory calculations (even third order becomes prohibitively difficult), it raises fundamental questions concerning the high energy/low energy (or short range/long range) dichotomy underlying the whole theory. The double-partitioning technique described by Barrett is partly vitiated by this effect. There, low-lying two-particle states are systematically excluded from the reaction-matrix ladder and reintroduced as long-range configuration-mixing factors in the perturbation expansion. Ideally, one could have hoped to find some range of "cutoff" energies such that all two-body states important for short-range correlations lay at much higher energies, while all two-body states important for long-range correlations lay at lower energies. The double-partitioning results would then be unchanged on moving the cutoff through this range. Unfortunately, the tensor correlations seem to fill in this "energy gap". It has been suggested at this conference that a triple-partitioning technique may be needed, involving some special way of treating the tensor correlations at intermediate energies. I would like to point out a second possibility, involving double partitioning "from the top down" - the existing method works from the bottom up, by taking low-lying states from the reaction-matrix ladder and transferring them to the low orders of perturbation theory.

One could imagine instead first partitioning the A-particle Hilbert space into a large, finite model space (perhaps all A nucleons anywhere below the N-th oscillator shell) and an excluded space. The effective interaction in the model space would be treated by a cluster approach, and would have a two-body part given by the reaction matrix with a high Pauli cutoff. The tensor correlations scattering outside the model space would then introduce a three-body term in the effective interaction. One would then proceed to partition the model space again, in the standard way, into a smaller model space of the closed-core-plus-few-valence-particles type, and an excluded space, and then do ordinary effective-interaction calculations with a two-body plus three-body "bare" force and an exact cutoff on all intermediate-state sums. But one way or another, the Vary-Sauer-Wong tensor effect must be handled - we cannot simply ignore it.

In the light of the irreducible imprecision inherent in the perturbation theory approach, it would perhaps be wise to adopt a more qualitative attitude and to search for understanding in terms of simple physical concepts which can be associated (pictorially) with certain terms in the perturbation series. As emphasized by Harvey, a major part of the renormalization of effective operators appears to be associated with self-consistent-field effects. Köhler and Zucker have independently suggested that our favourite two-body renormalization, core polarization, may in fact simply reflect a change in the self-consistent field due to the addition of two valence particles, and in that sense is hardly a legitimate two-body effect. Similarly, we are all familiar with the picture of single-particle energies changing smoothly with nucleon number as particles are added to a shell. However, all our calculations are done with fixed single-particle energies. Perhaps we should reinterpret some apparent two-body terms as arising from this smooth change in the single-particle energies. Bertsch has shown how Pauli-blocking effects can change the effective two-body force towards the end of a shell, something which shows up in our theory as an effective three-body force. Given that exact numerical precision is out of our reach, it is possible that new, qualitative, points of view such as these could be useful in refining our understanding both of effective interaction theory and of nature.

Sooner or later, we are forced to confront experiment, and here we must thank Schiffer and Petrovich for keeping us honest by showing some genuine experimental data. The message of the experimental data seems to be that things are in fact much simpler than our theories and calculations would suggest. Talmi showed that excellent agreement with experiment is frequently achieved using very simple configurations, and pure two-body forces, while Schiffer demonstrated again that the two-valence-nucleon data seems to be compatible with a simple universal force in simple configurations (though much more thought clearly needs to be devoted to the fact that he uses centroids of simple-configuration strengths rather than observed spectra). We have come a long way since a good fit to experimental data was the only criterion for the quality of a calculation, but we should not let the pendulum swing too far the other way - working with one hand on the data is still a reasonable modus operandi. On the other hand, as S.Y. Lee pointed out here and as Gerry Brown has been saying for some time, we should be careful not to agree with experiment too soon. Knowing from semi-phenomenological studies that deformed 4p2h states can push the ground state of ^{18}O down by an MeV, we should not be happy with calculations which get the correct ground-state energy without including the contribution of these states.

I feel this is the appropriate place to comment on what I call the "new phenomenology", the use of density-dependent forces to explain nuclear structure. Negele gave a convincing demonstration that such forces arise from the attempt to find a local representation of the reaction matrix, and that there is no reason to believe that one set of parameters in such a representation will be appropriate for calculations both of nuclear bulk properties and of nuclear spectra. Too often, I feel, the Skyrme force of today plays the role of the gaussian-plus-Rosenfeld-exchange-mixture of twenty years ago. I do not wish to belittle the work done in this field by Zamick and others. We can clearly gain much insight into nuclear structure by careful and judicious studies of this kind. But this work is rather far from microscopic effective-interaction theory. As an illustration of the difference, let me mention the calculation of breathing mode energies. These come much too low with conventional forces, acceptably high with density-dependent forces. But a careful check shows that the density-dependent force, used exactly like a conventional force, produces very much the same low monopole energies. The big repulsive push comes from renormalization factors (like $\frac{20}{9}$ times the repulsive $\rho^{2/3}$ term), and the challenge is to understand in microscopic terms these dynamic density-dependent effects. The density-dependent studies can highlight specific physical features, but do not in themselves give an adequate microscopic explanation.

One other point where our theory has not been adequately confronted with experiment is in the appearance of many-body effective forces and operators. The theory unambiguously predicts such effects as soon as the number of valence particles exceeds one or two, while the experimental data, as analyzed by shell model phenomenology, generally seems to require little, if any, of such effects. It may well be that what appear formally as many-body effects are more physically interpreted as smooth changes in core fields, single-particle energies or two-body interactions under the influence of additional valence particles. We need to think much more about such effects.

We have also heard presented some suggested new methods for approaching the theory of nuclear structure. Green's function techniques have remained popular with many people, though too often they seem to be useful mainly for formal manipulations. When one gets down to calculations, these tend to appear difficult, to be based on questionable ansatzes, and to be of the nucleus-by-nucleus type - relationships between neighbouring nuclei seem to be lost. I am yet to be convinced that they represent a viable alternative.

French's very entertaining presentation of the statistical approach contained some very scary implications. The impressive agreement of the eigenvalues of an 839-dimensional shell-model matrix with the simple gaussian distribution of the (random) statistical theory is hardly encouraging to one who makes his living among the lowest four 0^+ states! But of course it is precisely in the low-energy tail of the statistical distributions that one expects to find significant fluctuations, and these are our bread-and-butter effects. So perhaps we need not feel too threatened by French and Co., though it may well be healthy to bear in mind that we really need a theory of deviations from smooth statistical behaviour.

The methods described by Kümmel are most impressive, and strike me as very promising. He gives up entirely the short-range/long-range dichotomy, with the solution of Bethe-Goldstone and Bethe-Fadeev type equations forming an explicit part of his procedure, rather than a preliminary stage. It may thus be dangerous to draw conclusions about the grouping of diagrams according to steps in the Kümmel iteration scheme, since groupings appropriate to short-range correlations and those appropriate to long-range configuration-mixing correlations will probably be mixed together. However, the tensor correlations should be automatically included and do not constitute an additional complication. This method shares with other competitors to perturbation theory the feature that each nucleus is a new problem, with no simple connection between the one-body part of ^{17}O and that of ^{18}O, for instance. It is clearly a very interesting approach, deserving much wider attention, and I will eagerly be awaiting the promised results on nuclear spectra.

To summarize my summary, I would say that there are definitely major obstacles in the way of doing really convincing calculations. We have a theory, we are aware of the weakness of our computations, and we should certainly invest the effort needed to plug the more obvious holes. But we should recognize the lack of precision inherent in our inability to go to high orders in perturbation theory and adopt more qualitative methods of extracting information and gaining understanding. New ideas, new viewpoints are much needed. We should also try to withstand the temptations of seductive ^{18}O, and pay some attention to other systems. Gerry Brown has frequently commented that ^{18}O has too many special problems to serve as a typical nucleus. He has also remarked that it is difficult to construct a theory which does not fit the spectrum of ^{18}O. With full recognition of the difficulties involved, and an eye on the broader features of nuclei, we can still keep the show on the road.

30,–

Lecture Notes in Physics

Bisher erschienen/Already published

Vol. 1: J. C. Erdmann, Wärmeleitung in Kristallen, theoretische Grundlagen und fortgeschrittenene experimentelle Methoden. 1969. DM 22,–

Vol. 2: K. Hepp, Théorie de la renormalisation. 1969. DM 20,–

Vol. 3: A. Martin, Scattering Theory: Unitarity, Analyticity and Crossing. 1969. DM 18,–

Vol. 4: G. Ludwig, Deutung des Begriffs physikalische Theorie und axiomatische Grundlegung der Hilbertraumstruktur der Quantenmechanik durch Hauptsätze des Messens. 1970. Vergriffen.

Vol. 5: M. Schaaf, The Reduction of the Product of Two Irreducible Unitary Representations of the Proper Orthochronous Quantummechanical Poincaré Group. 1970. DM 18,–

Vol. 6: Group Representations in Mathematics and Physics. Edited by V. Bargmann. 1970. DM 27,–

Vol. 7: R. Balescu, J. L. Lebowitz, I. Prigogine, P. Résibois, Z. W. Salsburg, Lectures in Statistical Physics. 1971. DM 20,–

Vol. 8: Proceedings of the Second International Conference on Numerical Methods in Fluid Dynamics. Edited by M. Holt. 1971. Out of print.

Vol. 9: D. W. Robinson, The Thermodynamic Pressure in Quantum Statistical Mechanics. 1971. DM 18,–

Vol. 10: J. M. Stewart, Non-Equilibrium Relativistic Kinetic Theory. 1971. DM 18,–

Vol. 11: O. Steinmann, Perturbation Expansions in Axiomatic Field Theory. 1971. DM 18,–

Vol. 12: Statistical Models and Turbulence. Edited by C. Van Atta and M. Rosenblatt. Reprint of the First Edition 1975. DM 28,–

Vol. 13: M. Ryan, Hamiltonian Cosmology. 1972. DM 20,–

Vol. 14: Methods of Local and Global Differential Geometry in General Relativity. Edited by D. Farnsworth, J. Fink, J. Porter and A. Thompson. 1972. DM 20,–

Vol. 15: M. Fierz, Vorlesungen zur Entwicklungsgeschichte der Mechanik. 1972. DM 18,–

Vol. 16: H.-O. Georgii, Phasenübergang 1. Art bei Gittergasmodellen. 1972. DM 20,–

Vol. 17: Strong Interaction Physics. Edited by W. Rühl and A. Vancura. 1973. DM 32,–

Vol. 18: Proceedings of the Third International Conference on Numerical Methods in Fluid Mechanics, Vol. I. Edited by H. Cabannes and R. Temam. 1973. DM 20,–

Vol. 19: Proceedings of the Third International Conference on Numerical Methods in Fluid Mechanics, Vol. II. Edited by H. Cabannes and R. Temam. 1973. DM 29,–

Vol. 20: Statistical Mechanics and Mathematical Problems. Edited by A. Lenard. 1973. DM 24,–

Vol. 21: Optimization and Stability Problems in Continuum Mechanics. Edited by P. K. C. Wang. 1973. DM 18,–

Vol. 22: Proceedings of the Europhysics Study Conference on Intermediate Processes in Nuclear Reactions. Edited by N. Cindro, P. Kulišić and Th. Mayer-Kuckuk. 1973. DM 29,–

Vol. 23: Nuclear Structure Physics. Proceedings of the Minerva Symposium on Physics. Edited by U. Smilansky, I. Talmi, and H. A. Weidenmüller. 1973. DM 29,–

Vol. 24: R. F. Snipes, Statistical Mechanical Theory of the Electrolytic Transport of Non-electrolytes. 1973. DM 22,–

Vol. 25: Constructive Quantum Field Theory. The 1973 "Ettore Majorana" International School of Mathematical Physics. Edited by G. Velo and A. Wightman. 1973. DM 29,–

Vol. 26: A. Hubert, Theorie der Domänenwände in geordneten Medien. 1974. DM 28,–

Vol. 27: R. Kh. Zeytounian, Notes sur les Ecoulements Rotationnels de Fluides Parfaits. 1974. DM 28,–

Vol. 28: Lectures in Statistical Physics. Edited by W. C. Schieve and J. S. Turner. 1974. DM 24,–

Vol. 29: Foundations of Quantum Mechanics and Ordered Linear Spaces. Advanced Study Institute Held in Marburg 1973. Edited by A. Hartkämper and H. Neumann. 1974. DM 26,–

Vol. 30: Polarization Nuclear Physics. Proceedings of a Meeting held at Ebermannstadt October 1–5, 1973. Edited by D. Fick. 1974. DM 24,–

Vol. 31: Transport Phenomena. Sitges International School of Statistical Mechanics, June 1974. Edited by G. Kirczenow and J. Marro. DM 39,–

Vol. 32: Particles, Quantum Fields and Statistical Mechanics. Proceedings of the 1973 Summer Institute in Theoretical Physics held at the Centro de Investigacion y de Estudios Avanzados del IPN – Mexico City. Edited by M. Alexanian and A. Zepeda. 1975. DM 18,–

Vol. 33: Classical and Quantum Mechanical Aspects of Heavy Ion Collisions. Symposium held at the Max-Planck-Institut für Kernphysik, Heidelberg, Germany, October 2–5, 1974. Edited by H. L. Harney, P. Braun-Munzinger and C. K. Gelbke. 1975. DM 28,–

Vol. 34: One-Dimensional Conductors, GPS Summer School Proceedings, 1974. Edited by H. G. Schuster. 1975. DM 32,–

Vol. 35: Proceedings of the Fourth International Conference on Numerical Methods in Fluid Dynamics. June 24–28, 1974, University of Colorado. Edited by R. D. Richtmyer. 1975. DM 37,–

Vol. 36: R. Gatignol, Théorie Cinétique des Gaz à Répartition Discrète de Vitesses. 1975. DM 23,–

Vol. 37: Trends in Elementary Particle Theory. Proceedings 1974. Edited by H. Rollnik and K. Dietz. 1975. DM 37,–

Vol. 38: Dynamical Systems, Theory and Applications. Proceedings 1974. Edited by J. Moser. 1975. DM 46,–

Vol. 39: International Symposium on Mathematical Problems in Theoretical Physics. Proceedings 1975. Edited by H. Araki. 1975. DM 44,–

Vol. 40: Effective Interactions and Operators in Nuclei. Proceedings 1975. Edited by B. R. Barrett. 1975. DM 30,–

SPRINGER TRACTS IN MODERN PHYSICS

Ergebnisse der exakten Naturwissenschaften

Springer-Verlag
Berlin
Heidelberg
New York

Volume 66

30 figures. III, 173 pages. 1973
Cloth DM 78,—; US $31.90
ISBN 3-540-06189-4

Quantum Statistics

in Optics and Solid-State Physics

R. Graham: Statistical Theory of Instabilities in Stationary Nonequilibrium Systems with Applications to Lasers and Nonlinear Optics.
F. Haake: Statistical Treatment of Open Systems by Generalized Master Equations.

Volume 67

III, 69 pages. 1973
Cloth DM 38,—; US $15.50
ISBN 3-540-06216-5

S. Ferrara, R. Gatto, A. F. Grillo:

Conformal Algebra in Space-Time

and Operator Product Expansion

Introduction to the Conformal Group in Space-Time. Broken Conformal Symmetry. Restrictions from Conformal Covariance on Equal-Time Commutators. Manifestly Conformal Covariant Structure of Space-Time. Conformal Invariant Vacuum Expectation Values. Operator Products and Conformal Invariance on the Light-Cone. Consequences of Exact Conformal Symmetry on Operator Product Expansions. Conclusions and Outlook.

Volume 68

77 figures. 48 tables. III, 205 pages. 1973
Cloth DM 88,—; US $35.90
ISBN 3-540-06341-2

Solid-State Physics

D. Schmid: Nuclear Magnetic Double Resonance — Principles and Applications in Solid-State Physics.
D. Bäuerle: Vibrational Spectra of Electron and Hydrogen Centers in Ionic Crystals.
J. Behringer: Factor Group Analysis Revisited and Unified.

Volume 69

13 figures. III, 121 pages. 1973
Cloth DM 78,—; US $31.90
ISBN 3-540-06376-5

Astrophysics

G. Börner: On the Properties of Matter in Neutron Stars.
J. Stewart, M. Walker: Black Holes: the Outside Story.

Prices are subject to change without notice
■ Prospectus with Classified Index of Authors and Titles Volumes 36—74 on request

Volume 70

II, 135 pages. 1974
Cloth DM 77,—; US $31.50
ISBN 3-540-06630-6

Quantum Optics

G. S. Agarwal: Quantum Statistical Theories of Spontaneous Emission and their Relation to Other Approaches.

Volume 71

116 figures. III, 245 pages. 1974
Cloth DM 98,—; US $40.00
ISBN 3-540-06641-1

Nuclear Physics

H. Überall: Study of Nuclear Structure by Muon Capture.
P. Singer: Emission of Particles Following Muon Capture in Intermediate and Heavy Nuclei.
J. S. Levinger: The Two and Three Body Problem.

Volume 72

32 figures. II, 145 pages. 1974
Cloth DM 78,—; US $31.90
ISBN 3-540-06742-6

D. Langbein:

Theory of Van der Waals Attraction

Introduction. Pair Interactions. Multiplet Interactions. Macroscopic Particles. Retardation. Retarded Dispersion Energy. Schrödinger Formalism. Electrons and Photons.

Volume 73

110 figures. VI, 303 pages. 1975
Cloth DM 97,—; US $39.60
ISBN 3-540-06943-7

Excitons at High Density

Editors: H. Haken, S. Nikitine
Biexcitons. Electron-Hole Droplets. Biexcitons and Droplets. Special Optical Properties of Excitons at High Density. Laser Action of Excitons. Excitonic Polaritons at Higher Densities.

Volume 74

75 figures. III, 153 pages. 1974
Cloth DM 78,—; US $31.90
ISBN 3-540-06946-1

Solid-State Physics

G. Bauer: Determination of Electron Temperatures and of Hot Electron Distribution Functions in Semiconductors.
G. Borstel, H. J. Falge, A. Otto: Surface and Bulk Phonon-Polaritons Observed by Attenuated Total Reflection.